Flash Points
of Organic and Organometallic Compounds

Flash Points of Organic and Organometallic Compounds

Richard M. Stephenson
Professor of Chemical Engineering
University of Connecticut
Storrs, Connecticut

Elsevier
New York • Amsterdam • London

Elsevier Science Publishing Co., Inc.
52 Vanderbilt Avenue, New York, New York 10017

Distributors outside the United States and Canada:

Elsevier Applied Science Publishers Ltd.
Crown House, Linton Road, Barking, Essex IG11 8JU, England

Library of Congress Cataloging in Publication Data

Stephenson, Richard Montgomery, 1917-
    Flash points of organic and organometallic compounds.

    Bibliography: p.
    1. Inflammable liquids.   2. Organometallic compounds.
I. Title.
TP361.S74   1987   661'.0028   87-6774
ISBN 0-444-01239-7

Current printing (last digit)
10 9 8 7 6 5 4 3 2 1

Manufactured in the United States of America

Foreword

This book brings together and makes easily accessible data on flash points presently listed in the standard references, plus many values measured in American industrial laboratories and European data as represented by the Fluka catalog and by values given in Nabert and Schoen. It is a collection that should be useful to those involved in manufacturing, handling, and shipping the many organics and organometallics now in production.

The author would like to thank David Harvey, President, and Chuck Pouchert, of Aldrich Chemical Company, for permission to use their material and for providing a computer printout of some six thousand chemicals listed in the Aldrich catalog. Flash points for most of these chemicals were measured in the Aldrich laboratories and are not available elsewhere.

Thanks are due also to Dr. Marjan Bace of Elsevier Science Publishing Co., Inc., who recognized the need for a handbook on flash points and handled the publishing arrangements. Particular thanks go to my wife, Mary, who prepared the entire camera-ready manuscript.

Richard M. Stephenson
The University of Connecticut
Storrs, Connecticut

December, 1986

Introduction

Flash points are of particular importance to those involved with the handling and shipping of organics and organometallics. As each country has its own shipping regulations, this can be a real problem for international shippers. Many flash points have been determined by individual chemical manufacturers in answer to their own needs. For example, about 80% of the thousands of flash points listed in the Aldrich catalog[1] have been measured in their own laboratories. Similarly, the Swiss laboratory supply house Fluka Chemie AG[3] has measured many of the values in their catalog. Kirk–Othmer[4] lists many flash points measured in industrial laboratories, and Ernest Flick[2] has done a good job of searching industrial sales catalogs and providing many flash points in his *Solvents Handbook*.

Note, however, that all flash point data must be approached with caution. Because values are often simply copied from one source to another, the appearance of the same number in several references does not mean it is necessarily more reliable. Flash points may also be strongly affected by small traces of impurities. Although values given in this table are for purities greater than 95%, in some cases it is noted that the value is for commercial or technical material. Sometimes the purity of the material is not known, but it is assumed that most industrial measurements are for typical commercial-grade material

All flash point values in the table are in Kelvins. Those determined by the open-cup method are starred (for example, 294.15*). Flash points determined by the closed-cup method are unstarred. Unfortunately, some references do not state which method was used, so that some of the unstarred values in this table were undoubtedly determined by the open-cup method. No attempt has been made to specify the apparatus (such as Cleveland or ASTM) used in these measurements.

In preparing the table of flash point data which follows, a complete search was made of the references listed below, and the source for each value is indicated in parentheses.

REFERENCES

1. *Aldrich Catalog of Fine Chemicals*, Aldrich Chemical Co., Inc., Milwaukee, Wisconsin, 1986–1987.

2. Ernest W. Flick, *Industrial Solvents Handbook*, 3rd Ed., Noyes Data Corp., Park Ridge, New Jersey, 1985.

3. *Fluka Chemie AG, Catalog 1986–87*, CH-9470, Bucks/Switzerland.

4. *Kirk–Othmer Encyclopedia of Chemical Technology*, 3rd Ed., John Wiley and Sons, Inc., New York, 1978.

5. K. Nabert and G. Schoen, *Sicherheitstechnische Kennzahlen brennbarer Gase und Daempfe*, 2nd Ed., Deutscher Eichverlag GmbH, Braunschweig, West Germany.

6. National Fire Protection Association, *Fire Protection Guide on Hazardous Materials*, Quincy, Massachusetts, 1984.

7. N. Irving Sax, *Dangerous Properties of Industrial Materials*, Van Nostrand Co., New York, 1984.

Flash Points

B_2H_6

Boron hydride, CA 19287-45-7: 205.37 (7).

$B_{10}H_{14}$

Decaborane, CA 17702-41-9: 353.15 (6).

Cl_2H_2Si

Dichlorosilane, CA 4109-96-0: 235.93 (6).

Cl_2OS

Thionyl chloride, CA 7719-09-7: Nonflammable (1).

ClO_2S

Sulfuryl chloride, CA 7791-25-5: Nonflammable (1).

Cl_2S

Sulfur dichloride, CA 10545-99-0: Nonflammable (1).

Cl_2S_2

Sulfur monochloride, CA 10025-67-9: 391.48 (6,7); 403.15* (4); nonflammable (1).

Cl_3HSi

Trichlorosilane, CA 10025-78-2: 245.15 (4,7); below 253.15 (5); 259.26* (1,6); 266.15 (3).

Cl_3PS

Thiophosphoryl chloride, CA 3982-91-0: Nonflammable (1,7).

H_2S_2

Hydrogen disulfide, CA 13465-07-1: Below 295.15 (7).

H_3NO

Hydroxylamine, CA 7803-49-8: Explodes 402.59 (6,7).

H_4N_2

Hydrazine, CA 302-01-2: 310.93 (6); 310.93* (7); 325.15 (1,6); 325.15* (4).

1

H_6N_2O

Hydrazine hydrate, CA 7803-57-8:   325.15 (3); 345.15* (4); 347.04 (1).

$CClF_3O_2S$

Trifluoromethanesulfonyl chloride, CA 421-83-0: Nonflammable (1).

$CClNO_3S$

Chlorosulfonyl isocyanate, CA 1189-71-5: Nonflammable (1).

CCl_2O

Phosgene, CA 75-44-5:   277.15 (3).

$CHBrCl_2$

Bromodichloromethane, CA 75-27-4: Nonflammable (1).

$CHBr_2Cl$

Chlorodibromomethane, CA 124-48-1: Nonflammable (1).

$CHBr_3$

Bromoform, CA 75-25-2: Nonflammable (1).

$CHCl_3$

Chloroform, CA 67-66-3: Nonflammable (1).

CHF_3O_3S

Trifluoromethanesulfonic acid, CA 1493-13-6: Nonflammable (1).

CHN

Hydrogen cyanide, CA 74-90-8: Below 253.15 (5); 255.37 (4,6,7).

CH_2BrCl

Bromochloromethane, CA 74-97-5: Nonflammable (1,2,7).

CH_2BrNO_2

Bromonitromethane, CA 563-70-2: Above 383.15 (1).

CH_2Br_2

Dibromomethane, CA 74-95-3: Nonflammable (1,2).

CH_2Cl_2

Dichloromethane, CA 75-09-2: Nonflammable (1,2,4,6,7).

3

CH_2Cl_3OP

Chloromethylphosphonic dichloride, CA 1983-26-2: Above 383.15 (1).

CH_2Cl_4Si

(Chloromethyl)trichlorosilane, CA 1558-25-4: 342.15 (1).

CH_2I_2

Diiodomethane, CA 75-11-6: Above 383.15 (1).

CH_2N_2

Cyanamide, CA 420-04-2: 413.71 (7); 414.26 (5,6).

CH_2O

Formaldehyde, CA 50-00-0: With 10% methanol, 337.15 (3); as 37% solution, 327.15 (5); 329.26 (1); 358.15 (6,7); as 37% solution with 15% methanol, 323.15 (6,7).

CH_2O_2

Formic acid, CA 64-18-6: 315.15 (5); 332.15 (4); 332.04* (2); 341.15 (1); 342.04 (3,6,7); as 90% solution, 323.15 (6).

CH_3AsCl_2

Dichloromethylarsine, CA 593-89-5: Above 378.15 (7).

CH_3Br

Bromomethane, CA 74-83-9: Nonflammable (1,4,6,7).

CH_3Cl

Chloromethane, CA 74-87-3: 227.59 (2,6); 233.15 (1).

CH_3ClO_2S

Methanesulfonyl chloride, CA 124-63-0: 383.15 (1).

CH_3Cl_2OP

Methyl dichlorophosphite, CA 3279-26-3: 298.15 (3); 313.15 (1).
Methylphosphonic dichloride, CA 676-97-1: Above 383.15 (1).

$CH_3Cl_2O_2P$

Methyl dichlorophosphate, CA 677-24-7: Above 385.93 (1).

CH_3Cl_3Ge

Methyltrichlorogermane, CA 993-10-2: Above 385.93 (1).

4

CH_3Cl_3Si

Methyltrichlorosilane, CA 75-79-6: 258.15 (1); 263.71 (6); 281.15 (3).

CH_3Cl_3Sn

Methyltin trichloride, CA 993-16-8: 258.15 (1).

Ch_3I

Iodomethane, CA 74-88-4: Nonflammable (1).

CH_3NO

Formamide, CA 75-12-7: 427.59* (1,2,6,7); 448.15 (4).

CH_3NO_2

Nitromethane, CA 75-52-7: 308.15 (1,2,4,6,7); 309.15 (3,5); 317.55* (4).

CH_4

Methane, CA 74-82-3: 50.59 (7); 85 (4); 85.93 (3).

CH_4BF_3O

Boron trifluoride-methanol complex, CA 373-57-9: 284.15 (1).

CH_4Cl_2Si

Dichloromethylsilane, CA 75-54-7: 241.15 (1,7); 263.15 (3,6).

CH_4N_2O

Formic acid hydrazide, CA 624-84-0: Above 385.93 (1).

CH_4O

Methanol, CA 67-56-1: 284.15 (1,2,3,4,5,6,7); 288.75* (4).

CH_4O_3S

Methanesulfonic acid, CA 75-75-2: Above 385.93 (1).

CH_4S

Methanethiol, CA 74-93-1: 255.15 (7); 255.37* (2).

CH_5N

Methylamine, CA 74-89-5: As 30% solution, 273.45* (2); as 40% solution, 262.59 (1); 260.93 to 277.59* (2).

CH_6N_2

Methylhydrazine, CA 60-34-4: 264.82 (6); 274.15* (4); 287.15 (3); 294.26 (1); 296.15 (7).

CN_4O_8

Tetranitromethane, CA 509-14-8: Above 385.93 (1).

CS_2

Carbon disulfide, CA 75-15-0: 239.26 (1); 243.15 (2,3,4,6,7).

$C_2Br_2O_2$

Oxalyl bromide, CA 15219-34-8: Nonflammable (1).

C_2ClF_3

Chlorotrifluoroethylene, CA 79-38-9: 245.37 (7).

C_2ClNO_2

N-(Chlorocarbonyl)isocyanate, CA 27738-96-1: Above 385.93 (1).

$C_2Cl_2O_2$

Oxalyl chloride, CA 79-37-8: Nonflammable (1).

C_2Cl_4

1,1,2,2-Tetrachloroethylene, CA 127-18-4: Nonflammable (1,7).

C_2HBr_3O

Tribromoacetaldehyde, CA 115-17-3: 338.71 (1).

C_2HCl_3

Trichloroethylene, CA 79-01-6: Nonflammable (1,4).

C_2HCl_3O

Dichloroacetyl chloride, CA 79-36-7: 339.26 (6,7); nonflammable (1).

$C_2HCl_3O_2$

Trichloroacetic acid, CA 76-03-9: Above 385.93 (1); nonflammable (7).

C_2H_2BrClO

Bromoacetyl chloride, CA 22118-09-8: Nonflammable (1).

6

C_2H_2BrN

Bromoacetonitrile, CA 590-17-0: Above 385.93 (1).

$C_2H_2Br_2$

1,2-Dibromoethylene, CA 540-49-8: Nonflammable (1).

$C_2H_2Br_2O_2$

Dibromoacetic acid, CA 631-64-1: Above 385.93 (1).

$C_2H_2Br_3Cl_2O_2P$

2,2,2-Tribromoethyl dichlorophosphate, CA 53676-22-5: Above 385.93 (1).

$C_2H_2Br_4$

1,1,1,2-Tetrabromoethane, CA 630-16-0: Above 385.93 (1).
1,1,2,2-Tetrabromoethane, CA 79-27-6: Nonflammable (1).

C_2H_2ClN

Chloroacetonitrile, CA 107-14-2: 320.15 (1); 327.15 (3).

C_2H_2ClNO

Chloromethyl isocyanate, CA 7093-91-6: 292.59 (1).

C_2H_2ClNS

Chloromethyl thiocyanate, CA 3268-79-9: 357.04 (1).

$C_2H_2Cl_2$

1,1-Dichloroethylene, CA 75-35-4: 245.15 (4,6); 250.37 (1); 255.37 (2); 255.37* (7); 257.15* (4) 263.15 (3,5).

1,2-Dichloroethylene, cis, CA 156-59-2: 277.04 (7); 279.15 (1,2,3,4,5).

1,2-Dichloroethylene, trans, CA 156-60-5: 275.38 (7); 277.15 (2,4); 279.15 (1,3,6).

1,2-Dichloroethylene, mixed isomers, CA 540-59-0: 275.37 (6); 279.26 (1).

$C_2H_2Cl_2O$

Chloroacetyl chloride, CA 79-04-9: Nonflammable (1,7).
2,2-Dichloroacetaldehyde, CA 79-02-7: 333.15 (7).

$C_2H_2Cl_2O_2$

Dichloroacetic acid, CA 79-43-6: Above 385.93 (1).

$C_2H_2Cl_2O_3S$

Chlorosulfonylacetyl chloride, CA 4025-77-8: Above 385.93 (1).

$C_2H_2Cl_5O_2P$

2,2,2-Trichloroethyl phosphorodichloridate, CA 18868-46-7: Above 385.93 (1).

C_2H_2FN

Fluoroacetonitrile, CA 503-20-8: 259.26 (1).

$C_2H_2F_2O_2$

Difluoroacetic acid, CA 381-73-7: 351.15 (1,4).

C_2H_2IN

Iodoacetonitrile, CA 624-75-9: 359.26 (1).

C_2H_3Br

Vinyl bromide, CA 593-60-2: Nonflammable (1).

C_2H_3BrO

Acetyl bromide, CA 506-96-7: 275.15 (3); above 385.93 (1).

$C_2H_3BrO_2$

Bromoacetic acid, CA 79-08-3: Above 385.93 (1).

C_2H_3Cl

Vinyl chloride, CA 75-01-4: 195.4* (4); 265.15* (7).

C_2H_3ClO

Acetyl chloride, CA 75-36-5: 269.15 (5); 277.59 (1,3,4,6,7).
2-Chloroacetaldehyde, CA 107-20-0: 360.93 (7).

C_2H_3ClOS

Methyl chlorothioformate, CA 18369-83-0: 310.93 (2).

$C_2H_3ClO_2$

Chloroacetic acid, 79-11-8: 399.15 (5,7); 423.15 (6).
Methyl chloroformate, CA 79-22-1: 283.15 (3); 285.37 (7); 290.93 (1,4);
297.55* (4).

$C_2H_3Cl_2NO_2$

1,1-Dichloro-1-nitroethane, CA 594-72-9: 348.71* (2,6,7); 349.15 (5).

$C_2H_3Cl_3$

1,1,1-Trichloroethane, CA 71-55-6: Nonflammable (1,2,4,6).

1,1,2-Trichloroethane, CA 79-00-5: Nonflammable (1,2,4).

$C_2H_3Cl_3O$

2,2,2-Trichloroethanol, CA 115-20-8: Above 385.93 (1).

$C_2H_3Cl_3Si$

Trichlorovinylsilane, CA 97-94-5: 263.71 (1,7); 283.15 (3); 294.26* (6).

$C_2H_3F_3O$

2,2,2-Trifluoroethanol, CA 75-89-8: 302.59 (1); 305.37* (2); 308.15 (3); 314.15* (4).

$C_2H_3F_3O_3S$

Methyl trifluoromethanesulfonate, CA 333-27-7: 311.48 (1).

C_2H_3N

Acetonitrile, CA 75-05-8: 275.15 (3,5); 278.71 (1,4); 278.71* (2,4,6,7).

C_2H_3NO

Methyl isocyanate, CA 624-83-9: Below 258.15 (7); 266.15 (1,6); 266.15* (4).

C_2H_3NS

Methyl isothiocyanate, CA 556-61-6: 305.37 (1); 310.15 (3).

Methyl thiocyanate, CA 556-64-9: 311.48 (1).

C_2H_4

Ethylene, CA 74-85-1: 137.04 (2).

$C_2H_4Br_2$

1,2-Dibromoethane, CA 106-93-4: Nonflammable (1,4,7).

C_2H_4ClN

2-Chloroacrylonitrile, CA 920-37-6: 281.15 (7).

$C_2H_4ClNO_2$

1-Chloro-1-nitroethane, CA 598-92-5: 329.15 (5); 329.26* (2,6).

$C_2H_4ClO_3P$

2-Chloro-2-oxo-1,3,2-dioxaphospholane, CA 6609-64-9: 286.15 (3).

$C_2H_4Cl_2$

1,1-Dichloroethane, CA 75-34-3: 261.15 (4); 263.15 (5); 267.59 (1,6); 267.59* (7); 278.15 (3).

1,2-Dichloroethane, CA 107-06-2: 286.15 (3,4,5,6,7); 288.71 (1); 290.15 (4); 294.15* (2,4).

$C_2H_4Cl_2O$

Bis(chloromethyl) ether, CA 542-88-1: Below 265.93 (7).

2,2-Dichloroethanol, CA 598-38-9: 351.48 (1).

Dichloromethyl methyl ether, CA 4885-02-3: 294.15 (3); 315.37 (1).

$C_2H_4Cl_2O_2S$

2-Chloro-1-ethanesulfonyl chloride, CA 1622-32-8: 383.15 (1).

$C_2H_4F_3N$

2,2,2-Trifluoroethylamine, CA 753-90-2: 256.48 (1).

C_2H_4O

Acetaldehyde, CA 75-07-0: 233.15 (1,6); 235.15 (4,7); 246.15 (3).

Ethylene oxide, CA 75-21-8: 244.26 (6); 253.15 (3,7); below 255.37* (2,4).

C_2H_4OS

Thioacetic acid, CA 507-09-5: 263.15 (3); 284.26 (1).

$C_2H_4O_2$

Acetic acid, CA 64-19-7: 313.15 (1,3,5,6); 315.93 (7); 317.59* (2); 330.15* (4).

Methyl formate, CA 107-31-3: 240.93 (1,2,4); 254.15 (3,6,7); as 99% solution, 246.48 (1).

$C_2H_4O_2S$

Mercaptoacetic acid, CA 68-11-1: Above 385.93 (1).

$C_2H_4O_3$

Peroxyacetic acid, CA 79-21-0: As 40% solution, 313.71* (7).

$C_2H_4O_3S$

Glycol sulfite, CA 3741-38-6: 352.59 (1).

$C_2H_4O_4$

Glyoxylic acid monohydrate, CA 563-96-2: Above 385.93 (1).

C_2H_4S

Ethylene sulfide, CA 420-12-2: 239.15 (3): 283.15 (1).

$C_2H_5AlCl_2$

Dichloroethylaluminum, CA 563-43-9: May ignite spontaneously in air (6).

C_2H_5Br

Bromoethane, CA 74-96-4: Below 253.15 (5,7); 253.15 (4); nonflammable (1,2,6).

C_2H_5BrO

2-Bromoethanol, CA 540-51-2: 313.71 (1).

Bromoethyl methyl ether, CA 13057-17-5: 299.82 (1).

C_2H_5Cl

Chloroethane, CA 75-00-3: 223.15 (3,4,6,7); 230.37* (2,4).

C_2H_5ClO

2-Chloroethanol, CA 107-07-3: 328.15 (3,5); 330.35 (4); 333.15 (1,6); 333.15* (2,7).

Chloromethyl methyl ether, CA 107-30-2: 288.71 (1).

$C_2H_5ClO_2S$

Ethanesulfonyl chloride, CA 594-44-5: Above 385.93 (1).

C_2H_5ClS

Chloromethyl methyl sulfide, CA 2373-51-5: 290.15 (1,3).

$C_2H_5Cl_2OP$

Ethyl dichlorophosphite, CA 1498-42-6: 277.59 (1).

Ethyl phosphonic dichloride, CA 1066-50-8: Above 385.93 (1).

$C_2H_5Cl_2OPS$

Ethyl dichlorothiophosphate, CA 1498-64-2: Above 385.93 (1).

$C_2H_5Cl_2O_2P$

Ethyl dichlorophosphate, CA 1498-51-7: Above 385.93 (1).

$C_2H_5Cl_3Si$

  Chloromethyldichloromethylsilane, CA 1558-33-4: 325.93 (1).

  Ethyltrichlorosilane, CA 115-21-9: 263.15 (1); 295.15 (3,5); 295.37* (6,7).

C_2H_5FO

  2-Fluoroethanol, CA 371-62-0: 304.26 (1); 307.15 (3).

$C_2H_5F_3N_2$

  2,2,2-Trifluoroethylhydrazine, CA 5042-30-8: As 70% solution, 315.93 (1).

C_2H_5I

  Iodoethane, CA 75-03-6: 305.15 (3); nonflammable (1).

C_2H_5IO

  2-Iodoethanol, CA 624-76-0: 338.71 (1).

C_2H_5N

  Ethyleneimine, CA 151-56-4: 260.15 (5); 262.04 (2,4,6,7).

C_2H_5NO

  Acetaldoxime, CA 107-29-9: 295.38 (7); 309.15 (3); 313.15 (1).

  N-Methylformamid, CA 123-39-7: Below 295.15 (7); 372.04 (1).

$C_2H_5NO_2$

  Ethyl nitrite, CA 109-95-5: 238.15 (6,7).

  Nitroethane, CA 79-24-3: 301.15 (3,5,6); 303.71 (1,2,4); 314.26* (2,4,7).

$C_2H_5NO_3$

  2-Nitroethanol, CA 625-48-9: Above 385.93 (1).

  Ethyl nitrate, CA 625-58-1: 283.15 (5,6,7).

C_2H_5OTl

  Thallous ethoxide, CA 20398-06-5: Above 385.93 (1).

C_2H_6

  Ethane, CA 74-84-0: 138.15 (2,4); 143.15 (7).

$C_2H_6BF_3O$

  Boron trifluoride dimethyl etherate, CA 353-42-4: 308.71 (1).

12

$C_2H_6BF_3S$

   Boron trifluoride methyl sulfide complex, CA 353-43-5: 257.59 (1).

$C_2H_6ClNO_2S$

   Dimethylsulfamoyl chloride, CA 13360-57-1: 367.59 (1).

$C_2H_6ClO_2PS$

   Dimethyl chlorothiophosphate, CA 2524-03-0: 378.15 (1).

$C_2H_6Cl_2Si$

   Dichlorodimethylsilane, CA 75-78-5: 257.04 (1); 264.15 (7); 268.15 (3).

   Ethyldichlorosilane, CA 1789-58-8: 272.04 (6).

$C_2H_6N_2O$

   N-Nitrosodimethylamine, CA 62-75-9: 334.26 (1).

C_2H_6O

   Dimethyl ether, CA 115-10-6: 232.04 (2,7).

   Ethanol, CA 64-17-5: 285.15 to 287.15 (3,4,5,6,7); 289.26* (2); 292.05* (4); as 5% solution, 333.15; as 10% solution, 320.15; as 20% solution, 308.65; as 30% solution, 302.15; as 40% solution, 298.65; as 50% solution, 297.15; as 60% solution, 295.65; as 70% solution, 294.15; as 80% solution, 292.65; as 90% solution, 290.65; as 95% solution, 289.15 (5).

C_2H_6OS

   Dimethyl sulfoxide, CA 67-68-5: 355.15 (3); 368.15 (1,5); 368.15* (2,4,6,7).

   2-Mercaptoethanol, CA 60-24-2: 347.04 (1,5); 347.04* (6,7); 349.82* (2); 350.15 (3).

$C_2H_6O_2$

   Ethylene glycol, CA 107-21-1: 383.15 to 389.15 (1,2,4,5,6,7); 388.75 to 391.48* (2,4).

$C_2H_6O_2S$

   Methyl sulfone, CA 67-71-0: 416.48 (1).

$C_2H_6O_2S_2$

   Methyl methanethiosulfonate, CA 2949-92-0: 360.93 (1).

$C_2H_6O_3S$

   Dimethyl sulfite, CA 616-42-2: 303.71 (1); 311.15 (3).

   Ethane sulfonic acid, CA 594-45-6: Above 385.93 (1).   *(continues)*

$C_2H_6O_3S$ *(continued)*

   Methyl methanesulfonate, CA 66-27-3: 377.59 (1).

$C_2H_6O_4S$

   Dimethyl sulfate, CA 77-78-1: 356.48 (1,3,5); 356.48* (6,7); 389.15* (4).
   2-Ethylsulfonylethanol, CA 107-36-8: 460.93* (2).

C_2H_6S

   Dimethyl sulfide, CA 75-18-3: 236.48 (1); 239.15 (7); 258.15 (3).
   Ethanethiol, CA 75-08-1: Below 255.15 (5,6); 255.93 (1,7); 255.37* (1),
   263.15 (3).

$C_2H_6S_2$

   Dimethyl disulfide, CA 624-92-0: 280.15 (7); 288.15 (3); 297.59 (1).
   1,2-Ethanedithiol, CA 540-63-6: 318.15 (3); 323.15 (1).

$C_2H_7BBr_2S$

   Dibromoborane-methyl sulfide complex, CA 55671-55-1: 299.82 (1).

$C_2H_7BCl_2$

   Dichloroborane-methyl sulfide complex, CA 63642-42-0: 297.04 (1).

$C_2H_7BO_3$

   Boric acid, ethyl ester, CA 51845-86-4: 284.26 (7).

C_2H_7ClSi

   Chlorodimethyl silane, CA 1066-35-9: 244.26 (1); 250.15 (3).

C_2H_7N

   Dimethylamine, CA 124-40-3: As 25% solution, 279.4* (2); as 40% solution,
   255.15 (3,7); 253.15 to 264.26* (2); 288.71 (1); as 60% solution, below
   253.15* (2).
   Ethylamine, CA 75-04-7: 226.48 (3); 236.15 (3); 256.48 (1,7); below
   266.48* (2); as 70% solution, 253.15 (3); 255.93 (1).

C_2H_7NO

   2-Aminoethanol, CA 141-43-5: 358.15 (4,5); 366.48 (1,2,3); 366.48* (2,7).

$C_2H_7O_3P$

   Dimethyl phosphite, CA 868-85-9: 302.59 (1); 369.26* (2).

14

C_2H_8BBrS

Monobromoborane-methyl sulfide complex, CA 55652-52-3: 283.15 (1).

C_2H_8BClS

Monochloroborane-methyl sulfide complex, CA 63348-81-2: 297.59 (1).

$C_2H_8N_2$

1,1-Dimethylhydrazine, CA 57-14-7: 255.15 (5); 258.15 (6,7); 258.15* (4); 263.15 (3); 274.26 (1).

1,2-Dimethylhydrazine, CA 540-73-8: Below 296.15 (7).

Ethylenediamine, CA 107-15-3: 304.15 (3); 307.04 (1,5,6); 313.15 (2,4); as 78% solution, 316.48* (2); as 76% solution, 338.71* (6).

$C_2H_8N_2O$

2-Hydroxyethylhydrazine, CA 109-84-2: 347.04 (1); 379.82 (7).

C_2H_9BS

Borane-methyl sulfide complex, CA 13292-87-0: 272.15 (3); 291.48 (1).

$C_2H_{10}BN$

Borane-dimethylamine complex, CA 74-94-2: 316.48 (1).

$C_2H_{10}B_2$

1,1-Dimethyldiborane, CA 16924-32-6: Below 263.15 (7).

1,2-Dimethyldiborane, CA 17156-88-6: Below 218.15 (7).

$C_3Cl_2N_2$

Dichloromalononitrile, CA 13063-43-9: Above 385.93 (1).

$C_3Cl_3NO_2$

Trichloroacetyl isocyanate, CA 3019-71-4: 338.71 (1).

C_3Cl_6O

Hexachloroacetone, CA 116-16-5: Nonflammable (1).

C_3F_6O

Hexafluoroacetone, CA 684-16-2: Nonflammable (7).

C_3HCl_5

Pentachlorocyclopropane, CA 6262-51-7: 373.71 (1).

C_3H_2BrNS

2-Bromothiazole, CA 3034-53-5:   336.48 (1); 340.15 (3).

C_3H_2ClN

2-Chloroacrylonitrile, CA 920-37-6:   279.82 (1); 283.15 (3).

$C_3H_2ClNO_2$

Chloroacetyl isocyanate, CA 4461-30-7:   334.82 (1).

$C_3H_2Cl_2O_2$

Malonyl dichloride, CA 1663-67-8:   320.37 (1).

$C_3H_2Cl_2O_3$

4,5-Dichloro-1,3-dioxolan-2-one, CA 3967-55-3:   Above 385.93 (1).

$C_3H_2F_6O$

1,1,1,3,3,3-Hexafluoro-2-propanol, CA 920-66-1:   Nonflammable (1).

$C_3H_2N_2$

Malononitrile, CA 109-77-3:   385.37 (1); 403.15* (7).

$C_3H_2N_2O_2$

Methylene diisocyanate, CA 4747-90-4:   358.15* (6).

$C_3H_2O_2$

Propiolic acid, CA 471-25-0:   332.04 (1): 337.15 (3).

$C_3H_2O_3$

Vinylene carbonate, CA 872-36-6:   345.93 (1).

C_3H_3Br

Propargyl bromide, CA 106-96-7:   260.93 (7); 283.15 (6); 291.15 (3,5); 291.48* (7).

$C_3H_3Br_2ClO$

2,3-Dibromopropionyl chloride, CA 18791-02-1:   339.82 (1).

$C_3H_3Br_2N$

2,3-Dibromopropionitrile ±, CA 4554-16-9:   350.93 (1).

C_3H_3Cl

Propargyl chloride, CA 624-65-7: 272.15 (3): below 288.15 (7); 292.04 (1).

$C_3H_3ClF_2O_2$

Methyl 2-chloro-2,2-difluoroacetate, CA 1514-87-0: 292.59 (1).

C_3H_3ClO

Acryloyl chloride, CA 814-68-6: 287.15 (3); 289.26 (1).

$C_3H_3ClO_2$

Vinyl chloroformate, CA 5130-24-5: 268.71 (1).

$C_3H_3ClO_3$

Chloroethylene carbonate, CA 3967-54-2: Above 385.93 (1).
Methyl oxalyl chloride, CA 5781-53-3: 319.82 (1).

$C_3H_3Cl_3$

1,2,3-Trichloropropylene, CA 96-19-5: 355.37* (7).

$C_3H_3Cl_3O$

2,3-Dichloropropionyl chloride, CA 7623-13-4: 383.15 (1).

1,1,1-Trichloroacetone, CA 921-03-9: 352.59 (1).

$C_3H_3Cl_3O_2$

Methyl trichloroacetate, CA 598-99-2: 345.93 (1).

$C_3H_3F_3O$

1,1,1-Trifluoroacetone, CA 421-50-0: 242.59 (1); 295.15 (3).

$C_3H_3F_3O_2$

Methyl trifluoroacetate, CA 431-47-0: 253.15 (3); 265.93 (1).

$C_3H_3F_5O$

2,2,3,3,3-Pentafluoro-1-propanol, CA 422-05-9: Nonflammable (1).

C_3H_3N

Acrylonitrile, CA 107-13-1: 268.15 (5); 268.15* (4); 272.15 (3,7); 273.15 (1); 273.15* (6).

C_3H_3NO

Isoxazole, CA 288-14-2: 282.04 (1); 285.15 (3).

Oxazole, CA 288-42-6: 292.04 (1).

Pyruvonitrile, CA 631-57-2: 287.59 (1).

$C_3H_3NO_2$

Cyanoacetic acid, CA 372-09-8: 380.15 (4); 380.93 (1).

C_3H_3NS

Thiazole, CA 288-47-1: 295.37 (1); 299.15 (3).

C_3H_4BrClO

2-Bromopropionyl chloride ±, CA 7148-74-5: 324.82 (1,3).

3-Bromopropionyl chloride, CA 15486-96-1: 352.59 (1).

$C_3H_4BrClO_2$

2-Bromoethyl chloroformate, CA 4801-27-8: 334.82 (1).

C_3H_4BrN

3-Bromopropionitrile, CA 2417-90-5: 370.37 (1).

$C_3H_4Br_2$

1,3-Dibromo-1-propene, mixed isomers, CA 627-15-6: 299.82 (1).

2,3-Dibromopropene, CA 513-31-5: 354.26 (1).

$C_3H_4Br_2O$

2-Bromopropionyl bromide, CA 563-76-8: 314.15 (3); above 385.93 (1).

2,3-Dibromo-2-propen-1-ol, mixed isomers, CA 7228-11-7: 377.15 (3).

C_3H_4ClN

2-Chloropropionitrile ±, CA 1617-17-0: 306.48 (1).

3-Chloropropionitrile, CA 542-76-7: 348.71 (1,6,7); 356.15 (3).

C_3H_4ClNO

2-Chloroethyl isocyanate, CA 1943-83-5: 329.82 (1).

$C_3H_4Cl_2$

1,1-Dichloro-1-propene, CA 563-58-6: 273.71 (1); 308.15 (3).

1,3-Dichloropropene, mixed isomers, CA 542-75-6: 300.93 (1); 308.15 (3,6,7).

2,3-Dichloro-1-propene, CA 78-88-6: 283.15 (1,7); 288.15 (6); 295.15 (3).

$C_3H_4Cl_2O$

2-Chloropropionyl chloride, CA 7623-09-8: 304.26 (1).

3-Chloropropionyl chloride, CA 625-36-5: 334.82 (1).

1,1-Dichloroacetone, CA 513-88-2: 297.59 (1); 316.15 (3).

1,3-Dichloroacetone, CA 534-07-6: 362.59 (1).

$C_3H_4Cl_2O_2$

1-Chloroethyl chloroformate, CA 50893-53-3: 313.71 (1).

2-Chloroethyl chloroformate, CA 627-11-2: 343.15 (1).

2,2-Dichloropropionic acid, CA 75-99-0: Above 385.93 (1).

Methyl dichloroacetate, CA 116-54-1: 337.15 (3); 353.15 (1).

$C_3H_4F_3NO$

N-Methyltrifluoroacetamide, CA 815-06-5: 347.04 (1).

$C_3H_4N_2$

Imidazole, CA 288-32-4: 418.15 (1).

$C_3H_4N_2O$

2-Cyanoacetamide, CA 107-91-5: 488.15 (1).

C_3H_4O

Acrolein, CA 107-02-8: 247.15 (3,4,6); 248.15 (7); 254.26 (1); 255.15* (4).

Methoxyacetylene, CA 6443-91-0: Below 253.15 (7).

Propargyl alcohol, CA 107-19-7: 306.15* (7); 309.15 (1,6); 309.15* (4,6).

$C_3H_4O_2$

Acrylic acid, CA 79-10-7: 319.15 (3); 323.15* (2,6); 327.59 (1); 327.59* (7); 341.15* (4).

beta-Propiolactone, CA 57-57-8: 343.15 (1); 347.15 (2,5,6).

Vinyl formate, CA 692-45-5: Below 273.15 (7).

$C_3H_4O_3$

Ethylene carbonate, CA 96-49-1: 416.48* (6,7); 433.15 (1).

Pyruvic acid, CA 127-17-3: 357.04 (1).

$C_3H_4S_3$

Ethylene trithiocarbonate, CA 822-38-8: 436.48 (1).

C_3H_5Br

Allyl bromide, CA 106-95-6: 270.93 (1); 254.82 (7); 272.15 (2,3,5,6,7).

1-Bromo-1-propene, mixed isomers, CA 590-14-7: 277.59 (1).

2-Bromopropene, CA 557-93-7: 277.59 (1).

Cyclopropyl bromide, CA 4333-56-6: 266.48 (1); 318.15 (3).

C_3H_5BrO

Epibromohydrin, CA 3132-64-7: Below 267.59 (7); 329.26 (1); 332.15 (3).

$C_3H_5BrO_2$

Bromomethyl acetate, CA 96-32-2: 330.37 (1).

2-Bromopropionic acid ±, CA 598-72-1: 373.15 (1).

3-Bromopropionic acid, CA 590-92-1: 338.71 (1).

Methyl bromoacetate, CA 96-32-2: 335.93 (1); 337.15 (3).

$C_3H_5Br_2Cl$

1,2-Dibromo-3-chloropropane, CA 96-12-8: 349.82* (7).

C_3H_5Cl

Allyl chloride, CA 107-05-1: 241.48 (3,4,6,7); 244.26 (1); 277.15 (2).

1-Chloro-1-propene, CA 590-21-6: Below 243.59 (7); below 267.04 (6,7).

2-Chloro-1-propene, CA 557-98-2: Below 238.71 (1); below 253.15 (3,5,6).

C_3H_5ClO

Chloroacetone, CA 78-95-5: 300.93 (1); 308.15 (3).

Epichlorohydrin, CA 106-89-8: 294.15 (7); 301.15 (5); 304.26* (6); 305.15 (3); 307.04 (1); 313.75* (2,4).

Propionyl chloride, CA 79-03-8: 285.15 (1,3,5,6,7).

C_3H_5ClOS

Ethyl chlorothioformate, CA 2941-64-2: 303.71 (1); 324.82 (2).

$C_3H_5ClO_2$

2-Chloropropionic acid, CA 598-78-7: 374.82 (1); 380.37 (6,7).

3-Chloropropionic acid, CA 107-94-8: Above 385.93 (1).

Ethyl chloroformate, CA 541-41-3: 275.37 (1,7); 289.15 (3,5,6); 291.45 (4); 300.95* (4).

Methoxyacetyl chloride, CA 38870-89-2: 302.04 (1).

Methyl chloroacetate, CA 96-34-4: 320.15 (3,5); 324.82 (1); 330.37* (6).

$C_3H_5Cl_2NO_2$

1,1-Dichloro-1-nitropropane, CA 595-44-8: 339.15 (5); 339.26* (2,6).

N,N-Dichlorourethane, CA 13698-16-3: 369.82 (1).

$C_3H_5Cl_3$

1,2,3-Trichloropropane, CA 96-18-4: 347.15 (2,3,5); 355.37 (1); 355.37* (6,7).

$C_3H_5Cl_3O$

1,1,1-Trichloro-2-propanol, CA 76-00-6: 355.37 (1).

$C_3H_5Cl_3Si$

Allyltrichlorosilane, CA 107-37-9: 304.26 (1); 308.15* (6,7).

C_3H_5FO

Epifluorohydrin, 90%, CA 503-09-3: 277.59 (1).

Fluoroacetone, CA 430-51-3: 280.37 (1).

$C_3H_5F_3O_3S$

Ethyl trifluoromethanesulfonate, CA 425-75-2: 308.71 (1).

C_3H_5I

Allyl iodide, CA 556-56-9: 289.15 (3); 291.48 (1).

C_3H_5N

Propargylamine, CA 2450-71-7: 277.04 (1); 279.15 (3).

Propionitrile, CA 107-12-0: 275.15 (5,6,7); 279.26 (1); 282.15 (3).

C_3H_5NO

Ethyl isocyanate, CA 109-90-0: 263.15 (3); 266.48 (1).

3-Hydroxypropionitrile, CA 109-78-4: Above 385.93 (1,4); 402.59* (6,7).

Lactonitrile ±, CA 78-97-7: 350.15 (1,2,3,4,5,6,7).

Methoxyacetonitrile, CA 1738-36-9: 305.15 (1,3).

C_3H_5NS

Ethyl isothiocyanate, CA 542-85-8: 297.15 (3); 305.37 (1).

(Methylthio)acetonitrile, CA 35120-10-6: 340.93 (1).

$C_3H_5N_3$

3-Aminopyrazole, CA 1820-80-0: Above 385.93 (1).

$C_3H_5N_3O_9$

Nitroglycerine, CA 55-63-0: Explodes (6).

C_3H_6

Propylene, CA 115-07-1: 165.37 (3,7).

C_3H_6BrCl

1-Bromo-3-chloropropane, CA 109-70-6: Nonflammable (1,3); 318.15 (3).
2-Bromo-1-chloropropane, CA 3017-95-6: Above 385.93 (1).
Propylene chlorobromide, mixed isomers: Nonflammable (2).

$C_3H_6BrNO_2$

2-Bromo-2-nitropropane, CA 5447-97-2: 334.16 (1).

$C_3H_6Br_2$

1,2-Dibromopropane ±, CA 78-75-1: Nonflammable (2).
1,3-Dibromopropane, CA 109-64-8: Nonflammable (2); 327.59 (1).
2,2-Dibromopropane, CA 594-16-1: Above 385.93 (1).

$C_3H_6Br_2O$

1,3-Dibromo-2-propanol, CA 96-21-9: 319.82 (1).
2,3-Dibromo-1-propanol, CA 96-13-9: Above 385.93 (1).

C_3H_6ClI

1-Chloro-3-iodopropane, CA 6940-76-7: Above 385.93 (1).

C_3H_6ClNO

Dimethylcarbamoyl chloride, CA 79-44-7: 341.48 (1).

C_3H_6ClNS

Dimethylthiocarbamoyl chloride, CA 16420-13-6: 371.48 (1).

$C_3H_6ClNO_2$

1-Chloro-1-nitropropane, CA 600-25-9: 335.15 (5); 335.37* (2,6,7).
2-Chloro-2-nitropropane, CA 594-71-8: 330.15 (5); 330.37* (2,6,7).

$C_3H_6Cl_2$

1,1-Dichloropropane, CA 78-99-9: 294.15 (7).
1,2-Dichloropropane, CA 78-87-5: 277.59 (1); 288.15 (3,5,6,7); 294.15* (2).
1,3-Dichloropropane, CA 142-28-9: 293.15 (3); 294.15 (7); 305.37 (1).
2,2-Dichloropropane, CA 594-20-7: 268.15 (1); 293.15 (3).

$C_3H_6Cl_2O$

2,2-Dichloroethyl methyl ether, CA 34862-07-2: 307.04 (1).

2,3-Dichloro-1-propanol, CA 616-23-9: 364.15* (4); 366.48 (6).

1,3-Dichloro-2-propanol, CA 96-23-1: 347.15 (2,5); 347.15* (6,7); 358.15 (1,3).

$C_3H_6Cl_2O_2S$

3-Chloropropanesulfonyl chloride, CA 1633-82-5: Above 385.93 (1).

$C_3H_6Cl_2Si$

Dichloromethylvinylsilane, CA 124-70-9: 272.15 (3); 277.59 (1).

$C_3H_6F_2O$

1,3-Difluoro-2-propanol, CA 453-13-4: 315.37 (1).

$C_3H_6I_2$

1,3-Diiodopropane, CA 627-31-6: Above 385.93 (1).

$C_3H_6N_2$

Dimethyl cyanamide, CA 1467-79-4: 331.48 (1); 338.71 (7); 344.15 (5,6).

C_3H_6O

Acetone, CA 67-64-1: Below 253.15 (5); 253.15 (2,6); 255.38 (1,4,7); 254.15 (3); 254.26* (2); 257.15* (4).

Allyl alcohol, CA 107-18-6: 294.15 (3,4,5,6,7); 295.37 (1,2); 305.37* (2).

Methyl vinyl ether, CA 107-25-5: 217.04* (2); 217.15 (4); 293.59 (7).

Propionaldehyde, CA 123-38-6: 243.15 (6); 246.48 (1); below 255.37* (2); 263.71 to 265.93* (7); 264.15 (3).

Propylene oxide, CA 75-56-9: 235.93 (1,2,3,6); 235.93* (2,7); 238.15 (4).

Trimethylene oxide, CA 503-30-0: 244.82 (1); 253.15 (3).

$C_3H_6O_2$

Acetol, technical, CA 116-09-6: 329.26 (1).

1,3-Dioxolane, CA 646-06-0: 270.15 (3); as 99% solution, 270.37 (1); 275.15 (1,5,7); 274.82* (6).

Ethyl formate, CA 109-94-4: 253.15 to 254.15 (1,3,4,5,6,7).

Glycidol, CA 556-52-5: 339.15 (3); 354.26 (1).

Methyl acetate, CA 79-20-9: 258.15* (2); 263.15 (1,4,5,6,7); 289.15 (3).

Propionic acid, CA 79-09-4: 324.82 (1,6); 327.15 (3,4); 329.82* (2).

$C_2H_6O_2S$

3-Mercaptopropionic acid, CA 107-96-0: 367.04 (1).

(Methylthio)acetic acid, CA 2444-37-3: 383.15 (3).

Methyl thioglycolate, CA 2365-48-2: 303.15 (1); 329.15 (3).

Methyl vinyl sulfone, CA 3680-02-2: Above 385.93 (1).

Thiolactic acid, CA 79-42-5: 360.93 (1).

$C_3H_6O_3$

Dimethyl carbonate, CA 616-38-6: 289.15 (3,4); 291.48 (1); 292.04* (6,7); 294.85* (4).

Glyceraldehyde, CA 453-17-8: Above 385.93 (1).

Lactic acid, CA 50-21-5: Above 385.93 (1).

Methoxyacetic acid, CA 625-45-6: Above 385.93 (1).

1,3,5-Trioxane, CA 110-88-3: 318.15 (1,5); 318.15* (6,7).

$C_3H_6O_3S$

1,3-Propane sulfone, CA 1120-71-4: Above 385.93 (1).

C_3H_6S

Allyl mercaptan, CA 870-23-5: 263.15 (7); 291.15 (3); 294.26 (1).

Propylene sulfide, CA 1072-43-1: 263.15 (3); 283.15 (1).

Trimethylene sulfide, CA 287-27-4: 253.15 (3); 270.37 (1).

C_3H_7Br

1-Bromopropane, CA 106-94-5: Below 267.59 (7); 287.15 (3); 298.71 (1).

2-Bromopropane, CA 75-26-3: Nonflammable (2); 252.15 (3); below 263.15 (7); 292.59 (1).

C_3H_7BrO

2-Bromoethyl methyl ether, CA 6482-24-2: 301.48 (1).

3-Bromo-1-propanol, CA 627-18-9: 338.15 (3); 367.04 (1).

1-Bromo-2-propanol, CA 19686-73-6: 316.15 (3); as 70% solution, 327.59 (1).

$C_3H_7BrO_2$

3-Bromo-1,2-propanediol, CA 4704-77-2: 383.15 (1).

C_3H_7Cl

1-Chloropropane, CA 540-54-5: Below 253.15 (5); 255.15 (3); 291.48 (1).

2-Chloropropane, CA 75-29-6: 238.15 (1); 241.15 (6,7); 252.15 (3).

24

C_3H_7ClO

Chloroethyl methyl ether, CA 627-42-9: 283.15 (3); 288.15 (1).

Chloromethyl ethyl ether, CA 3188-13-4: Below 254.15 (7); 292.59 (1).

2-Chloro-1-propanol, CA 78-89-7: 324.82 (6,7); for (S)-(+)-isomer, 317.59 (1).

3-Chloro-1-propanol, CA 627-30-5: 346.48 (1).

1-Chloro-2-propanol, CA 127-00-4: 324.82 (1,3,4,5); 324.82* (2,6).

$C_3H_7ClO_2$

3-Chloro-1,2-propanediol, CA 96-24-2: 332.04 (1); 410.93* (2,4).

$C_3H_7ClO_2S$

1-Propanesulfonyl chloride, CA 10147-36-1: 370.15 (3).

2-Propanesulfonyl chloride, CA 10147-37-2: 359.26 (1).

$C_3H_7ClO_3S$

2-Chloroethyl methanesulfonate, CA 3570-58-9: Above 385.93 (1).

C_3H_7ClS

2-Chloroethyl methyl sulfide, CA 542-81-4: 315.37 (1,3).

3-Chloro-1-propanethiol, CA 17481-19-5: 316.48 (1).

$C_3H_7Cl_2OP$

Propylphosphonic dichloride, CA 4708-04-7: Above 385.93 (1).

$C_3H_7Cl_3Si$

Propyltrichlorosilane, CA 141-57-1: 309.82 (6); 310.93 (7).

C_3H_7I

1-Iodopropane, CA 107-08-4: 317.59 (1); 341.15 (3).

2-Iodopropane, CA 75-30-9: 275.15 (3); 315.37 (1).

C_3H_7N

Allylamine, CA 107-11-9: 244.26 (1,6,7); 274.15 (3).

Azetidine, CA 503-29-7: As 90% solution, 252.59 (1).

Cyclopropylamine, CA 765-30-0: 247.59 (1); 274.15 (3,7).

2-Methylaziridine, CA 75-55-8: 258.15 (1); 263.15 (7).

Propyleneimine, CA 75-55-8: 263.15 (3).

Trimethyleneimine, CA 503-29-7: 297.15 (3).

C_3H_7NO

N,N-Dimethylformamide, CA 68-12-2: 330.93 (1,4,5,6,7); 340.15* (2,4); 342.15 (3); as 90% solution, 347.04; as 70% solution, 374.82; as 60% solution, nonflammable (2).

N-Methylacetamide, CA 79-16-3: 381.48 (1).

C_3H_7NOS

O-Ethyl thiocarbamate, CA 625-57-0: 383.15 (1).

$C_3H_7NO_2$

Isopropyl nitrite, CA 541-42-4: Below 283.15 (7).

Methyl N-hydroxyacetimidate, CA 41599-36-4: 320.37 (1).

1-Nitropropane, CA 108-03-2: 307.04 (1,7); 308.75 (2,3,4,6); 322.04* (2,4).

2-Nitropropane, CA 79-46-9: 297.04 (6); 299.15 (3); 300.93 (2,4,7); 310.95* (2,4); 312.15* (4).

Urethane, CA 51-79-6: 365.37 (1).

$C_3H_7NO_3$

Isopropyl nitrate, CA 1712-64-7: 284.15 (7); 285.93 (1,3).

2-Nitro-1-propanol, CA 2902-96-7: 373.15 (1).

Propyl nitrate, CA 627-13-4: 293.15 (6,7).

C_3H_7NS

N,N-Dimethylthioformamide, CA 758-16-7: 372.59 (1).

Thiazolidine, CA 504-78-9: 329.26 (1).

C_3H_8

Propane, CA 74-98-6: 168.71 (2,7); 169 (4).

$C_3H_8BrClSi$

(Bromomethyl)chlorodimethylsilane, CA 16532-02-8: 314.26 (1); 323.15 (3).

$C_3H_8Cl_2Si$

Chloro(chloromethyl)dimethylsilane, CA 1719-57-9: 294.26 (1); 295.15 (3).

$C_3H_8N_2O_2$

Ethyl carbazate, CA 4114-31-2: 359.26 (1).

C_3H_8O

Ethyl methyl ether, CA 540-67-0: 235.93 (6,7).

1-Propanol, CA 71-23-8: 288.15 (1,2,7); 295.15 (5); 297.15 (3,6); 298.15 (4); 302.15* (4); 305.37* (2); as 80% solution, 302.15;   *(continues)*

26

C₃H₈O *(continued)*

as 60% solution, 304.15; as 40% solution, 305.15; as 20% solution, 307.15; as 10% solution, 314.15 (5).

2-Propanol, CA 67-63-0: 285.15 (3,4,5,6,7); 288.75* (4); 290.35* (4); 294.26* (2); as 88% solution, 287.04 (6); 291.45 (4); 297.04* (2); as 80% solution, 290.15; as 60% solution, 292.15; as 40% solution, 293.15; as 20% solution, 302.15; as 10% solution, 312.15 (5).

C₃H₈OS

2-(Methylthio)ethanol, CA 5271-38-5: 343.15 (1).

C₃H₈OS₂

2,3-Dimercapto-1-propanol, CA 59-52-9: Above 385.93 (1).

Methyl methylsulfinylmethyl sulfide, CA 33577-16-1: Above 385.93 (1).

C₃H₈O₂

Dimethoxymethane, CA 109-87-5: 240.93* (6); below 253.15 (5); 255.15 (1,3,7); 255.37* (2).

2-Methoxyethanol, CA 109-86-4: 312.15 (2,4,5,6); 319.26 (1); 319.26* (2,7); 316.45* (4); 325.15 (3).

1,2-Propanediol ±, CA 57-55-6: 372.15 (3,5,6); 372.15* (2,4,7); 374.15 (4); 377.59 (2); 380.37 (1).

1,3-Propanediol, CA 504-63-2: 352.59 (1).

C₃H₈O₂S

3-Mercapto-1,2-propanediol, CA 96-27-5: Above 385.93 (1).

C₃H₈O₃

Glycerol, CA 56-81-5: 433.15 (1,2,5,7); 449.85* (2,4); 472.15 (4,6).

C₃H₈O₃S

Ethyl methanesulfonate, CA 62-50-0: 373.15 (1).

C₃H₈S

Ethyl methyl sulfide, CA 624-89-5: 258.15 (1); 273.15 (3).

1-Propanethiol, CA 107-03-9: 252.59 (1,7); 258.15* (2); 263.15 (3).

2-Propanethiol, CA 75-33-2: Below 238.71 (1,7); 238.71* (2); 263.15 (3).

C₃H₈S₂

Bis(methylthio)methane, CA 1618-26-4: 317.04 (1,3).

1,2-Propanedithiol, CA 814-67-5: 335.15 (3).

1,3-Propanedithiol, CA 109-80-8: 332.04 (1); 335.15 (3).

$C_3H_9BO_3$

Trimethyl borate, CA 121-43-7: 264.82 (1); 270.15 (3).

$C_3H_9B_3O_6$

Trimethoxyboroxine, CA 102-24-9: 283.15 (1).

C_3H_9BrGe

Bromotrimethylgermane, CA 1066-37-1: 310.37 (1).

C_3H_9BrSi

Bromotrimethylsilane, CA 2857-97-8: 298.15 (3); 305.37 (1).

C_3H_9BrSn

Trimethyltin bromide, CA 1066-44-0: 383.15 (1).

C_3H_9ClGe

Chlorotrimethylgermane, CA 1529-47-1: 274.82 (1).

$C_3H_9ClO_3SSi$

Trimethylsilyl chlorosulfonate, CA 4353-77-9: 315.15 (3).

C_3H_9ClSi

Chlorotrimethylsilane, CA 75-77-4: 245.37 (1,6,7); 253.15 (7); 255.15 (3).

C_3H_9ClSn

Trimethyltin chloride, CA 1066-45-1: 370.37 (1).

C_3H_9ISi

Iodotrimethylsilane, CA 16029-98-4: 241.48 (1); 293.15 (3).

C_3H_9N

N-Ethylmethylamine, CA 624-78-2: 275.15 (3).

Isopropylamine, CA 75-31-0: 235.93* (6,7); 239.26 (2); 240.93 (1); 247.15 (3,4).

Propylamine, CA 107-10-8: 235.93 (1,3,4,6,7); 243.15 (2).

Trimethylamine, CA 75-50-3: 253.15 (3); 266.48 (1,7); as 25% solution, 276.48 (1); 269.26 to 283.15* (2).

C_3H_9NO

1-Amino-2-propanol ±, CA 78-96-6: 344.15 (5); 347.04 (1); 350.38 (6,7); 347.04* (2).

(continues)

28

C_3H_9NO *(continued)*

1-Amino-2-propanol, (R)-(-), CA 2799-16-8: 344.15 (3); 347.04 (1).

1-Amino-2-propanol, (S)-(+), CA 2799-17-9: 349.26 (1).

2-Amino-1-propanol ±, CA 6168-72-5: 357.04 (1).

2-Amino-1-propanol, (R)-(-), CA 35320-23-1: 357.04 (1).

2-Amino-1-propanol, (S)-(+), CA 2749-11-3: 335.93 (1).

3-Amino-1-propanol, CA 156-87-6: 352.59 (1); 352.59* (6,7).

2-Methoxyethylamine, CA 109-85-3: 282.59 (1,3); 288.71 (2).

2-(Methylamino)ethanol, CA 109-83-1: 344.26 (1); 345.93 (1); 347.04* (6,7); 347.15 (5); 363.15 (3).

$C_3H_9NO_2$

3-Amino-1,2-propanediol, CA 616-30-8: Above 385.93 (1).

C_3H_9NS

2-(Methylthio)ethylamine, CA 18542-42-2: 309.15 (3).

$C_3H_3N_3Si$

Azidotrimethylsilane, CA 4648-54-8: 296.48 (1).

$C_3H_9O_3P$

Dimethyl methylphosphonate, CA 756-79-6: 316.48 (1).

Trimethyl phosphite, CA 121-45-9: 300.93 (1,3); 310.93* (2); 327.59* (2,6,7).

$C_3H_9O_4P$

Trimethyl phosphate, CA 512-56-1: Nonflammable (1).

Tris(hydroxymethyl)phosphine oxide, CA 1067-12-5: Above 385.93 (1).

$C_3H_{10}N_2$

1,2-Diaminopropane, CA 78-90-0: 297.15 (7); 306.48 (1,5); 306.48* (6,7); 316.15* (4); 322.04* (2); 322.15 (3).

1,3-Diaminopropane, CA 109-76-2: 297.04* (6,7); 297.15 (3); 322.04 (1); 322.04* (7); 323.15* (4).

N-Methylethylenediamine, CA 109-81-9: 300.15 (3); 314.26 (1).

$C_3H_{10}N_2O$

1,3-Diamino-2-propanol, CA 616-29-5: Above 385.93 (1); 405.37 (6).

$C_3H_{10}O_3Si$

Trimethoxysilane, CA 2487-90-3: 268.71 (1); 282.15 (3).

$C_3H_{11}NOSi$

O-(Trimethylsilyl)hydroxylamine, CA 22737-36-6: 282.04 (1).

$C_4Cl_2F_4O_3$

Chlorodifluoroacetic anhydride, CA 2834-23-3: Nonflammable (1).

C_4Cl_4S

Tetrachlorothiophene, CA 6012-97-1: Above 385.93 (1).

C_4Cl_6

Hexachloro-1,3-butadiene, CA 87-68-3: Nonflammable (1).

$C_4Cl_6O_3$

Trichloroacetic anhydride, CA 4124-31-6: Nonflammable (1).

$C_4F_6O_3$

Trifluoroacetic anhydride, CA 407-25-0: Nonflammable (1).

C_4HBrO_3

Bromomaleic anhydride, CA 5926-51-2: Above 385.93 (1).

$C_4HCl_3N_2$

2,4,6-Trichloropyrimidine, CA 3764-01-0: Above 385.93 (1).

$C_4HF_7O_2$

Heptafluorobutyric acid, CA 375-22-4: Nonflammable (1).

$C_4H_2Br_2S$

2,5-Dibromothiophene, CA 3141-27-3: 372.59 (1).
3,4-Dibromothiophene, CA 3141-26-2: Above 385.93 (1).

$C_4H_2Cl_2N_2$

2,6-Dichloropyrazine, CA 4774-14-5: 383.15 (1).

$C_4H_2Cl_2O_2$

Fumaryl chloride, CA 627-63-4: 347.04 (1); 350.15 (3).

$C_4H_2Cl_2O_3$

Mucochloric acid, CA 87-56-9: 373.15 (1).

$C_4H_2Cl_2S$

2,5-Dichlorothiophene, CA 3172-52-9: 332.59 (1).

$C_4H_2Cl_4O_3$

Dichloroacetic anhydride, CA 4124-30-5: Above 385.93 (1).

$C_4H_2F_6O_2$

2,2,2-Trifluoroethyl trifluoroacetate, CA 407-38-5: 273.15 (1).

$C_4H_2O_3$

Maleic anhydride, CA 108-31-6: 374.82 (4,6,7); 376.48 (1,5); 383.15* (4).

C_4H_3BrO

3-Bromofuran, technical, CA 22037-28-1: 276.48 (1).

C_4H_3BrS

2-Bromothiophene, CA 1003-09-4: 325.15 (3); 333.15 (1).

3-Bromothiophene, CA 872-31-1: 325.15 (3); 329.82 (1).

$C_4H_3ClN_2$

Chloropyrazine, CA 14508-49-7: 330.93 (1).

$C_4H_3ClO_2S_2$

2-Thiophenesulfonyl chloride, CA 16629-19-9: Above 385.93 (1).

C_4H_3ClS

2-Chlorothiophene, CA 96-43-5: 295.93 (1,3).

$C_4H_3Cl_3O_2$

3-Hydroxy-4,4,4-trichlorobutyric acid *beta*-lactone, CA 5895-35-2: Above 385.93 (1).

$C_4H_3F_5O_2$

Methyl pentafluoropropionate, CA 378-75-6: 278.71 (1).

$C_4H_3F_7O$

2,2,3,3,4,4,4-Heptafluoro-1-butanol, CA 375-01-9: 293.15 (1).

C_4H_3IS

2-Iodothiophene, CA 3437-95-4: 344.26 (1).

$C_4H_3NO_2S$

2-Nitrothiophene, technical, CA 609-40-5:   As 85% solution, 367.04 (1).

C_4H_4

1-Buten-3-yne, CA 689-97-4:   Below 224.26 (7).

C_4H_4BrN

3-Bromo-2-methylacrylonitrile, mixed isomers: 343.71 (1).

$C_4H_4Cl_2$

2,3-Dichloro-1,3-butadiene, CA 1653-19-6:   283.15 (6).
1,4-Dichloro-2-butyne, CA 821-10-3:   433.15 (1).
3,4-Dichlorocyclobutene, *cis*, CA 2957-95-1:   328.15 (3).

$C_4H_4Cl_2O_2$

Succinyl chloride, CA 543-20-4:   349.82 (1).

$C_4H_4Cl_2O_3$

Chloroacetic anhydride, technical, CA 541-88-8:   Above 385.93 (1).

$C_4H_4Cl_2O_4$

Ethylene, bis(chloroformate), CA 124-05-0:   399.15 (4); 407.15* (4).

$C_4H_4Cl_7O_3P$

Bis(2,2,2-trichloroethyl) phosphorochloridate, CA 17672-53-6:   Above 385.93 (1).

$C_4H_4F_6O$

2,2,2-Trifluoroethyl ether, CA 333-36-8:   274.82 (1).

$C_4H_4I_2O_3$

Iodoacetic anhydride, CA 54907-61-8:   Above 385.93 (1).

$C_4H_4N_2$

Pyrazine, CA 290-37-9:   328.71 (1).
Pyridazine, CA 289-80-5:   358.15 (1).
Pyrimidine, CA 289-95-2:   304.26 (1).
Succinonitrile, CA 110-61-2:   405.37 (6,7).

32

C_4H_4O

3-Butyn-2-one, CA 1423-60-5: 272.04 (1).

Furan, CA 110-00-9: 237.59 (1,2,3,4,7).

$C_4H_4O_2$

Diketene, CA 674-82-8: 307.04 (6); 307.04* (7); 309.82 (1).

2(5H)-Furanone, CA 497-23-4: 374.26 (1).

Methyl propiolate, CA 922-67-8: 283.15 (1); 289.15 (3).

C_4H_4S

Thiophene, CA 110-02-1: 264.15 (5); 266.48* (2,4); 267.15 (7); 272.04 (1,3,4,6).

$C_4H_5BrO_2$

alpha-Bromo-gamma-butyrolactone, CA 5061-21-2: Above 385.93 (1).

C_4H_5Cl

1-Chloro-1,3-butadiene, CA 627-22-5: 253.15 (7).

2-Chloro-1,3-butadiene, CA 126-99-8: 253.15 (6,7); 253.15* (4).

C_4H_5ClO

Crotonyl chloride, technical, CA 10487-71-5: As 90% solution, 308.15 (1).

Cyclopropanecarboxylic acid chloride, CA 4023-34-1: 296.43 (1); 304.15 (3).

Methacryloyl chloride, CA 920-46-7: 295.15 (3); as 90% solution, 275.93 (1).

$C_4H_5ClO_2$

Allyl chloroformate, CA 2937-50-0: 300.95* (4); 304.26 (1,3,4,5,6,7).

$C_4H_5ClO_3$

Acetoxyacetyl chloride, CA 13831-31-7: 344.26 (1).

Ethyl oxalyl chloride, CA 4755-77-5: 314.82 (1).

$C_4H_5Cl_3O_2$

Ethyl trichloroacetate, CA 515-84-4: 338.15 (1); 346.15 (3).

$C_4H_5F_3OS$

S-Ethyl trifluorothioacetate, CA 383-64-2: 277.04 (1); 293.15 (3).

$C_4H_5F_3O_2$

Ethyl trifluoroacetate, CA 383-63-1: 256.15 (3); 272.04 (1).

C_4H_5N

Allyl cyanide, CA 109-75-1: 292.15 (3,7); 297.04 (1).

Crotononitrile, mixed isomers, CA 4786-20-3: 289.15 (3,7); 293.15 (1).

Cyclopropyl cyanide, CA 5500-21-0: 305.93 (1); 313.15 (3).

Methacrylonitrile, CA 126-98-7: 272.15 (3); 274.26 (6); 285.37 (1,7).

Pyrrole, CA 109-97-7: 306.48 (1); 309.15 (3); 312.15 (2,4,5,6,7).

C_4H_5NO

Allyl isocyanate, CA 1476-23-9: 316.48 (1).

5-Methylisoxazole, CA 5765-44-6: 301.15 (3); 302.59 (1).

4-Methyloxazole, CA 693-93-6: 287.59 (1).

$C_4H_5NO_2$

Ethyl cyanoformate, CA 623-49-4: 297.59 (1).

Methyl cyanoacetate, CA 105-34-0: 383.15 (1,3).

Methyl isocyanoacetate, CA 39687-95-1: 357.15 (3); 362.59 (1).

$C_4H_5NO_2S$

Ethoxycarbonyl isothiocyanate, CA 16182-04-0: 323.71 (1).

C_4H_5NS

Allyl isothiocyanate, CA 57-06-7: 302.15 (3); 319.26 (1,5,6,7).

5-Methylisothiazole, CA 693-97-0: 312.59 (1).

4-Methylthiazole, CA 693-95-8: 305.37 (1).

5-Methylthiazole, CA 3581-89-3: 315.37 (1).

C_4H_6

1,2-Butadiene, CA 590-19-2: Below 255.37 (7).

1,3-Butadiene, CA 106-99-0: 197.04 (2,7).

2-Butyne, CA 503-17-3: Below 238.71 (1); 248.15 (3).

Cyclobutene, CA 822-35-5: Below 263.71 (7).

C_4H_6BrClO

4-Bromobutyryl chloride, CA 927-58-2: 364.26 (1).

C_4H_6BrN

4-Bromobutyronitrile, CA 5332-06-9: 377.04 (1).

$C_4H_6Br_2$

1,4-Dibromo-2-butene, CA 821-06-7: Above 385.93 (1).

34

$C_4H_6Br_2O$

2-Bromobutyryl bromide ±, CA 26074-52-2:   Above 385.93 (1).

2-Bromoisobutyryl bromide, CA 20769-85-1:   Above 385.93 (1).

C_4H_6ClN

4-Chlorobutyronitrile, CA 628-20-6:   358.15 (1); 364.15 (3).

3-Chloro-2-methylpropionitrile, technical, CA 7659-45-2:   As 85% solution, 334.26 (1).

C_4H_6ClNO

3-Chloropropyl isocyanate, CA 13010-19-0:   334.82 (1).

$C_4H_6Cl_2$

3-Chloro-2-chloromethyl-1-propene, CA 1871-57-4:   309.82 (1).

1,3-Dichloro-2-butene, mixed isomers, CA 926-57-8:   300.15 (5,6); 307.04 (1).

1,4-Dichloro-2-butene, technical, CA 764-41-0:   300.15 (3); 332.59 (1).

1,4-Dichloro-2-butene, *cis*, CA 1476-11-5:   328.71 (1).

1,4-Dichloro-2-butene, *trans*, CA 110-57-6:   As 85% solution, 327.04 (1).

3,4-Dichloro-1-butene, CA 760-23-6:   301.48 (1); 318.15 (6).

$C_4H_6Cl_2O$

4-Chlorobutyryl chloride, CA 4635-59-0:   345.93 (1).

$C_4H_6Cl_2O_2$

Ethyl dichloroacetate, CA 535-15-9:   335.15 (3).

Methyl 2,3-dichloropropionate, CA 3674-09-7:   315.37 (1); 336.15 (3).

$C_4H_6Cl_2O_3$

Methyl 2,2-dichloro-2-methoxyacetate, CA 17640-25-4:   363.71 (1).

$C_4H_6Cl_3NSi$

3-Cyanopropyltrichlorosilane, CA 1071-27-8:   365.15 (3).

$C_4H_6Cl_5OP$

2,2,2-Trichloro-1,1-dimethylethyl dichlorophosphite, CA 39177-74-7: 383.15 (1).

$C_4H_6N_2$

1-Methylimidazole, CA 616-47-7:   365.37 (1).

4-Methylimidazole, CA 822-36-6:   Above 385.93 (1).

(continues)

$C_4H_6N_2$ *(continued)*

  3-Methylpyrazole, CA 1453-58-3: 366.48 (1).

  4-Methylpyrazole, CA 7554-65-6: 369.26 (1).

$C_4H_6N_2O_2$

  Ethyl diazoacetate, CA 623-73-4: 299.82 (1); 319.15 (3).

$C_4H_6N_2S$

  2-Amino-4-methylthiazole, CA 1603-91-4: Above 385.93 (1).

$C_4H_6N_2S_2$

  Dimethyl *N*-cyanodithioiminocarbonate, CA 10191-60-3: Above 385.93 (1).

C_4H_6O

  Butadiene monoxide, CA 930-22-3: Below 223.15 (6,7); 223.15 (1).

  2-Butyn-1-ol, CA 764-01-2: 324.82 (1).

  3-Butyn-1-ol, CA 927-74-2: 309.26 (1).

  3-Butyn-2-ol ±, CA 65337-13-5: 307.59 (1).

  Crotonaldehyde, CA 123-73-9: 282.04 (1); 286.15 (3,5,6,7); 286.15* (4).

  Cyclobutanone, CA 1191-95-3: 283.15 (1); 300.15 (3).

  2,3-Dihydrofuran, CA 1191-99-7: 248.71 (1); 253.15 (3).

  2,5-Dihydrofuran, CA 1708-29-8: 256.48 (1); 274.15 (3).

  Methacrolein, CA 78-85-3: 258.15 (1,5); 274.82* (4,6,7).

  Methyl propargyl ether, CA 627-41-8: 254.82 (1).

  Methyl vinyl ketone, CA 78-94-4: 266.48 (1,3,5,6,7).

  Vinyl ether, CA 109-93-3: Below 243.15 (6,7).

C_4H_6OS

  Tetrahydrothiophene-3-one, CA 1003-04-9: 350.37 (1); 351.15 (3).

  gamma-Thiobutyrolactone, CA 1003-10-7: 358.71 (1).

$C_4H_6O_2$

  Allyl formate, CA 1838-59-1: Below 227.60 (7).

  1,3-Butadiene diepoxide ±, CA 298-18-0: As 90% solution, 263.15 (3); as 97% solution, 318.71 (1).

  2,3-Butanedione, CA 431-03-8: 279.15 (5); 280.15 (3); 298.82 (6); 299.82 (1,7).

  2-Butyne-1,4-diol, CA 110-65-6: 425.15* (4); 425.37 (1).

  beta-Butyrolactone ±, CA 36536-46-6: 333.15 (1).

  gamma-Butyrolactone, CA 96-48-0: 371.48 (1,2); 371.48* (4,6,7).

  Crotonic acid, CA 3724-65-0: 360.93 (1,3); 360.93* (6,7).

(continues)

$C_4H_6O_2$ *(continued)*

Cyclopropanecarboxylic acid, CA 1759-53-1: 344.82 (1).

Methacrylic acid, CA 79-41-4: 341.15 (5); 350.15 (1,3); 350.15* (4,6,7).

Methyl acrylate, CA 96-33-3: 270.15 (3,4,5); 270.15* (6,7); 279.82 (1); 283.15* (2,4).

Vinyl acetate, CA 108-05-4: 265.15 (3,5,6,7); 244.15* (4); 265.15* (4); 266.48 (1).

Vinylacetic acid, CA 625-38-7: 338.71 (1).

$C_4H_6O_2S$

Acetyl sulfide, CA 3232-39-1: 299.82 (1).

Butadiene sulfone, CA 77-79-2: Above 385.93 (1).

Vinyl sulfone, CA 77-77-0: 375.93 (1).

$C_4H_6O_3$

Acetic anhydride, CA 108-24-7: 322.15 (3,5,6); 327.59 (1,7).

1,3-Dioxan-2-one, CA 2453-03-4: Above 385.93 (1).

2-Ketobutyric acid, CA 600-18-0: 354.82 (1).

Methyl pyruvate, CA 600-22-6: 312.59 (1); 320.15 (3).

Propylene carbonate, CA 108-32-7: 408.15 (5); 408.15* (6,7); 405.37 (1).

$C_4H_6O_4$

Dimethyl oxalate, CA 553-90-2: 328.15 (3); 348.15 (1).

Ethylene glycol diformate, CA 629-15-2: 366.15 (5); 366.48* (4,6,7).

$C_4H_6O_5$

Dimethyl dicarbonate, CA 4525-33-1: 353.15 (1,3).

$C_4H_6O_6$

Tartaric acid ±, CA 133-37-9: 483.15* (5,6).

C_4H_7Br

2-Bromo-2-butene, mixed isomers, CA 13294-71-8: 273.15 (3); 274.26 (1).

4-Bromo-1-butene, CA 5162-44-7: 274.15 (3); 282.59 (1).

(Bromomethyl)cyclopropane, CA 7051-34-5: 314.82 (1).

1-Bromo-2-methylpropene, CA 3017-69-4: 280.93 (1).

Crotyl bromide, *trans*, CA 29576-14-5: As 80% solution, 284.26 (1); as 85% solution, 286.15 (3).

Cyclobutyl bromide, CA 4399-47-7: 295.37 (1,3).

C_4H_7BrO

1-Bromo-2-butanone, CA 816-40-0: As 90% solution, 341.48 (1).

4-Bromo-1,2-epoxybutane, CA 61847-07-7: 332.15 (3).

1-Bromo-2-ethoxyethylene, cis, CA 23521-49-5: As 95% solution, 325.15 (3); as 90% solution, 320.93 (1).

$C_4H_7BrO_2$

2-Bromobutyric acid, CA 80-58-0: Above 385.93 (1).

4-Bromobutyric acid, CA 2623-87-2: Above 385.93 (1).

2-Bromoethyl acetate, CA 927-68-4: 344.26 (1).

2-Bromoethyl-1,3-dioxolane, CA 4360-63-8: 335.93 (1).

Ethyl bromoacetate, CA 105-36-2: 320.93 (1,6,7); 331.15 (3).

Methyl 2-bromopropionate ±, CA 5445-17-0: 324.82 (1); 348.15 (3).

Methyl 3-bromopropionate, CA 3395-91-3: 348.15 (1,3).

C_4H_7Cl

2-Chloro-2-butene, mixed isomers, CA 4461-41-0: 253.71 (1); 248.15 (7).

3-Chloro-1-butene, CA 563-52-0: 246.15 (3,7); 253.15 (1,6).

Chlorobutene, mixed isomers, CA 39437-98-4: Below 294.15 (5).

(Chloromethyl)cyclopropane, CA 5911-08-0: 271.48 (1).

1-Chloro-2-methylpropene, CA 513-37-1: 262.15 (3); 272.04 (1).

3-Chloro-2-methylpropene, CA 563-47-3: 254.15 (7); 261.15 (5,6); 263.15 (1,3).

Crotyl chloride, CA 591-97-9: 253.71 (6); as 70% solution, 258.15 (1).

Crotyl chloride, mixed isomers, CA 4894-61-5: As 60% solution, 263.15 (3).

C_4H_7ClO

Butyryl chloride, CA 141-75-3: 291.15 (3); 294.82 (1).

3-Chloro-2-butanone, CA 4091-39-8: 294.26 (1); 301.15 (3).

2-Chloroethyl vinyl ether, CA 110-75-8: 289.26 (1); 299.82* (6,7); 300.15 (5); 305.37* (2).

Isobutyryl chloride, CA 79-30-1: 274.26 (1); 281.15 (3).

C_4H_7ClOS

Isopropyl chlorothioformate, CA 13889-93-5: 327.59 (2).

Propyl chlorothioformate, CA 13889-92-4: 335.93 (2,6).

$C_4H_7ClO_2$

4-Chlorobutyric acid, CA 627-00-9: Above 385.93 (1).

2-Chloroethyl acetate, CA 542-58-5: 339.26 (6).

(continues)

38

$C_4H_7ClO_2$ *(continued)*

Ethyl chloroacetate, CA 105-39-5: 310.93 (7); 327.15 (3,5); 337.04* (6); 338.71 (1).

Isopropyl chloroformate, CA 108-23-6: 261.48 (1); 296.45 (4); 300.95* (4).

Methyl 2-chloropropionate, CA 17639-93-9: 305.15 (3); 311.48 (1).

Methyl 2-chloropropionate, (R)-(+), CA 77287-29-7: 310.37 (1).

Methyl 2-chloropropionate, (S)-(-), CA 73246-45-4: 311.48 (1).

Propyl chloroformate, CA 109-61-5: 302.04 (1); 307.55* (4).

$C_4H_7Cl_2O_4P$

Dimethyl-2,2-dichlorovinyl phosphate, CA 62-73-7: 449.82* (6).

$C_4H_7FO_2$

Ethyl fluoroacetate, CA 459-72-3: 303.15 (1); 304.15 (3).

$C_4H_7F_3O_2$

1-Ethoxy-2,2,2-trifluoroethanol, CA 433-27-2: 312.59 (1).

$C_4H_7IO_2$

Ethyl iodoacetate, CA 623-48-3: 349.82 (1).

C_4H_7N

Butyronitrile, CA 109-74-0: 289.82 (1); 291.15 (3); 297.59* (6); 299.26* (7); 302.15* (4).

Isobutyronitrile, CA 78-82-0: 277.04 (1); 281.15 (3,6,7).

N-Methylpropargylamine, CA 35161-71-8: 273.71 (1).

3-Pyrroline, technical, CA 109-96-6: As 75% solution, 254.82 (1).

C_4H_7NO

Acetone cyanohydrin, CA 75-86-5: 336.15 (4); 337.04 (1); 347.04 (6,7); 347.15* (4); 348.15 (3,5).

Isopropyl isocyanate, CA 1795-48-8: 270.37 (1,3).

3-Methoxypropionitrile, CA 110-67-8: 334.26 (1); 338.15 (5); 338.15* (6,7); 341.15 (3).

2-Methyl-2-oxazoline, CA 1120-64-5: 293.15 (1).

Propyl isocyanate, CA 110-78-1: 268.15 (3); 299.82 (1).

2-Pyrrolidinone, CA 616-45-5: 383.15 (1); 402.55* (4,6); 416.15 (5).

$C_4H_7NO_2$

3-Methyl-2-oxazolidinone, CA 19836-78-3: Above 385.93 (1).

$C_4H_7NO_4$

Ethyl nitroacetate, CA 626-35-7: 365.37 (1).

C_4H_7NS

2-Methyl-2-thiazoline, CA 2346-00-1: 310.37 (1).

Propyl isothiocyanate, CA 628-30-8: 258.15 (1).

$C_4H_7NS_2$

2-(Methylthio)-2-thiazoline, CA 19975-56-5: 370.93 (1).

C_4H_8

1-Butene, CA 106-98-9: 193.15 (2,7).

2-Butene, *cis*, CA 590-18-1: 199.82 (2,7).

2-Butene, *trans*, CA 624-64-6: 199.82 (2,7).

Cyclobutane, CA 287-23-0: Below 283.15 (7).

Isobutylene, CA 115-11-7: 197.04 (2).

C_4H_8BrCl

1-Bromo-4-chlorobutane, CA 6940-78-9: 333.15 (1).

1-Bromo-3-chloro-2-methylpropane ±, CA 6974-77-2: Above 385.93 (1).

$C_4H_8Br_2$

1,2-Dibromobutane, CA 533-98-2: Above 385.93 (1).

1,3-Dibromobutane ±, CA 107-80-2: Nonflammable (1).

1,4-Dibromobutane, CA 110-52-1: Above 385.93 (1).

1,2-Dibromo-2-methylpropane, CA 594-34-3: 383.15 (1).

$C_4H_8Br_2O$

1,4-Dibromo-2-butanol, technical, CA 19398-47-1: As 85% solution, above 385.93 (1).

1,2-Dibromoethyl vinyl ether ±, CA 2983-26-8: 383.15 (1).

C_4H_8ClNO

N-(2-Chloroethyl)acetamide, CA 7355-58-0: 383.15 (1).

$C_4H_8ClO_3P$

2-Chloro-4,5-dimethyl-1,3,2-dioxaphospholane-2-oxide, CA 891-48-3: 329.26 (1).

$C_4H_8Cl_2$

1,3-Dichlorobutane, CA 1190-22-3: 303.71 (1).

(continues)

40

$C_4H_8Cl_2$ *(continued)*

 1,4-Dichlorobutane, CA 110-56-5: 313.15 (1); 325.15 (3,5,6).

 2,3-Dichlorobutane, CA 7581-97-7: 291.48 (1); 293.15 (5); 363.15* (6).

 Dichlorobutane, mixed 1,2 and 1,3 isomers, CA 26761-81-9: 294.15 (7).

 1,2-Dichloro-2-methylpropane, CA 594-37-6: 288.71 (1).

$C_4H_8Cl_2O$

 2-Chloroethyl ether, CA 111-44-4: 328.15 (1,2,3,5,6,7); 352.15* (2).

 Ethyl, 1,2-dichloroethyl ether, CA 623-46-1: 315.15 (3).

$C_4H_8Cl_2S$

 Bis(2-chloroethyl)sulfide, CA 505-60-2: 378.15 (7).

$C_4H_8I_2$

 1,4-Diiodobutane, CA 628-21-7: Nonflammable (1).

$C_4H_8N_2$

 Dimethylaminoacetonitrile, CA 926-64-7: Below 296.15 (7); 308.15 (3); 309.82 (1).

 N-Methyl-*beta*-alaninenitrile, CA 693-05-0: Above 385.93 (1).

$C_4H_8N_2O$

 1-Nitrosopyrrolidine, CA 930-55-2: 356.48 (1).

C_4H_8O

 2-Butanone, CA 78-93-3: 264.26 (6); 267.15 (4); 267.15* (4,7), 269.82 (1,2); 272.15 (3,4,5).

 3-Buten-1-ol, CA 627-27-0: 305.37 (1); 306.15 (3); 310.93 (5,6).

 3-Buten-2-ol ±, CA 598-32-3: 289.82 (1).

 Butyraldehyde, CA 123-72-8: 250.93 (6); 262.04 (1); 263.71* (2,4); 268.15 (3); 266.48 (7).

 Crotyl alcohol, mixed isomers, CA 6117-91-5: 300.37 (6); 306.15 (5,7); 307.15 (3); 310.37 (1); 318.15* (2).

 Cyclobutanol, CA 2919-23-5: 294.26 (1).

 Cyclopropyl carbinol, CA 2516-33-8: 308.15 (1); 313.15 (3).

 Cyclopropyl methyl ether, CA 540-47-6: Below 283.15 (7).

 1,2-Epoxybutane, CA 106-88-7: 251.48 (6); 258.15 (3,5,7); 260.93 (1).

 2,3-Epoxybutane, *cis*, CA 1758-33-4: 251.48 (1).

 2,3-Epoxybutane, *trans*, CA 21490-63-1: 246.48 (1).

 2,3-Epoxybutane, mixed isomers, CA 3266-23-7: 258.15 (6).

 Ethyl vinyl ether, CA 109-92-2: 227.59 (1); below 227.59 (6,7); 255.15 (3).

(continues)

C_4H_8O *(continued)*

Isobutyraldehyde, CA 78-84-2: 233.15 (1,7); 248.15 (3); 254.82 (6); 262.55* (2,4).

2-Methoxypropene, CA 116-11-0: 243.71 (1); 248.15 (3).

2-Methyl-2-propen-1-ol, CA 513-42-8: 305.15 (3); 306.48 (1,6).

Tetrahydrofuran, CA 109-99-9: 253.15 (5); 256.15 (1,3,7); 259.15 (4,6).

C_4H_8OS

3-Methoxythietane: 277.59 (1).

3-(Methylthio)propionaldehyde, CA 3268-49-3: 334.26 (1,5,6); 333.15 (3).

Tetramethylene sulfoxide, CA 1600-44-8: Above 385.93 (1).

1,4-Thioxane, CA 15980-15-1: 315.37 (1,5,6,7).

$C_4H_8O_2$

Acetyl methyl carbinol, CA 513-86-0: 323.15 (1).

2-Butene-1,4-diol, *cis*, CA 110-64-5: 401.48 (1); 401.48* (1,4,6).

Butyric acid, CA 107-92-6: 339.15 (4); 344.82 (3,5,6,7); 349.82 (1); 349.82 (2).

1,3-Dioxane, CA 505-22-6: 274.15 (7); 278.15 (3); 288.15 (1).

1,4-Dioxane, CA 123-91-1: 284.15 (3,5); 285.37 (1,4,6,7); 296.45* (4).

Ethyl acetate, CA 141-78-6: 268.15* (2); 269.15 (4,5,6); 269.82 (1); 271.15 (3).

Glycidyl methyl ether ±, CA 930-37-0: 292.04 (1).

3-Hydroxybutyraldehyde, CA 107-89-1: 338.71* (6,7); 356.15 (5).

3-Hydroxytetrahydrofuran, CA 453-20-3: 354.26 (1).

Isobutyric acid, CA 79-31-2: 328.71 (1,3,6); 328.71* (7); 350.15 (5).

Isopropyl formate, CA 625-55-8: 267.59 (5,6,7).

Methoxyacetone, CA 5878-19-3: 298.15 (1).

2-Methyl-1,3-dioxolane, CA 497-26-7: 270.93 (1); 278.15 (3).

Methyl propionate, CA 554-12-1: 271.15 (3,4,5,6,7); 279.26 (1).

Propyl formate, CA 110-74-7: 269.26 (1); 270.37 (5,6,7).

$C_4H_8O_2S$

Ethyl 2-mercaptoacetate, CA 623-51-8: 320.93 (1); 330.15 (3).

Ethyl vinyl sulfone, CA 1889-59-4: Above 385.93 (1).

Methyl 3-mercaptopropionate, CA 2935-90-2: 333.15 (1).

Methyl (methylthio)acetate, CA 16630-66-3: 331.15 (3).

Sulfolane, CA 126-33-0: 438.71 (1,2,7); 449.82* (6).

$C_4H_8O_3$

Ethoxyacetic acid, CA 627-03-2: 370.93 (1).

(continues)

$C_4H_8O_3$ *(continued)*

Ethylene glycol monoacetate, technical, CA 542-59-6: As 50% solution, 361.48 (1); 374.82* (2,6,7); 375.15 (5).

Glycerol formal, CA 86687-05-0: 349.82 (1).

3-Hydroxybutyric acid, technical, CA 300-85-6: Above 385.93 (1).

2-Methoxy-1,3-dioxolane, CA 19693-75-5: 304.26 (1); 307.15 (3).

Methyl lactate, (R)-(+), CA 17392-83-5: 322.59 (1).

Methyl lactate, (S)-(-), CA 27871-49-4: 322.59 (1).

Methyl lactate ±, CA 547-64-8: 322.15 to 324.85 (2,3,5,6).

Methyl methoxyacetate, CA 6290-49-9: 308.71 (1); 315.15 (3).

$C_4H_8O_3S$

1,4-Butane sultone, CA 1633-83-6: Above 385.93 (1).

4,5-Dimethyl-2-oxo-1,3,2-dioxathiolane, CA 4440-90-8: 342.59 (1).

Methyl (methylsulfinyl)acetate, CA 52147-67-8: Above 385.93 (1).

C_4H_8S

Allyl methyl sulfide, CA 10152-76-8: 291.48 (1).

Tetrahydrothiophene, CA 110-01-0: 285.15 (3); 285.93 (1); 286.48* (2).

$C_4H_8S_2$

1,3-Dithiane, CA 505-23-7: 363.71 (1).

Ethyl dithioacetate, CA 870-73-5: 316.15 (3).

C_4H_9Br

1-Bromobutane, CA 109-65-9: 286.15 (5); 291.48 (6); 291.48* (7); 295.15 (3); 297.04 (1,2).

2-Bromobutane, CA 78-76-2: 294.26 (1,2,3,7).

1-Bromo-2-methylpropane, CA 78-77-3: Below 273.15 (5); 291.48 (1); 295.15 (7); 305.37 (2).

2-Bromo-2-methylpropane, CA 507-19-7: 255.15 (7); 289.15 (3); 291.48 (1); nonflammable (2).

C_4H_9BrO

2-Bromoethyl ethyl ether, technical, CA 592-55-2: 278.15 (7); 294.26 (1).

$C_4H_9BrO_2$

2-Bromo-1,1-dimethoxyethane, CA 7252-83-7: 324.15 (3); 327.04 (1).

C_4H_9Cl

1-Chlorobutane, CA 109-69-3: 261.15 (3,5,7); 263.71 (6); 263.71* (7); 266.48 (1); 266.48* (2).   *(continues)*

C$_4$H$_9$Cl *(continued)*

2-Chlorobutane, CA 78-86-4: 258.15 (1); 263.15 (3,7).

1-Chloro-2-methylpropane, CA 513-36-0: 267.15 (7); below 273.15 (5); below 294.26 (6); 294.26 (1).

2-Chloro-2-methylpropane, CA 507-20-0: 263.15 (3); below 273.15 (5,6); 273.15 (7); 291.48 (1).

C$_4$H$_9$ClO

4-Chloro-1-butanol, technical, CA 928-51-8: As 85% solution, 309.26; as 90+% solution, 323.15 (3).

2-Chloroethyl ethyl ether, CA 628-34-2: 279.15 (3); 288.71 (1).

1-Chloro-2-methyl-2-propanol, CA 558-42-9: 311.15 (3).

C$_4$H$_9$ClO$_2$

1-Chloro-2,2-dimethoxyethane, CA 97-97-2: 302.04 (1); 308.15 (3); 317.04 (6).

2-(2-Chloroethoxy)ethanol, CA 628-89-7: 363.71 (1); 380.15 (5); 380.37* (6,7).

2-Methoxyethoxymethyl chloride, CA 3970-21-6: 327.15 (3); above 385.93 (1).

C$_4$H$_9$ClO$_2$S

1-Butanesulfonyl chloride, CA 2386-60-9: 352.59 (1).

C$_4$H$_9$ClS

2-Chloroethyl ethyl sulfide, CA 693-07-2: 325.37 (1).

C$_4$H$_9$Cl$_3$Si

Butyltrichlorosilane, CA 7521-80-4: 327.59* (6,7); 328.15 (5).

C$_4$H$_9$Cl$_3$Sn

Butyltin trichloride, CA 1118-46-3: 354.26 (1).

C$_4$H$_9$F

2-Fluoro-2-methylpropane, CA 353-61-7: 260.93 (1).

C$_4$H$_9$F$_3$O$_3$SSi

Trimethylsilyl trifluoromethanesulfonate, CA 27607-77-8: 298.71 (1).

C$_4$H$_9$I

1-Iodobutane, CA 542-69-8: 263.15 (3); 306.48 (1).

2-Iodobutane, CA 513-48-4: 263.15 (7); 297.04 (1,3).

(continues)

44

C_4H_9I *(continued)*

1-Iodo-2-methylpropane, CA 513-38-2: 273.15 (3,7); 285.93 (1).

2-Iodo-2-methylpropane, CA 558-17-8: 263.15 (3); 280.93 (1).

C_4H_9N

N-Allyl methylamine, CA 627-37-2: 265.15 (3).

(Aminomethyl)cyclopropane, CA 2516-47-4: 275.15 (3).

Cyclobutylamine, CA 2516-34-9: 268.71 (1); 291.15 (3).

Pyrrolidine, CA 123-75-1: 275.93 (1,3,5,6,7).

C_4H_9NO

1-Aziridineethanol, CA 1072-52-2: 340.37 (1).

2-Butanone oxime, CA 96-29-7: 333.15 (3); 342.04 to 349.82 (6).

Butyraldoxime, CA 110-69-0: 330.93 (5,6,7).

N,N-Dimethylacetamide, CA 127-19-5: 336.15 (4); 339.15 (3); 343.15 (1); 343.15* (2,6); 350.37* (7).

N-Ethylacetamide, CA 625-50-3: 383.15 (6,7).

N-Methylpropionamide, CA 1187-58-2: 342.15 (3); 378.71 (1).

Morpholine, CA 110-91-8: 308.15 (4); 308.71 (1); 309.82* (6); 310.93* (2,7); 311.15 (3,5); 311.15* (4).

2-Pyrrolidinol, CA 40499-83-0: 383.15 (1).

C_4H_9NOSi

Trimethylsilyl isocyanate, CA 1118-02-1: 268.15 (3); 270.37 (1).

$C_4H_9NO_2$

N-Acetylethanolamine, technical, CA 142-26-7: 449.82 (1); 452.59* (6,7).

Butyl nitrite, CA 544-16-1: 259.82 (1); 260.93 (7).

tert-Butyl nitrite, technical, CA 540-80-7: As 90% solution, 259.82 (1); 263.15 (3).

Ethyl *N*-hydroxyacetimidate, CA 10576-12-2: 328.15 (3); 349.82 (1).

2-Methyl-2-nitropropane, CA 594-70-7: 292.59 (1).

1-Nitrobutane, CA 317.15 (3); 320.93 (1).

$C_4H_9NO_2S$

Homocysteine ±, CA 6027-13-0: Above 385.93 (1).

$C_4H_9NO_3$

Butyl nitrate, CA 928-45-0: 309.26 (6).

Isobutyl nitrate, CA 543-29-3: 294.26 (1).

3-Nitro-2-butanol, mixed isomers, CA 6270-16-2: 364.26 (1).

C_4H_9NS

Thiomorpholine, CA 123-90-0: 333.15 (1); above 371.15 (3).

C_4H_9NSSi

Trimethylsilyl isothiocyanate, CA 2290-65-5: 302.15 (3); 308.15 (1).

C_4H_9NSi

Trimethylsilyl cyanide, CA 7677-24-9: 274.26 (1); 288.15 (3).

$C_4H_9O_5P$

Trimethyl phosphonoformate, CA 31142-23-1: Above 385.93 (1).

C_4H_{10}

Butane, CA 106-97-8: 199 (4); 199.82 (2); 213.15 (7).
Isobutane, CA 75-28-5: 190 (4); 190.37 (2).

$C_4H_{10}BF_3O$

Boron trifluoride etherate, CA 109-63-7: 332.15 (3); 337.04* (6); 338.71 (1).

$C_4H_{10}BrMgo$

Magnesium bromide etherate, CA 29858-07-9: 308.15 (1).

$C_4H_{10}ClO_2P$

Diethyl chlorophosphite, CA 589-57-1: 274.26 (1).

$C_4H_{10}ClO_2PS$

Diethyl chlorothiophosphate, CA 2524-04-1: Above 385.93 (1).

$C_4H_{10}ClO_3P$

Diethyl chlorophosphate, CA 814-49-3: 334.26 (1).

$C_4H_{10}Cl_2Si$

Bis(chloromethyl)dimethylsilane, CA 2817-46-6: 319.26 (1).
Dichlorodiethylsilane, CA 1719-53-5: 294.15 (3); 297.15 (7); 301.48 (1).
(Dichloromethyl)trimethylsilane, CA 5926-38-5: 298.71 (1).

$C_4H_{10}N_2$

Piperazine, CA 110-85-0: 350.93* (2); 354.26* (6); 358.15* (4); 360.93* (7); 380.15* (4); 382.59 (1).
Piperazine hexahydrate, CA 142-63-2: 360.93 (1).

46

$C_4H_{10}N_2O$

N-Acetylethylenediamine, CA 1001-53-2: Above 385.93 (1).

4-Aminomorpholine, CA 4319-49-7: 331.48 (1).

$C_4H_{10}O$

1-Butanol, CA 71-36-3: 307.15 (3); 308.15 (1,4,5,7); 309.82 (6); 319.26* (2).

2-Butanol ±, CA 15892-23-6: 263.15 (7); 296.15 (3); 297.15 (2,4,5,6); 299.82 (1); 299.82* (2); 304.15* (4).

2-Butanol, (R)-(-), CA 14898-79-4: 299.82 (1).

2-Butanol, (S)-(+), CA 4221-99-2: 299.82 (1).

Diethyl ether, CA 60-29-7: 224.15 (4); 228.15 (4,6,7); 233.15 (1,2,3).

2-Methyl-1-propanol, CA 78-83-1: 300.15 (5); 301.15 (3,4,6,7); 310.37 (1); 312.59* (2).

2-Methyl-2-propanol, CA 75-65-0: 282.05 (2,4); 283.15 (7); 284.15 (1,3,4,5,6); 288.71* (2).

Methyl propyl ether, CA 557-17-5: Below 253.15 (5,6).

$C_4H_{10}OS$

Ethyl 2-hydroxyethyl sulfide, CA 110-77-0: Above 385.93 (1).

3-Mercapto-2-butanol, technical, CA 54812-86-1: 334.82 (1).

$C_4H_{10}OS_2$

2-Mercaptoethyl ether, CA 2150-02-9: 363.15 (3); 371.48 (1).

$C_4H_{10}O_2$

Acetaldehyde dimethyl acetal, CA 534-15-6: 255.93 (1); 274.15 (7).

1,2-Butanediol, CA 584-03-2: 313.15 (6); 363.15 (5,6,7); 366.48 (1).

1,3-Butanediol ±, CA 107-88-0: 382.15 (5); 394.26 (1,6,7); 394.26* (2).

1,3-Butanediol, (R)-(-), CA 6290-03-5: 394.26 (1).

1,3-Butanediol, (S)-(+), CA 24621-61-2: 394.26 (1).

1,4-Butanediol, CA 110-63-4: 394.26 (7); 394.26* (4,6); above 385.93 (1).

2,3-Butanediol, CA 513-85-9: 358.15 (1,5,7); 358.15* (2,6).

2,3-Butanediol, (2S,3S)-(+), CA 19132-06-0: 358.15 (1).

2,3-Butanediol, (2R,3R)-(-), CA 24347-58-8: 358.15 (1).

tert-Butyl hydroperoxide, CA 75-91-2: As 90% solution, 313.71 (1); as 70% solution, 320.15 (3); 335.93 (1).

2-Ethoxyethanol, CA 110-80-5: 313.15 (5); 315.15 (2,3,4); 316.48 (6); 317.59 (1); 319.15 to 322.15* (2,4).

Ethylene glycol dimethyl ether, CA 110-71-4: 271.48 (6); 273.15 (1); 277.55* (2,4); 278.15 (3).

1-Methoxy-2-propanol ±, CA 107-98-2: 304.82 (2); 306.15 (3,4); 309.15* (4); 310.95 (7); 310.95* (2,4,6); 311.15 (5).

$C_4H_{10}O_2S$

3-Methylthio-1,2-propanediol, CA 22551-26-4: Above 385.93 (1).

2,2'-Thiodiethanol, CA 111-48-8: 383.15 (1); 433.15* (2,4,6,7).

$C_4H_{10}O_2S_2$

Dithiothreitol, CA 27565-41-9: Above 385.93 (1).

1,4-Dithiothreitol, *levo*, CA 16096-97-2: Above 385.93 (1).

2-Hydroxyethyl disulfide, CA 1892-29-1: Above 385.93 (1).

$C_4H_{10}O_3$

1,2,4-Butanetriol, CA 3068-00-6: Above 385.93 (1); 439.82* (2).

Diethylene glycol, CA 111-46-6: 397.04 (2,6,7); 411.15 (4); 416.45 (1); 416.45* (2,4); 419.15 (5).

3-Methoxy-1,2-propanediol ±, CA 36887-04-4: Above 385.93 (1).

Trimethylolmethane, technical, CA 4704-94-3: 362.59 (1).

Trimethyl orthoformate, CA 149-73-5: 283.15 (3); 288.71 (1,7).

$C_4H_{10}O_3S$

Diethyl sulfite, CA 623-81-4: 327.15 (1,3).

(Ethylsulfonyl)ethanol, CA 513-12-2: 461.48 (7).

Propyl methanesulfonate, CA 1912-31-8: 382.04 (1).

$C_4H_{10}O_3Si$

Diethyl silicate, commercial, CA 18954-71-7: 316.15 (4).

$C_4H_{10}O_4S$

Diethyl sulfate, CA 64-67-5: 351.48 (1); 377.59 (5,6,7); 386.15* (4).

$C_4H_{10}S$

1-Butanethiol, CA 109-79-5: 263.15 (3); 274.82 (6,7); 285.93 (1); 285.93* (2).

Ethyl sulfide, CA 352-93-2: 263.71 (1,3,7).

1-Methyl-1-propanethiol, CA 513-53-1: 249.82 (6); 263.15 (3); 269.26* (2); 294.26 (1).

2-Methyl-1-propanethiol, CA 513-44-0: 263.71 (1,2,3).

2-Methyl-2-propanethiol, CA 77-66-1: Below 244.26 (6,7); 248.71 (1); 263.15 (3); 265.37* (2).

$C_4H_{10}S_2$

1,4-Butanedithiol, CA 1191-08-8: 342.15 (1); 353.15 (3).

2,3-Butanedithiol, CA 4532-64-3: 325.37 (1).

Ethyl disulfide, CA 110-81-6: 313.15 (1).

$C_4H_{10}S_3$

2-Mercaptoethyl sulfide, CA 3570-55-6:   363.71 (1).

Tris(methylthio)methane, CA 5418-86-0:   359.15 (3); 368. (1).

$C_4H_{11}BrSi$

Bromomethyltrimethylsilane, CA 18243-41-9:   289.26 (1).

$C_4H_{11}ClSi$

Chloromethyltrimethylsilane, CA 2344-80-1:   269.15 (3); 270.37 (1).

$C_4H_{11}ISi$

(Iodomethyl)trimethylsilane, CA 4206-67-1:   302.15 (3); 304.82 (1).

$C_4H_{11}N$

1-Aminobutane, CA 109-73-9:   258.71 (1); 260.93 (6); 260.93* (7); 266.15 (3); 269.26 (2); 318.15* (2).

2-Aminobutane ±, CA 33966-50-6:   253.71 (1); 264.15 (3,5,6,7); 265.37 (2); 266.48* (2).

2-Aminobutane, (R)-(-), CA 13250-12-9:   253.71 (1).

2-Aminobutane, (S)-(+), CA 513-49-5:   253.71 (1).

tert-Butylamine, CA 75-64-9:   263.15 (3); 264.26 (1,4).

Diethylamine, CA 109-89-7:   237.15 (3); 244.26 (1); 247.15 (4); 250.37 (6); 252.04 (2); 255.15 (7).

N,N-Dimethylethylamine, CA 598-56-1:   237.04 (1); 245.15 (3).

Isobutylamine, CA 78-81-9:   252.59 (1); 264.15 (3,5,6,7); below 266.48* (2).

N-Isopropylmethylamine, CA 4747-21-1:   253.15 (3).

$C_4H_{11}NO$

2-Amino-1-butanol ±, CA 13054-87-0:   347.04* (2,6,7); 347.15 (5); 357.59 (1); 368.15 (3).

2-Amino-1-butanol, (R)-(-), CA 5856-63-3:   355.37 (1); 368.15 (3).

2-Amino-1-butanol, (S)-(+), CA 5856-62-2:   352.59 (1).

4-Amino-1-butanol, CA 13325-10-5:   380.93 (1).

1-Amino-2-ethoxyethane, CA 110-76-9:   294.26 (2).

1-Amino-3-methoxypropane, CA 5332-73-0:   305.15 (3,5,6); 305.37* (7).

2-Amino-1-methoxypropane ±, CA 37143-54-7:   282.04 (1); 288.71 (2).

2-Amino-2-methyl-1-propanol, CA 124-68-5:   340.37 (1,6); 340.37* (7); 344.15 (5); 356.48 (2).

N,N-Diethylhydroxylamine, CA 3710-84-7:   318.15 (1).

N,N-Dimethylethanolamine, CA 108-01-0:   304.15 (3,5); 313.15 (2); 313.71 (1); 313.71* (6,7); 314.15* (4).

2-(Ethylamino)ethanol, CA 110-73-6:   344.26 (1,3,5); 344.26* (4,6,7); 345.37 (2).

$C_4H_{11}NO_2$

Diethanolamine, CA 111-42-2: 410.15* (4); 410.93 (1,5); 410.93* (2); 424.82* (7); 445.37* (6).

Diethanolamine hydrochloride, CA 14426-21-2: Above 385.93 (1).

2,2-Dimethoxyethylamine, CA 22483-09-6: 317.15 (3); 326.48 (1).

$C_4H_{11}O_2PS_2$

Diethyl dithiophosphate, technical, CA 298-06-6: As 90% solution, 355.93 (1).

$C_4H_{11}O_3P$

Diethyl phosphite, CA 762-04-9: 338.15 (3); 363.71 (1); 363.71* (2).

$C_4H_{11}O_4P$

Phosphoric acid, monobutyl ester, CA 1623-15-0: 383.15* (7).

$C_4H_{12}ClN_2OP$

Bis(dimethylamino)phosphorochloridate, technical, CA 1605-65-8: As 90% solution, above 385.93 (1).

$C_4H_{12}ClN_2P$

Bis(dimethylamino)chlorophosphine, CA 3348-44-5: 288.71 (1).

$C_4H_{12}FN$

Tetramethylammonium fluoride tetrahydrate, CA 17787-40-5: Above 385.93 (1).

$C_4H_{12}N_2$

1,3-Diaminobutane, CA 590-88-5: 324.82 (7); 324.82* (6).

1,4-Diaminobutane, CA 110-60-1: 324.82 (1); 335.15 (3).

1,2-Diamino-2-methylpropane, CA 811-93-8: 297.04 (1).

N,N-Dimethylethylenediamine, CA 108-00-9: 284.15 (7); 285.15 (3); 297.04 (1).

N,N'-Dimethylethylenediamine, technical, CA 110-70-3: As 85% solution, 295.93 (1); 299.15 (3).

N-Ethylethylenediamine, CA 110-72-5: 283.15 (1); 307.15 (3).

N-Methyl-1,3-propanediamine, CA 6291-84-5: 308.71 (1); 313.15 (3).

Tetramethylhydrazine, CA 6415-12-9: 300.15 (3).

$C_4H_{12}N_2O$

2-(2-Aminoethylamino)ethanol, CA 111-41-1: 375.38 (7); 383.15 (1); 402.59 (2,5,6); 408.15 (6)

$C_4H_{12}OSi$

Methoxytrimethylsilane, CA 1825-61-2: 242.59 (1); 253.15 (3).

Trimethylsilylmethanol, CA 3219-63-4: 329.15 (3).

$C_4H_{12}O_3SSi$

Trimethylsilyl methanesulfonate, CA 10090-05-8: 303.15 (3).

$C_4H_{12}O_3Si$

Methyltrimethoxysilane, CA 1185-55-3: 284.26 (1); 294.15 (4).

$C_4H_{12}O_4Si$

Tetramethoxysilane, CA 681-84-5: 293.71 (1,3); 318.15 (4).

$C_4H_{12}Pb$

Tetramethyl lead, CA 75-74-1: Below 294.15 (3); 310.93 (6,7).

$C_4H_{12}SSi$

(Methylthio)trimethylsilane, CA 3908-55-2: 295.15 (3).

$C_4H_{12}Si$

Dimethylethylsilane, CA 758-21-4: Below 238.71 (1).

Tetramethylsilane, CA 75-76-3: 245.93 (1); 253.15 (3).

$C_4H_{12}Sn$

Tetramethyltin, CA 594-27-4: 260.37 (1).

$C_4H_{13}NSi$

Trimethylsilylmethylamine, CA 18166-02-4: 282.15 (3).

$C_4H_{13}N_3$

Diethylenetriamine, CA 111-40-0: 367.59 (1); 370.93* (6); 371.15 (4); 372.04 (2); 374.82* (2,4,7); 375.15 (5).

$C_4H_{14}BN$

Borane-diethylamine complex, CA 2670-68-0: 359.82 (1).

$C_4H_{14}OSi_2$

1,1,3,3-Tetramethyldisiloxane, technical, CA 3277-26-7: 251.15 (3); 263.15 (1).

$C_4H_{15}NSi_2$

1,1,3,3-Tetramethyldisilazane, CA 15933-59-2: 270.15 (3).

C_4NiO_4

Nickel carbonyl, CA 13463-39-3: Below 253.15 (5,6,7).

C_5ClF_4N

3-Chloro-2,4,5,6-tetrafluoropyridine, CA 1735-84-8: Above 385.93 (1).

$C_5Cl_2F_3N$

3,5-Dichloro-2,4,6-trifluoropyridine, CA 1737-93-5: Above 385.95 (1).

C_5Cl_6

Hexachlorocyclopentadiene, CA 77-47-4: Nonflammable (1,7).

C_5F_5N

Pentafluoropyridine, CA 700-16-3: 297.04 (1); 298.15 (3).

C_5FeO_5

Iron carbonyl, CA 13463-40-6: 255.15 (3); 258.15 (1,5,6,7).

$C_5H_2Cl_2N_2O_2$

2,6-Dichloro-3-nitropyridine, technical, CA 16013-85-7: Above 385.93 (1).

$C_5H_2F_6O_2$

1,1,1,5,5,5-Hexafluoro-2,4-pentanedione, CA 1522-22-1: 305.15 (3); nonflammable (1).

C_5H_3BrOS

5-Bromo-2-thiophenecarboxaldehyde, technical, CA 4701-17-1: 372.04 (1).

C_5H_3ClOS

2-Thiophenecarbonyl chloride, CA 5271-67-0: 363.71 (1).

$C_5H_3ClO_2$

2-Furoyl chloride, CA 527-69-5: 358.15 (1).

$C_5H_3F_2N$

2,6-Difluoropyridine, CA 1513-65-1: 305.15 (4); 358.15 (1).

$C_5H_3F_6NO_2$

  N-Methylbis(trifluoroacetamide), CA 685-27-8: 318.15 (1,3).

$C_5H_3F_7O_2$

  Methyl heptafluorobutyrate, CA 356-24-1: Above 385.93 (1).

C_5H_3NO

  2-Furonitrile, CA 617-90-3: 308.15 (1).

$C_5H_3NO_4$

  5-Nitro-2-furaldehyde, CA 698-63-5: 306.48 (1).

C_5H_3NS

  2-Thiophenecarbonitrile, CA 1003-31-2: 326.48 (1).

C_5H_4BrN

  2-Bromopyridine, CA 109-04-6: 327.59 (1); 346.15 (3).

  3-Bromopyridine, CA 626-55-1: 324.82 (1); 330.15 (3).

C_5H_4ClN

  2-Chloropyridine, CA 109-09-1: 338.15 (1).

  3-Chloropyridine, CA 626-60-8: 338.71 (1).

$C_5H_4Cl_2N_2$

  2,4-Dichloro-6-methylpyrimidine, CA 5424-21-5: Above 385.93 (1).

$C_5H_4Cl_2N_2S$

  4,6-Dichloro-2-(methylthio)pyrimidine, CA 6299-25-8: Above 385.93 (1).

$C_5H_4Cl_2O_2$

  Itaconyl chloride, CA 1931-60-8: Above 385.93 (1).

$C_5H_4Cl_2S$

  2-Chloro-5-chloromethylthiophene, CA 23784-96-5: 373.71 (1).

C_5H_4FN

  2-Fluoropyridine, CA 372-48-5: 297.15 (3,4); 301.48 (1).

  3-Fluoropyridine, CA 372-47-4: 286.48 (1.4).

C_5H_4OS

  2-Thiophenecarboxaldehyde, CA 98-03-3: 327.15 (3); 350.93 (1).

  3-Thiophenecarboxaldehyde, CA 498-62-4: 327.15 (3); 347.04 (1).

$C_5H_4O_2$

  4-Cyclopentene-1,3-dione, CA 930-60-9: 357.04 (1).

  2-Furaldehyde, CA 98-01-1: 333.15 (1,3,4,5,6,7); 334.85 (2,4).

  3-Furaldehyde, CA 498-60-2: 321.48 (1).

  2H-Pyran-2-one, CA 504-31-4: 367.59 (1).

  4H-Pyran-4-one, CA 108-97-4: 374.26 (1).

$C_5H_4O_3$

  Citraconic anhydride, CA 616-02-4: 374.82 (1).

$C_5H_5ClN_2S$

  4-Chloro-2-methylthiopyrimidine, CA 49844-90-8: Above 385.93 (1).

$C_5H_5F_3O_2$

  2,2,2-Trifluoroethyl acrylate, CA 407-47-6: 285.37 (1).

  1,1,1-Trifluoro-2,4-pentanedione, CA 367-57-7: 297.15 (3); 299.26 (1).

$C_5H_5F_5O_2$

  Ethyl pentafluoropropionate, CA 426-65-3: 274.82 (1).

C_5H_5N

  Pyridine, CA 110-86-1: 290.15 (3,5); 293.15 (1,6,7).

C_6H_5NO

  Pyrrole-2-carboxaldehyde, CA 1003-29-8: 379.82 (1).

C_5H_5NOS

  2-Acetylthiazole, CA 24295-03-2: 351.48 (1).

$C_5H_5NO_2$

  1-Cyanovinyl acetate, CA 3061-65-2: 337.04 (1); 342.15 (3).

C_5H_6

  1,3-Cyclopentadiene, CA 542-92-7: 298.15 (7).

  2-Methyl-1-buten-3-yne, CA 78-80-8: 265.93* (6); 266.48 (1,7).

$C_5H_6Br_2$

1,2-Dibromocyclopentene, technical, CA 75415-78-0: 383.15 (1).

$C_5H_6Cl_2O_2$

Glutaryl dichloride, CA 2873-74-7: 379.82 (1).

$C_5H_6Cl_4O_2$

2,2,2-Trichloro-1,1-dimethylethyl chloroformate, CA 66270-36-8: Nonflammable (1).

$C_5H_6N_2$

2-Aminopyridine, CA 504-29-0: 365.37 (1).

Glutaronitrile, CA 544-13-8: Above 385.93 (1).

2-Methylpyrazine, CA 109-08-0: 323.15 (1,3,5); 323.15* (6,7).

3-Methylpyridazine, CA 1632-76-4: 360.93 (1).

4-Methylpyrimidine, CA 3438-46-8: 313.71 (1).

1-Vinylimidazole, CA 1072-63-5: 354.26 (1).

C_5H_6O

2-Cyclopentenone, CA 930-30-3: 315.37 (1); 336.15 (3).

2-Methylfuran, CA 534-22-5: 243.15 (2,6,7); 250.93 (1).

C_5H_6OS

Furfuryl mercaptan, CA 98-02-2: 318.15 (1).

2-Methyl-3-furanthiol, technical, CA 28588-74-1: 309.82 (1).

2-Thiophenemethanol, CA 636-72-6: Above 385.93 (1).

$C_5H_6O_2$

alpha-Angelicalactone, CA 591-12-8: 341.48 (1).

5,6-Dihydro-2H-pyran-2-one, CA 3393-45-1: 319.26 (1).

Ethyl propiolate, CA 623-47-2: 296.48 (1); 298.15 (3).

3-Furanmethanol, CA 4412-91-3: 311.48 (1).

Furfuryl alcohol, CA 98-00-0: 338.15 (1,2,4); 348.15 (3,5); 348.15* (6,7);
350.15 (2).

2-Methoxyfuran, CA 25414-22-6: 283.15 (1).

Methyl 2-butynoate, CA 23326-27-4: 318.15 (1).

alpha-Methylene-gamma-butyrolactone, CA 547-65-9: 310.37 (1); 363.15 (3).

gamma-Methylene-gamma-butyrolactone, CA 10008-73-8: 363.15 (3).

$C_5H_6O_3$

Glutaric anhydride, CA 108-55-4: Above 385.93 (1).   (continues)

$C_5H_6O_3$ *(continued)*

Methylsuccinic anhydride, CA 4100-80-5: Above 385.93 (1).

C_5H_6S

2-Methylthiophene, CA 554-14-3: 280.37 (1); 281.15 (3,7).

3-Methylthiophene, CA 616-44-4: 284.26 (1); 285.15 (3).

$C_5H_7BrO_2$

alpha-Bromo-*gamma*-valerolactone, mixed isomers, CA 25966-39-6: Above 385.93 (1).

Methyl 2-(bromomethyl)acrylate, CA 4224-69-5: 351.48 (1).

Methyl 2-bromo-2-butenoate, mixed isomers, CA 17642-18-1: 336.15 (3).

Methyl 4-bromo-2-butenoate, mixed isomers, CA 1117-71-1: 335.15 (3); as 90% solution, 364.82 (1).

$C_5H_7BrO_3$

Ethyl bromopyruvate, technical, CA 70-23-5: 372.04 (1).

C_5H_7Cl

1-Chloro-1-cyclopentene, CA 930-29-0: 282.59 (1).

3-Chloro-3-methyl-1-butyne, CA 1111-97-3: 277.59 (1).

5-Chloro-1-pentyne, CA 14267-92-6: 288.71 (1).

C_5H_7ClO

2-Chlorocyclopentanone, CA 694-28-0: 350.37 (1).

Cyclobutanecarboxylic acid chloride, CA 5006-22-4: 310.37 (1).

3,3-Dimethylacryloyl chloride, CA 3350-78-5: 324.26 (1).

$C_5H_7ClO_2$

2-Chloroallyl acetate, CA 2916-14-5: 321.48 (1).

3-Chloro-2,4-pentanedione, CA 1694-29-7: 285.37 (1); 325.15 (3).

$C_5H_7ClO_3$

3-Carbomethoxypropionyl chloride, CA 1490-25-1: 347.04 (1).

Ethyl malonyl chloride, technical, CA 33142-21-1: 340.93 (1).

Methyl 2-chloroacetoacetate, CA 4755-81-1: 344.82 (1).

Methyl 4-chloroacetoacetate, CA 32807-28-6: 324.15 (3); 375.93 (1).

C_5H_7N

2-Methyl-2-butenenitrile, CA 4403-61-6: 256.15 (3).

2-Methyl-3-butenenitrile, CA 16529-56-9: 284.15 (3).

(continues)

C_5H_7N *(continued)*

 1-Methylpyrrole, CA 96-54-8: 288.15 (1); 289.15 (3,6,7).

 2-Pentenenitrile, CA 13284-42-9: 297.15 (3).

 3-Pentenenitrile, CA 4635-87-4: 314.15 (3).

C_5H_7NO

 3,5-Dimethylisooxazole, CA 300-87-8: 304.26 (1).

 3-Ethoxyacrylonitrile, CA 61310-53-0: 323.15 (3); 354.82 (1).

 Furfurylamine, CA 617-89-0: 310.15 (3,5); 310.37* (6,7); 318.15 (1).

$C_5H_7NO_2$

 Ethyl cyanoacetate, CA 105-56-5: 383.15 (1,4,6,7).

 Ethyl isocyanoacetate, CA 2999-46-4: 357.59 (1,3).

$C_5H_7NO_2S$

 3-Ethyl-2-thioxo-4-oxazolidinone, CA 10574-66-0: Above 385.93 (1).

$C_5H_7NO_3$

 4-Acetoxy-2-azetidinone, CA 28562-53-0: Above 385.93 (1).

 Ethyl isocyanatoacetate, CA 2949-22-6: 345.93 (1).

$C_5H_7NO_6$

 Dimethyl nitromalonate, CA 5437-67-2: 383.15 (1).

C_5H_7NS

 4,5-Dimethylthiazole, CA 3581-91-7: 324.26 (1).

 2-Thiophenemethylamine, CA 27757-85-3: 347.04 (1).

$C_5H_7N_3$

 2-Hydrazinopyridine, CA 4930-98-7: Above 385.93 (1).

C_5H_8

 Cyclopentene, CA 142-29-0: 180.26 (7); 235.93 (2); 244.15 (3,6); below
 238.71 (1).

 Isoprene, CA 78-79-5: 219.26 (1,3,6,7); 225.15 (2,4).

 3-Methyl-1,2-butadiene, CA 598-25-4: 248.15 (3); 260.93 (1).

 Methylenecyclobutane, CA 1120-56-5: Below 238.71 (1).

 1,3-Pentadiene, *cis*, 1574-41-0: 258.15 (3); 244.26 (1).

 1,3-Pentadiene, *trans*, CA 2004-70-8: 258.15 (3); 244.26 (1).

 1,3-Pentadiene, mixed isomers, CA 504-60-9: 230.15 (7); 244.26 (2,6); as
 90% solution, below 238.71 (1).

(continues)

C_5H_8 *(continued)*

1,4-Pentadiene, CA 591-93-5:   253.15 (3); 277.59 (1); below 273.15 (7).

1-Pentyne, CA 627-19-0:   Below 238.71 (1); below 253.15 (5,6); 253.15 (3,7).

2-Pentyne, CA 627-21-4:   242.59 (1).

C_5H_8BN

Borane-pyridine complex, CA 110-51-0:   294.26 (1,3).

C_5H_8BrN

5-Bromovaleronitrile, CA 5414-21-1:   Above 385.93 (1).

$C_5H_8Br_2O_2$

Ethyl 2,3-dibromopropionate, CA 3674-13-3:   364.82 (1).

Methyl 3-bromo-2-(bromomethyl)propionate, CA 22262-60-8:   Above 385.93 (1).

C_5H_8ClN

5-Chlorovaleronitrile, CA 6280-87-1:   371.48 (1).

$C_5H_8Cl_2$

1,1-Dichlorocyclopentane, CA 31038-06-9:   311.48 (1).

$C_5H_8Cl_2O$

3-Chloropivaloyl chloride, CA 4300-97-4:   335.93 (1).

5-Chlorovaleroyl chloride, CA 1575-61-7:   363.71 (1); 364.15 (3).

$C_5H_8Cl_2O_2$

3,3-Dichloropivalic acid, CA 64855-18-1:   Above 385.93 (1).

$C_5H_8Cl_4P_2$

1,2-Bis(dichlorophosphino)cyclopentane, ± *trans*, CA 88293-05-4:   327.59 (1).

$C_5H_8F_4O$

2-Methyl-3,3,4,4-tetrafluoro-2-butanol, technical, CA 29553-26-2: 347.04 (1).

$C_5H_8N_2$

3-Dimethylaminoacrylonitrile, CA 2407-68-3:   Above 385.93 (1).

1,2-Dimethylimidazole, CA 1739-84-0:   365.37 (1).

1-Pyrrolidinecarbonitrile, CA 1530-88-7:   380.37 (1).

58

$C_5H_8N_2O$

4-Morpholinecarbonitrile, CA 1530-89-8: 377.59 (1).

C_5H_8O

Allyl vinyl ether, CA 3917-15-5: Below 293.15* (6,7).

Cyclopentanone, CA 120-92-3: 299.26 (5,6,7); 303.71 (1,3).

Cyclopentene oxide, CA 285-67-6: 283.15 (1).

Cyclopropyl methyl ketone, CA 765-43-5: 286.15 (3,7); 294.26 (1).

4,5-Dihydro-2-methylfuran, CA 1487-15-6: 260.93 (1).

3,4-Dihydro-2H-pyran, CA 110-87-2: 255.37 (6,7); 257.59 (1,3,5,7).

2-Ethylacrolein, CA 922-63-4: 272.15 (3); 274.26 (1).

Ethyl ethynyl carbinol, CA 4187-86-4: 302.59 (1).

Ethyl vinyl ketone, CA 1629-58-9: 266.48 (1); 263.15 (3).

1-Methoxy-1,3-butadiene, CA 3036-66-6: 267.59 (1).

2-Methyl-2-butenal, trans, CA 497-03-0: 291.48 (1); 294.15 (3).

2-Methyl-3-butyn-2-ol, CA 115-19-5: Below 294.15 (7); 295.15 (3); 298.15 (1); 298.15* (4,6); 303.93* (2).

Methyl isopropenyl ketone, CA 814-78-8: 294.15 (7).

2-Pentenal, trans, CA 1576-87-0: 295.93 (1).

3-Penten-2-one, technical, CA 625-33-2: 294.26 (1).

3-Pentyn-1-ol, CA 10229-10-4: 327.59 (1).

4-Pentyn-2-ol ±, CA 2117-11-5: 310.37 (1).

$C_5H_8O_2$

Allyl acetate, CA 591-87-7: 279.82 (1); 284.15 (3); 295.37 (7); 295.37* (6).

Cyclobutanecarboxylic acid, CA 3721-95-7: 357.04 (1).

Ethyl acrylate, CA 140-88-5: 281.15 (3); 282.15 (5); 283.15* (2,4,6); 288.71 (1,4,7).

Isopropenyl acetate, CA 108-22-5: 277.15 (3); 288.71 (6); 288.71* (7); 294.04 (1).

4-Methoxy-3-buten-2-one, technical, CA 4652-27-1: 336.48 (1).

alpha-Methyl-gamma-butyrolactone, CA 1679-47-6: 345.93 (1).

Methyl crotonate, CA 623-43-8: 277.59 (1); 272.15 (3).

Methyl cyclopropanecarboxylate, CA 2868-37-3: 290.93 (1).

1-Methylcyclopropanecarboxylic acid, CA 6914-76-7: 357.59 (1).

2-Methylcyclopropanecarboxylic acid, CA 29555-02-0: 360.93 (1).

Methyl methacrylate, CA 80-62-6: 283.15 (1,3,5); 283.15* (6,7); 285.93* (2,4).

'2-Methyltetrahydrofuran-3-one, CA 3188-00-9: 312.59 (1).

2,3-Pentanedione, CA 600-14-6: 292.04 (1,3).

2,4-Pentanedione, CA 123-54-6: 307.15 (1,5,6); 311.15 (3); 313.71* (7).

(continues)

$C_5H_8O_2$ *(continued)*

2-Pentenoic acid, *trans*, CA 13991-37-2: 375.37 (1).

4-Pentenoic acid, CA 591-80-0: 362.59 (1).

Tetrahydro-4*H*-pyran-4-one, CA 29943-42-8: 328.15 (3); 329.82 (1).

delta-Valerolactone, CA 542-28-9: 369.15 (3); 373.15 (1).

gamma-Valerolactone, CA 108-29-2: 354.26 (1); 369.15 (3); 369.26* (7).

Vinyl propionate, CA 105-38-4: 274.26* (6,7).

$C_5H_8O_3$

Ethyl pyruvate, CA 617-35-6: 318.71 (1,3).

2-Hydroxyethyl acrylate, CA 818-61-1: 340.93 (6); 372.59 (1); 377.59* (6).

Levulinic acid, CA 123-76-2: 410.93 (1).

Methyl acetoacetate, CA 105-45-3: 335.15 (3,5); 343.15 (1); 349.82 (4,6,7).

$C_5H_8O_4$

Dimethyl malonate, CA 108-59-8: 363.15 (1,4); 358.15 (3).

Methylene diacetate, CA 628-51-3: 347.04 (1).

C_5H_9Br

4-Bromo-2-methyl-2-butene, CA 870-63-3: 305.93 (1); 281.15 (3).

5-Bromo-1-pentene, CA 1119-51-3: 303.71 (1,3).

Cyclopentyl bromide, CA 137-43-9: 308.15 (1); 315.15 (3).

C_5H_9BrO

Tetrahydrofurfuryl bromide, CA 1192-30-9: 336.48 (1).

$C_5H_9BrO_2$

2-(2-Bromoethyl)-1,3-dioxolane, CA 18742-02-4: 338.15 (3); 358.15 (1).

2-Bromo-3-methylbutyric acid ±, CA 565-74-2: 380.37 (1).

2-Bromovaleric acid, CA 584-93-0: Above 385.93 (1).

5-Bromovaleric acid, CA 2067-33-6: Above 385.93 (1).

Ethyl 2-bromopropionate, CA 535-11-5: 324.82 (1); 339.15 (3).

Ethyl 3-bromopropionate, CA 539-74-2: 343.15 (3); 352.59 (1).

Isopropyl bromoacetate, CA 29921-57-1: 383.15 (1).

Methyl 2-bromobutyrate ±, CA 69043-96-5: 341.48 (1).

C_5H_9Cl

Cyclopentyl chloride, CA 930-28-9: 288.15 (1); 289.15 (3,7).

C_5H_9ClO

1-Chloro-3-pentanone, technical, CA 32830-97-0: 324.82 (1).

5-Chloro-2-pentanone, technical, CA 5891-21-4: 308.71 (1); 340.15 (3).

Isovaleryl chloride, CA 108-12-3: 292.04 (1); 304.15 (3).

Tetrahydrofurfuryl chloride, CA 3003-84-7: 318.15 (3); 320.93 (1).

Trimethylacetyl chloride, CA 3282-30-2: 282.04 (1); 292.15 (3).

Valeryl chloride, CA 638-29-9: 296.48 (1); 309.15 (3).

C_5H_9ClOS

Butyl chlorothioformate, CA 13889-94-6: 347.04 (2).

tert-Butyl chlorothioformate, CA 13889-95-7: 319.26 (2).

3-Chloropropyl thioacetate, technical, CA 13012-54-9: 350.93 (1).

$C_5H_9ClO_2$

Butyl chloroformate, CA 592-34-7: 298.15 (1); 311.15 (3); 319.15 (4); 325.35* (4).

sec-Butyl chloroformate, CA 17462-58-7: 308.75* (4); 311.15 (4).

3-Chloropivalic acid, CA 13511-38-1: Above 385.93 (1).

3-Chloropropyl acetate, CA 628-09-1: 340.93 (1).

5-Chlorovaleric acid, CA 1119-46-6: Above 385.93 (1).

Ethyl 2-chloropropionate ±, CA 535-13-7: 311.48 (1); 315.15 (3).

Ethyl 3-chloropropionate, CA 623-71-2: 327.59 (1); 334.15 (3).

Isobutyl chloroformate, CA 543-27-1: 300.93 (1); 307.55 (4); 309.15 (3); 312.65* (4).

Methyl 4-chlorobutyrate, CA 3153-37-5: 332.59 (1).

$C_5H_9Cl_3O_2Si$

Trimethylsilyl trichloroacetate, CA 25436-07-1: 301.15 (3).

$C_5H_9F_3O_2Si$

Trimethylsilyl trifluoroacetate, CA 400-53-3: 272.59 (1); 286.15 (3).

C_5H_9N

tert-Butyl isocyanide, CA 7188-38-7: 270.93 (1); 288.15 (3).

1-Dimethylamino-2-propyne, CA 7223-38-3: 265.37 (1); 289.15 (3).

1,1-Dimethylpropargylamine, 90% solution in water, CA 2978-58-7: 275.37 (1).

2-Methylbutyronitrile, (S)-(+), CA 25570-03-0: 297.59 (1).

1,2,3,6-Tetrahydropyridine, CA 694-05-3: 289.26 (1).

Trimethylacetonitrile, CA 630-18-2: 277.59 (1); 294.15 (3,7).

Valeronitrile, CA 110-59-8: 307.15 (3); 313.71 (1).

C_5H_9NO

Butyl isocyanate, CA 111-36-4: 284.15 (3); 293.15* (4); below 299.82* (6); 299.82 (1).

tert-Butyl isocyanate, CA 1609-86-5: 299.82 (1).

Cyclopentanone oxime, CA 1192-28-5: 365.37 (1).

N,N-Dimethylacrylamide, CA 2680-03-7: 343.15 (3); 344.82 (1).

3-Ethoxypropionitrile, CA 2141-62-0: 337.04 (1); 339.15 (3).

2-Ethyl-2-oxazoline, CA 10431-98-8: 302.59 (1).

3-Hydroxy-3-methylbutyronitrile, CA 13635-04-6: 370.15 (3).

1-Methyl-2-pyrrolidinone, CA 872-50-4: 359.26 (1); 368.15 (3); 368.71* (2,4,6,7).

5-Methyl-2-pyrrolidinone, CA 108-27-0: Above 385.93 (1).

N-Methyl-N-vinylacetamide, CA 3195-78-6: 331.15 (3); 332.04 (1).

1-Pyrrolidinecarboxaldehyde, CA 3760-54-1: 368.15 (1).

delta-Valerolactam, CA 675-20-7: Above 385.93 (1).

$C_5H_9NO_2$

4-Formylmorpholine, CA 4394-85-8: Above 385.93 (1).

Nitrocyclopentane, CA 2562-38-1: 340.37 (1).

$C_5H_9NO_4$

Methyl 4-nitrobutyrate, technical, CA 13013-02-0: 358.15 (1).

C_5H_9NS

Butyl isothiocyanate, CA 592-82-5: 339.26 (1).

tert-Butyl isothiocyanate, CA 590-42-1: 311.48 (1).

$C_5H_9N_3$

5-Amino-1-ethylpyrazole, CA 3528-58-3: Above 385.93 (1).

C_5H_{10}

Cyclopentane, CA 287-92-3: 235.93 (1,2); 253.15 (3); 266.15 (7).

Ethyl cyclopropane, CA 1191-96-4: Below 283.15 (7).

2-Methyl-1-butene, CA 563-46-2: Below 238.71 (1); 224.82 (2); 253.15 (3,7).

2-Methyl-2-butene, CA 513-35-9: 227.59 (1,2); 253.15 (3); 255.37* (7).

3-Methyl-1-butene, CA 563-45-1: 216.48 (1,2); below 253.15 (5); 266.15 (7); 267.15 (3).

1-Pentene, CA 109-67-1: 244.26 (1); 255.15 (3,7); 255.37* (6).

2-Pentene, mixed isomers, CA 109-68-2: 227.59 (1); 255.15 (3).

2-Pentene, cis, CA 627-20-3: 227.59 (2); 245.37 (1); below 253.15 (5,6); 255.15 (7).

2-Pentene, trans, CA 646-04-8: Below 253.15 (5.6).

$C_5H_{10}BrCl$

1-Bromo-5-chloropentane, CA 54512-75-3: 368.15 (1).

$C_5H_{10}Br_2$

1,4-Dibromopentane, CA 626-87-9: Above 385.93 (1).

1,5-Dibromopentane, CA 111-24-0: 352.59 (1).

$C_5H_{10}ClNO$

Diethylcarbamyl chloride, CA 88-10-8: 348.15 (1); 349.15 (3); 435.93 to 445.37* (6).

$C_5H_{10}Cl_2$

1,5-Dichloropentane, CA 628-76-2: 299.82 (1); above 299.82* (6,7); 314.15 (3).

Dichloropentane, mixed isomers, CA 30586-10-8: 309.26* (2); 314.15 (5); 314.26* (6,7).

$C_5H_{10}Cl_2O$

2,2-Bis(chloromethyl)-1-propanol, CA 5355-54-4: 377.59 (1).

$C_5H_{10}Cl_2O_2$

Bis(2-chloroethoxy)methane, CA 111-91-1: 383.15 (5); 383.15* (6,7).

$C_5H_{10}Cl_3O_3P$

Diethyl(trichloromethyl) phosphonate, CA 866-23-9: 365.15 (3).

$C_5H_{10}I_2$

1,5-Diiodopentane, CA 628-77-3: Above 385.93 (1).

$C_5H_{10}NO_3P$

Diethyl cyanophosphonate, CA 2942-58-7: 353.71 (1).

$C_5H_{10}N_2$

Diethylcyanamide, CA 617-83-4: 342.59 (1).

3-(Dimethylamino)propionitrile, CA 1738-25-6: Below 295.15 (7); 335.93 (1); 338.15 (5); 338.15* (6).

$C_5H_{10}N_2O$

1,3-Dimethyl-2-imidazolidinone, CA 80-73-9: 353.15 (1).

1-Piperazinecarboxaldehyde, CA 7755-92-2: 374.82 (1).

$C_5H_{10}O$

Allyl ethyl ether, CA 557-31-3: 252.59 (1); 263.15 (3).

Cyclobutanemethanol, CA 4415-82-1: 313.15 (1).

Cyclopentanol, CA 96-41-3: 324.26 (1,3,5,6).

Cyclopropylethanol, CA 765-42-4: 303.71 (1); 310.15 (3).

Ethyl 1-propenyl ether, CA 928-55-2: Above 265.93* (6,7); 293.15 (3).

Isopropyl vinyl ether, CA 926-65-8: 241.15 (6,7); below 253.15 (5).

Isovaleraldehyde, CA 590-86-3: 268.15 (3,7); 271.48 (1); 282.04* (6).

3-Methyl-2-butanone, CA 563-80-4: 270.15 (3); 279.26 (1).

2-Methyl-3-buten-2-ol, CA 115-18-4: 283.15 (3); 286.48 (1).

3-Methyl-2-buten-1-ol, CA 556-82-1: 316.48 (1).

3-Methyl-3-buten-1-ol, CA 763-32-6: 309.26 (1); 315.15 (3).

2-Methylbutyraldehyde, CA 96-17-3: 277.59 (1); 282.59* (6).

1-Methylcyclopropanemethanol, CA 2746-14-7: 307.04 (1).

2-Methylcyclopropanemethanol, mixed isomers: 312.59 (1).

2-Methyltetrahydrofuran, CA 96-47-9: 261.15 (3,5); 262.04 (1,6,7).

2-Pentanone, CA 107-87-9: 280.37 (1,2,4,5,6,7); 283.15 (3); 283.15* (2).

3-Pentanone, CA 96-22-0: 285.15 (5); 285.93 (1,3,4,7); 285.93* (6).

1-Penten-3-ol, CA 616-25-1: 298.15 (1); 301.15 (3).

2-Penten-1-ol, mixed isomers, CA 20273-24-9: 323.71 (1).

2-Penten-1-ol, *cis*, CA 1576-95-0; 321.48 (1).

3-Penten-2-ol, CA 1569-50-2: 300.93 (1); 309.15 (3).

4-Penten-1-ol, CA 821-09-0: Below 296.15 (7); 316.48 (1); 321.15 (3).

4-Penten-2-ol, CA 625-31-0: 298.71 (1); 303.15 (3).

Tetrahydropyran, CA 142-68-7: 253.15 (2,3,5,6); 257.59 (1).

Trimethylacetaldehyde, CA 630-19-3: 257.59 (1); 268.15 (3).

Valeraldehyde, CA 110-62-3: 277.15 (3); 285.15 (1,5,7); 285.37* (2,6).

$C_5H_{10}OS$

S-Methyl thiobutanoate, CA 2432-51-0: 307.59 (1).

3-(Methylthio)-2-butanone ±, CA 53475-15-3: 317.59 (1); 321.15 (3).

Thiopivalic acid, CA 55561-02-9: 297.04 (1).

$C_5H_{10}O_2$

3-Acetyl-1-propanol, CA 1071-73-4: 366.48 (1).

Acrolein dimethyl acetal, CA 6044-68-4: 270.37 (1); 292.15 (7).

Butyl formate, CA 592-84-7: 287.04 (1); 291.15 (3,5,6,7).

tert-Butyl formate, CA 762-75-4: 263.71 (1); 331.15 (3).

1,3-Cyclopentanediol, mixed isomers, CA 59719-74-3: Above 385.93 (1).

2,2-Dimethyl-1,3-dioxolane, CA 2916-31-6: 272.15 (7); 279.15 (3).

(continues)

64

$C_5H_{10}O_2$ *(continued)*

2,2-Dimethyl-3-hydroxypropionaldehyde, technical, CA 597-31-9: 356.48 (1).

3-Ethoxypropionaldehyde, CA 2806-85-1: 310.93* (7); 311.15 (5,6).

Ethyl propionate, CA 105-37-3: 285.37 (1,3,4,5,6,7).

3-Hydroxy-3-methyl-2-butanone, CA 115-22-0: 315.37 (1).

4-Hydroxy-3-methyl-2-butanone, technical, CA 3393-64-4: 354.82 (1).

Isobutyl formate, CA 542-55-2: 283.15 (1); below 294.15 (5,6,7).

Isopropyl acetate, CA 108-21-4: 274.82 (6); 277.15 (2,4,5,7); 279.15 (3,4); 279.15* (4); 289.82 (1).

Isovaleric acid, CA 503-74-2: 343.71 (1); 369.26* (2).

3-Methoxybutyraldehyde, CA 5281-76-5: 333.15 (5,6,7).

2-Methoxyethyl vinyl ether, CA 1663-35-0: 291.15 (5,7); 291.15* (6,7).

1-Methoxy-2-methylpropylene oxide ±, CA 26196-04-3: 279.26 (1); 291.15 (3).

2-Methoxytetrahydrofuran, CA 13436-45-8: 280.93 (1).

Methyl butyrate, CA 623-42-7: 284.82 (1); 287.15 (3,4,5,6,7).

2-Methylbutyric acid ±, CA 600-07-7: 347.04 (1); 356.15 (3).

2-Methylbutyric acid, (S)-(+), CA 1730-91-2: 347.04 (1).

4-Methyl-1,3-dioxane, CA 1120-97-4: 295.37 (1).

Methyl isobutyrate, CA 547-63-7: 276.48 (1); 285.15 (3); 286.15 (7).

3-Methyl-3-oxetanemethanol, CA 3143-02-0: 370.15 (3); 372.04 (1).

Propyl acetate, CA 109-60-4: 283.15 (5); 285.93 (1,2,6); 287.15 (3,4,7); 287.15* (2,4).

Tetrahydrofurfuryl alcohol, CA 97-99-4: 347.15 (5); 348.15 (3); 348.15* (6); 357.04 (1,2,4,7); 357.04* (2,4).

Tetrahydro-2*H*-pyran-2-ol, technical, CA 694-54-2: Above 385.93 (1).

Tetrahydro-4*H*-pyran-4-ol, CA 2081-44-9: 359.15 (3); 360.93 (1).

Trimethylacetic acid, CA 75-98-9: 337.04 (1,4).

Valeric acid, CA 109-52-4: 362.04 (1); 369.15 (3); 369.15* (2,6,7).

$C_5H_{10}O_2S$

Ethyl (methylthio)acetate, CA 4455-13-4: 332.59 (1).

Methyl 3-(methylthio)propionate, CA 13532-18-8: 345.37 (1).

3-Methylsulfolane, CA 872-93-5: Above 385.93 (1).

$C_5H_{10}O_3$

Diethyl carbonate, CA 105-58-8: 298.15 (3,5,6); 298.15* (7); 304.26 (1,2); 305.95 (4); 319.25* (4).

3-Ethoxypropionic acid, CA 4324-38-3: 380.37 (6); 380.37* (7).

Ethyl lactate ±, CA 97-64-3: 319.26 (5,6,7); 327.04 (2).

Ethyl lactate, (S)-(-), CA 687-47-8: 319.15 (3); 322.04 (1).

Ethyl methoxyacetate, CA 3938-96-3: 319.26 (1); 320.15 (3).

(continues)

$C_5H_{10}O_3$ *(continued)*

2-Methoxyethyl acetate, CA 110-49-6: 317.04 (1,5,6,7); 319.15 (3); 322.04 (2,6); 324.82* (2); 328.75* (4); 329.15 (4).

Methyl 2-hydroxybutyrate, CA 2110-78-3: 314.15 (3).

Methyl 3-hydroxybutyrate ±, CA 1487-49-6: 355.37* (6).

Methyl 3-hydroxybutyrate, (R)-(-), CA 3976-69-0: 344.82 (1); 346.15 (3).

Methyl 3-hydroxybutyrate, (S)-(+), CA 53562-86-0: 344.82 (1).

Methyl 2-hydroxyisobutyrate, CA 2110-78-3: 315.37 (1).

Methyl 3-hydroxy-2-methylpropionate, (S)-(+), CA 80657-57-4: 314.15 (3); 354.26 (1).

Methyl 3-hydroxy-2-methylpropionate, (R)-(-), CA 72657-23-9: 322.15 (3); 354.26 (1).

Pyruvic aldehyde, dimethyl acetal, CA 6342-56-9: 299.15 (3); 310.93 (1).

$C_5H_{10}O_4$

Glycerol monoacetate, CA 26446-35-5: 317.04 (6); 424.82* (7).

Methyl dimethoxyacetate, CA 89-91-8: 336.48 (1); 340.15 (3).

$C_5H_{10}O_5$

Ribulose hydrate, *dextro*, CA 488-84-6: Above 385.93 (1).

Xylulose, *levo*, CA 527-50-4: Above 385.93 (1).

$C_5H_{10}S$

Pentamethylene sulfide, CA 1613-51-0: 294.26 (1).

$C_5H_{10}S_2$

2-Methyl-1,3-dithiane, CA 6007-26-7: 348.15 (3).

$C_5H_{10}Si$

Ethynyltrimethylsilane, CA 1066-54-2: Below 238.71 (1); 247.15 (3).

$C_5H_{11}Br$

1-Bromo-2,2-dimethylpropane, CA 630-17-1: 279.82 (1); 282.15 (3).

1-Bromo-2-methylbutane, (S)-(+), CA 534-00-9: 295.37 (1).

1-Bromo-3-methylbutane, CA 107-82-4: 294.15 (7); 305.37 (1,2,3).

1-Bromopentane, CA 110-53-2: 304.26 (1); 305.37 (2,3,6,7).

2-Bromopentane ±, CA 107-81-3: 293.71 (1,7); 305.37 (7).

3-Bromopentane, CA 1809-10-5: 291.48 (1); 292.15 (3).

$C_5H_{11}BrO_2$

1-Bromo-2,2-dimethoxypropane, CA 126-38-5: 313.71 (1).

(continues)

$C_5H_{11}BrO_2$ *(continued)*

1-Bromo-2-(2-methoxyethoxy)ethane, CA 54149-17-6: 347.04 (1).

$C_5H_{11}BrO_2Si$

Trimethylsilyl bromoacetate, CA 18291-80-0: 301.48 (1); 320.15 (3).

$C_5H_{11}BrSi$

1-(Bromovinyl)trimethylsilane, CA 13683-41-5: 286.15 (3).

2-(Bromovinyl)trimethylsilane, CA 41309-43-7: 302.15 (3).

$C_5H_{11}Cl$

1-Chloro-3-methylbutane, CA 107-84-6: Below 294.15 (5,6,7).

2-Chloro-2-methylbutane, CA 594-36-5: 263.71 (1); 285.15 (7).

Neopentyl chloride, CA 753-89-9: 264.26 (1).

Pentyl chloride, CA 543-59-9: 276.15 (5); 284.26 (1); 285.15 (2,3); 285.37* (6,7).

Pentyl chloride, mixed isomers, CA 29656-63-1: 276.48* (6).

$C_5H_{11}ClO$

3-Chloro-2,2-dimethyl-1-propanol, CA 13401-56-4: 344.26 (1).

$C_5H_{11}ClSi$

Allylchlorodimethylsilane, CA 4028-23-3: 247.15 (3); 278.71 (1).

$C_5H_{11}Cl_3Si$

Pentyltrichlorosilane, CA 107-72-2: 303.71 (1); 335.93* (6).

$C_5H_{11}F$

1-Fluoropentane, CA 592-50-7: 250.37 (1).

$C_5H_{11}F_3O_3SSi$

Trimethylsilylmethyl trifluoromethanesulfonate, CA 64035-64-9: 315.15 (3).

$C_5H_{11}I$

1-Iodo-3-methylbutane, CA 541-28-6: 313.15 (3).

1-Iodopentane, CA 628-17-1: 316.15 (3); 324.26 (1).

Neopentyl iodide, CA 15501-33-4: 305.93 (1).

$C_5H_{11}N$

N-Allyldimethylamine, CA 2155-94-4: 296.15 (3).

Cyclopentylamine, CA 1003-03-8: 286.15 (3,7); 290.37 (1).   *(continues)*

$C_5H_{11}N$ *(continued)*

1-Methylpyrrolidine, CA 120-94-5: 251.48 (1); 259.26 (6); 276.15 (3,7).

Piperidine, CA 110-89-4: 276.15 (7); 277.59 (1); 289.15 (3,5,6).

Piperidine hydrochloride, CA 6091-44-7: Above 385.93 (1).

$C_5H_{11}NO$

N-tert-Butylformamide, CA 2425-74-2: 355.15 (3); 368.15 (1).

N,N-Diethylformamide, CA 617-84-5: 333.71 (1); 339.15 (3).

(Dimethylamino)acetone, CA 15364-56-4: 303.15 (1); 307.15 (3).

N,N-Dimethylpropionamide, CA 758-96-3: 335.37 (1).

4-Hydroxypiperidine, CA 5382-16-1: 380.93 (1).

4-Methylmorpholine, CA 109-02-4: 287.05 (4); 295.35* (4); 297.04 (1,6,7); 301.15 (3).

1-Methyl-3-pyrrolidinol, CA 13220-33-2: 343.71 (1).

2-Pyrrolidinemethanol, *(R)*-(-), CA 68832-13-3: 359.26 (1).

2-Pyrrolidinemethanol, *(S)*-(+), CA 23356-96-9: 359.26 (1).

Tetrahydrofurfurylamine, CA 4795-29-3: 316.15 (3); 318.71 (1).

$C_5H_{11}NO_2$

Butyl carbamate, CA 592-35-8: 382.04 (1).

2-Dimethylamino-1,3-dioxolane, CA 19449-26-4: 311.15 (3).

N-Ethylurethane, CA 623-78-9: 348.71 (1).

Isopentyl nitrite, CA 110-46-3: 253.15 (3); 283.15 (1); below 296.15 (7).

2-Methylaminomethyl-1,3-dioxolane, CA 57366-77-5: 331.48 (1).

1-Nitropentane, CA 628-05-7: 332.59 (1); 334.15 (3).

Pentyl nitrite, CA 463-04-7: 294.15 (6,7).

$C_5H_{11}NO_3$

N,N-Bis(2-hydroxyethyl)formamide, CA 25209-66-9: Above 385.93 (1).

Isopentyl nitrate, CA 543-87-3: 321.15 (3).

Isopentyl nitrate, mixed isomers: 309.26 (1).

3-Nitro-2-pentanol, mixed isomers, CA 5447-99-4: 363.71 (1).

Pentyl nitrate, CA 1002-16-0: 320.93* (6); 324.82* (7).

$C_5H_{11}NS$

2,2-Dimethylthiazolidine, CA 19351-18-9: 325.93 (1).

$C_5H_{11}NSi$

Trimethylsilylacetonitrile, CA 18293-53-3: 316.15 (3); 323.15 (1).

Trimethylsilylmethylisocyanide, CA 30718-17-3: Below 288.15 (3).

$C_5H_{11}N_2O_2P$

Ethyl dimethylamidocyanophosphate, CA 77-81-6: 350.93 (7).

$C_5H_{11}N_3Si$

2-Trimethylsilyl-1,2,3-triazole, CA 13518-80-4: 281.48 (1).

$C_5H_{11}O_4P$

Dimethyl (2-oxopropyl)phosphonate, CA 4202-14-6: 366.48 (1).

$C_5H_{11}O_5P$

Trimethyl phosphonoacetate, CA 5927-18-4: 343.15 (3); above 385.93 (1).

C_5H_{12}

2-Methylbutane, CA 78-78-4: 216.48 (1,2); 213.0 (4); below 222.04 (6,7); 222.15 (3).

Neopentane, CA 463-82-1: 198.0 (4); 208.15 (2).

Pentane, CA 109-66-0: 223.71 (1,2,3); below 233.15 (6,7); 233.15 (4); 233.15* (4).

$C_5H_{12}ClO_3P$

Diethyl chloromethylphosphonate, CA 3167-63-3: 359.82 (1).

$C_5H_{12}N_2$

1-Aminopiperidine, CA 2213-43-6: 309.26 (1).

Homopiperazine, CA 505-66-8: 337.59 (1).

1-Methylpiperazine, CA 109-01-3: 312.15 (3); 315.37 (1); 315.37* (6,7).

2-Methylpiperazine, CA 109-07-9: 285.93 (1); 346.15* (2).

$C_5H_{12}N_2O$

1,1,3,3-Tetramethylurea, CA 632-22-4: 338.71 (1); 348.15 (7).

$C_5H_{12}N_2O_2$

tert-Butyl carbazate, CA 870-46-2: 364.82 (1).

$C_5H_{12}O$

tert-Amyl alcohol, CA 75-85-4: 292.59 (4,5,6); 294.26 (1); 297.15* (4); 294.26* (2); 313.71 (7); 317.15 (3).

Butyl methyl ether, CA 628-28-4: 263.15 (1,3); below 265.37 (7).

tert-Butyl methyl ether, CA 1634-04-4: 248.15 (3); 263.15 (1).

Ethyl propyl ether, CA 628-32-0: Below 253.15 (5,6,7).

2-Methyl-1-butanol ±, CA 34713-94-5: 313.15 (5); 316.48 (1); 317.15 (3); 323.15* (4,6,7).   (continues)

C$_5$H$_{12}$O *(continued)*

2-Methyl-1-butanol, *(S)*-(-), CA 1565-80-6: 316.48 (1); 317.15 (3).

3-Methyl-1-butanol, CA 123-51-3: 316.15 (3,4,5,6,7); 318.71 (1); 324.82* (2); 329.15* (4).

3-Methyl-2-butanol, CA 598-75-4: 299.82 (1); 303.15 (5); 307.15 (3); 308.15* (4).

3-Methyl-2-butanol, *(S)*-(+), CA 1517-66-4: 299.82 (1).

Neopentyl alcohol, CA 75-84-3: 303.15 (5); 309.82 (1,4,6).

1-Pentanol, CA 71-41-0: 306.15 (4,6,7); 311.15 (3); 322.15 (5); 311.15* (4).

2-Pentanol, CA 6032-29-7: 307.15 (1,3,4,5,6); 313.71 (6); 313.71* (2,7); 315.15* (4).

3-Pentanol, CA 584-02-1: 292.04 (7); 303.15 (5); 312.15* (4); 313.71 (1,4,6).

Pentyl alcohol, mixed primary isomers, CA 30899-19-5: 320.93* (2); 330.37* (2).

C$_5$H$_{12}$O$_2$

Diethoxymethane, CA 462-95-3: 268.15 (1,3); below 294.15 (7).

1,1-Dimethoxypropane, CA 4744-10-9: 283.15 (7).

1,2-Dimethoxypropane, CA 7778-85-0: 273.71 (1).

2,2-Dimethoxypropane, CA 77-76-9: 262.04 (1); 263.15 (3); 266.15 (7).

2,2-Dimethyl-1,3-propanediol, CA 126-30-7: 380.37 (1); 402.15* (4,6); 424.82* (2).

3-Ethoxy-1-propanol, CA 111-35-3: 327.59 (1).

2-Isopropoxyethanol, CA 109-59-1: 306.48* (6); 316.15 (3).

3-Methoxy-1-butanol, CA 2517-43-3: 319.82 (1); 345.93* (2); 347.04* (6); 347.15 (5).

1,2-Pentanediol, CA 5343-92-0: 377.59 (1).

1,4-Pentanediol, CA 626-95-9: Above 385.93 (1).

1,5-Pentanediol, CA 111-29-5: 402.59 (1); 402.59* (6,7); 408.15 (5,7).

2,4-Pentanediol, mixed isomers, CA 625-69-4: 372.04* (2); 374.82 (1).

2,4-Pentanediol, *(2R,4R)*-(-), CA 42075-32-1: 374.82 (1).

2,4-Pentanediol, *(2S,4S)*-(+): 374.82 (1).

2-Propoxyethanol, CA 2807-30-9: 322.04 (2).

Propylene glycol, ethyl ether, CA 1569-02-4: 316.48* (2).

C$_5$H$_{12}$O$_2$S

3-Ethylthio-1,2-propanediol, CA 60763-78-2: Above 385.93 (1).

(Methylthio)acetaldehyde dimethyl acetal, CA 40015-15-4: 254.26 (1).

C$_5$H$_{12}$O$_2$Si

Trimethylsilyl acetate, CA 13411-48-8: 277.59 (1); 292.15 (3).

$C_5H_{12}O_3$

3-Ethoxy-1,2-propanediol, CA 1874-62-0: 383.15 (1).

2-(2-Methoxyethoxy)ethanol, CA 111-77-3: 357.04 (1); 359.85 (4); 362.04 (2); 366.15 (4,5); 366.48* (2,4,7); 369.26* (6).

1,1,2-Trimethoxyethane, CA 24332-20-5: 296.48 (1); 303.15 (3).

Trimethylorthoacetate, CA 1445-45-0: 288.15 (3); 289.82 (1).

$C_5H_{12}O_3Si$

Vinyltrimethoxysilane, CA 2768-02-7: 295.93 (1).

$C_5H_{12}O_4$

Tetramethyl orthocarbonate, CA 1850-14-2: 279.82 (1); 307.15 (3).

$C_5H_{12}S$

tert-Butyl methyl sulfide, CA 6163-64-0: 269.26 (1).

2-Methyl-1-butanethiol, CA 1878-18-8: 292.59 (1).

2-Methyl-2-butanethiol, technical, CA 1679-09-0: 272.04 (1); 272.04* (2).

3-Methyl-1-butanethiol, CA 541-31-1: 291.48 (1).

3-Methyl-2-butanethiol, CA 2084-18-6: 275.93* (6).

1-Pentanethiol, CA 110-66-7: 291.48 (1,3,5,7); 291.48* (6); 299.82* (2).

Pentyl mercaptans, mixed isomers: 291.15 (5); 291.48* (6).

$C_5H_{12}Si$

Vinyltrimethylsilane, CA 754-05-2: Below 238.71 (1); 249.15 (3).

$C_5H_{13}ClOSi$

(2-Chloroethoxy)trimethylsilane, CA 18157-17-0: 303.71 (1).

$C_5H_{13}ClSi$

Chlorodimethylisopropylsilane, CA 3634-56-8: 288.15 (1).

(1-Chloroethyl)trimethylsilane, CA 7787-87-3: 286.15 (3).

$C_5H_{13}N$

1-Aminopentane, CA 110-58-7: 272.15 (3,6); 277.59 (1); 280.15 (5); 284.26 (2).

2-Aminopentane ±, CA 63493-28-7: 308.15 (1,3); 266.48 (6).

N,N-Diethylmethylamine, CA 616-39-7: 249.26 (1); 289.15 (3).

N,N-Dimethylisopropylamine, CA 996-35-0: 249.15 (3).

1,2-Dimethylpropylamine ±, CA 598-74-3: 245.37 (1).

N-Ethylisopropylamine, CA 19961-27-4: 289.15 (3).

1-Ethylpropylamine, CA 616-24-0: 275.37 (1).   (continues)

$C_5H_{13}N$ *(continued)*

Isopentylamine, CA 107-85-7: 272.15 (3); 291.48 (1).

N-Methylbutylamine, CA 110-68-9: 272.15 (3); 274.82 (1,7); 285.93* (6).

2-Methylbutylamine, CA 96-15-1: 276.48 (1); 281.15 (3).

2-Methylbutylamine, (S)-(-), CA 20626-52-2: 276.48 (1); 285.15 (3).

Neopentylamine, CA 5813-64-9: 259.26 (1); 298.15 (3).

tert-Pentylamine, CA 594-39-8: 272.04 (1).

Pentylamine, mixed isomers: 280.37* (7); 291.48* (2).

$C_5H_{13}NO$

2-Amino-3-methyl-1-butanol ±, CA 16369-05-4: 363.15 (1).

2-Amino-3-methyl-1-butanol, (R)-(+), CA 4276-09-9: 351.15 (3); 363.15 (1).

2-Amino-3-methyl-1-butanol, (S)-(-), CA 2026-48-4: 364.26 (1).

2-Amino-1-pentanol ±, CA 4146-04-7: 368.15 (1).

5-Amino-1-pentanol, CA 2508-29-4: 338.71 (1).

1-Dimethylamino-2-propanol, CA 108-16-7: 300.93 (2); 305.37 (7); 308.15 (1); 308.15* (6); 314.15 (3); as 77% solution in water, 310.37 (2); as 70% solution in water, 313.15 (2).

2-Dimethylamino-2-propanol, CA 2475-27-6: 308.15 (5).

3-Dimethylamino-1-propanol, CA 3179-63-3: 309.26 (1); 314.15 (3).

3-Ethoxypropylamine, CA 6291-85-6: 305.93 (1,3).

2-(Isopropylamino)ethanol, CA 109-56-8: 351.15 (3).

2-(Propylamino)ethanol, CA 16369-21-4: 351.48 (1); 360.93 (2).

$C_5H_{13}NOS$

Methioninol, (S)-(-), CA 2899-37-8: Above 385.93 (1).

$C_5H_{13}NOSi$

N-(Trimethylsilyl)acetamide, CA 13435-12-6: 330.37 (1).

$C_5H_{13}NO_2$

2-Amino-2-ethyl-1,3-propanediol, CA 115-70-8: 347.59* (2); above 385.93 (1).

(N,N-Dimethylamino)dimethoxymethane, CA 4637-24-5: 278.15 (3); 280.37 (1).

3-Dimethylamino-1,2-propanediol ±, CA 623-57-4: 377.59 (1).

Methylaminoacetaldehyde dimethyl acetal, CA 122-07-6: 302.59 (1,3).

N-Methyldiethanolamine, CA 105-59-9: 388.71 (2); 399.82 (1); 399.82* (6).

$C_5H_{13}N_3$

1-Amino-4-methylpiperazine, CA 6928-85-4: 335.93 (1).

1,1,3,3-Tetramethylguanidine, CA 80-70-6: 333.15 (1).

$C_5H_{13}O_3P$

Diethyl methylphosphonate, CA 683-08-9: 348.71 (1).

$C_5H_{14}BNO$

Borane-4-methylmorpholine complex, CA 15648-16-5: 339.26 (1).

$C_5H_{14}N_2$

1,5-Diaminopentane, CA 462-94-2: 335.93 (1).

1-Dimethylamino-2-propylamine, CA 62689-51-4: 308.15 (3).

3-Dimethylaminopropylamine, CA 109-55-7: 288.71 (1); 304.15 (3); 308.15 (2,5); 310.93* (6,7); 318.15 (4).

2,2-Dimethyl-1,3-propanediamine, CA 7328-91-8: 320.37 (1).

N-Isopropylethylenediamine, CA 19522-67-9: 279.26 (1).

N-Propylethylenediamine, CA 111-39-7: 318.15 (3).

N,N,N',N'-Tetramethyldiaminomethane, CA 51-80-9: 260.37 (1); 262.15 (3).

N,N,N'-Trimethylethylenediamine, CA 142-25-6: 282.59 (1); 287.15 (3,7).

$C_5H_{14}N_2O$

1-Amino-2-(2-hydroxyethyl)aminopropane, CA 10138-74-6: 398.82* (6).

2-(3-Aminopropylamino)ethanol, CA 4461-39-6: 425.15 (3).

$C_5H_{14}OSi$

Ethoxytrimethylsilane, CA 1825-62-3: 254.26 (1); 272.15 (3).

(Methoxymethyl)trimethylsilane, CA 14704-14-4: 259.82 (1); 267.15 (3).

2-(Trimethylsilyl)ethanol, CA 2916-68-9: 323.71 (1); 329.15 (3).

$C_5H_{14}SSi$

(Ethylthio)trimethylsilane, CA 5573-62-6: 290.15 (3).

$C_5H_{14}Si$

Diethylmethylsilane, CA 760-32-7: 248.71 (1); 250.15 (3).

Dimethylisopropylsilane, CA 18209-61-5: 243.15 (1).

$C_5H_{15}NSi$

N,N-Dimethyltrimethylsilylamine, CA 2083-91-2: 253.71 (1); 293.15 (3).

$C_5H_{15}N_3$

N-(2-Aminoethyl)-1,3-propanediamine, CA 13531-52-7: 369.26 (1).

$C_5H_{15}O_3PSi$

Dimethyl trimethylsilyl phosphite, CA 36198-87-5: 305.15 (3).

C_6BrF_5

Bromopentafluorobenzene, CA 344-04-7: 360.92 (1).

C_6ClF_5

Chloropentafluorobenzene, CA 344-07-0: Nonflammable (1).

$C_6ClF_5O_2S$

Pentafluorobenzenesulfonyl chloride, CA 832-53-1: 363.71 (1).

$C_6Cl_3F_3$

1,3,5-Trichloro-2,4,6-trifluorobenzene, CA 319-88-0: 375.37 (1).

C_6Cl_6

Hexachlorobenzene, CA 118-74-1: 515.37 (7).

C_6F_5I

Iodopentafluorobenzene, CA 827-15-6: Nonflammable (1).

$C_6F_5NO_2$

Pentafluoronitrobenzene, CA 880-78-4: 363.71 (1).

C_6F_6

Hexafluorobenzene, CA 392-56-3: 283.15 (1,7).

$C_6F_{10}O_3$

Pentafluoropropionic anhydride, CA 356-42-3: Nonflammable (1).

C_6F_{12}

Perfluorocyclohexane, CA 355-68-0: Nonflammable (1).

$C_6F_{13}I$

Perfluorohexyl iodide, CA 355-43-1: Nonflammable (1).

C_6F_{14}

Perfluorohexane, CA 355-42-0: Nonflammable (1).

C_6HBrF_4

1-Bromo-2,3,5,6-tetrafluorobenzene, CA 1559-88-2: 327.04 (1).

$C_6HF_4NO_2$

2,3,4,6-Tetrafluoronitrobenzene, CA 314-41-0: 344.26 (1).

C_6HF_5

Pentafluorobenzene, CA 363-72-4: 287.04 (1).

C_6HF_5O

Pentafluorophenol, CA 771-61-9: 345.37 (1).

C_6HF_5S

Pentafluorothiophenol, CA 771-62-0: 324.82 (1).

$C_6H_2BrF_3$

1-Bromo-2,4,5-trifluorobenzene, CA 327-52-6: 328.71 (1).

1-Bromo-2,4,6-trifluorobenzene, CA 2367-76-2: Above 385.93 (1).

$C_6H_2Br_2F_2$

1,2-Dibromo-4,5-difluorobenzene, CA 64695-78-9: 317.59 (1).

$C_6H_2Cl_3NO_2$

1,2,3-Trichloro-4-nitrobenzene, CA 17700-09-3: Above 385.93 (1).

1,2,4-Trichloro-5-nitrobenzene, CA 89-69-0: Above 385.93 (1).

$C_6H_2Cl_4$

1,2,3,4-Tetrachlorobenzene, CA 634-66-2: Above 385.93 (1).

1,2,3,5-Tetrachlorobenzene, CA 634-90-2: Above 385.93 (1).

1,2,4,5-Tetrachlorobenzene, CA 95-94-3: Above 385.93 (1); 428.15 (6,7).

Tetrachlorobenzene, mixed isomers, CA 12408-10-5: 428.15 (5,6).

$C_6H_2FN_3O_3$

4-Fluoro-7-nitrobenzofurazan, CA 29270-56-2: Above 385.93 (1).

$C_6H_2F_3NO_2$

1,2,4-Trifluoro-5-nitrobenzene, CA 2105-61-5: 362.59 (1).

$C_6H_2F_4$

1,2,3,5-Tetrafluorobenzene, technical, CA 2367-82-0: As 85% solution, 277.59 (1).

1,2,4,5-Tetrafluorobenzene, CA 327-54-8: 277.15* (4); 289.26 (1).

$C_6H_2F_4O$

2,3,5,6-Tetrafluorophenol, CA 769-39-1: 352.59 (1).

$C_6H_2F_4S$

2,3,5,6-Tetrafluorothiophenol, CA 769-40-4: 321.48 (1).

$C_6H_2F_5N$

2,3,4,5,6-Pentafluoroaniline, CA 771-60-8: 347.04 (1).

C_6H_3BrFI

1-Bromo-3-fluoro-4-iodobenzene: 383.15 (1).

$C_6H_3BrF_2$

1-Bromo-2,4-difluorobenzene, CA 348-57-2: 324.82 (1).

1-Bromo-2,5-difluorobenzene, CA 399-94-0: 338.15 (1).

1-Bromo-2,6-difluorobenzene, CA 64248-56-2: 326.48 (1).

1-Bromo-3,4-difluorobenzene, CA 348-61-8: 306.48 (1).

1-Bromo-3,5-difluorobenzene, CA 461-96-1: 317.59 (1).

$C_6H_3Br_2F$

2,4-Dibromo-1-fluorobenzene, technical, CA 1435-53-6: As 70% solution,
365.37 (1).

$C_6H_3Br_3$

1,2,4-Tribromobenzene, CA 615-54-3: 383.15 (1).

$C_6H_3ClFNO_2$

3-Chloro-4-fluoronitrobenzene, CA 350-30-1: Above 385.93 (1).

$C_6H_3ClN_2O_4$

1-Chloro-2,4-dinitrobenzene, CA 97-00-7: 459.26 (1); 467.15* (5);
467.59 (7).

1-Chloro-3,4-dinitrobenzene, CA 610-40-2: Above 385.93 (1).

Chlorodinitrobenzene, mixed isomers, CA 25567-67-3: 467.59 (6).

$C_6H_3Cl_2FO$

2,6-Dichloro-4-fluorophenol, CA 392-71-2: 376.48 (1).

$C_6H_3Cl_2I$

1,2-Dichloro-3-iodobenzene, CA 2401-21-0: Above 385.93 (1).

1,3-Dichloro-4-iodobenzene, CA 29898-32-6: 383.15 (1).

1,4-Dichloro-2-iodobenzene, CA 29682-41-5: 367.04 (1).

$C_6H_3Cl_2NO_2$

2,3-Dichloronitrobenzene, CA 3209-22-1: 397.04 (1).

2,4-Dichloronitrobenzene, CA 611-06-3: Above 385.93 (1).

2,5-Dichloronitrobenzene, CA 89-61-2: Above 385.93 (1).

3,4-Dichloronitrobenzene, CA 99-54-7: 397.04 (1).

$C_6H_3Cl_3$

1,2,3-Trichlorobenzene, CA 87-61-6: 386.15 (6); 386.15* (2); 399.82 (1).

1,2,4-Trichlorobenzene, CA 120-82-1: 372.15 (4); 378.71 (6); 383.15 (4,5,7); 399.82 (1).

1,3,5-Trichlorobenzene, CA 108-70-3: 380.15 (5); 380.15* (2); 399.82 (1).

$C_6H_3Cl_3O$

2,3,6-Trichlorophenol, CA 933-75-5: 352.04 (1).

2,4,6-Trichlorophenol, CA 88-06-2: Nonflammable (1).

$C_6H_3Cl_3O_2S$

2,5-Dichlorobenzenesulfonyl chloride, CA 5402-73-3: Above 385.93 (1).

$C_6H_3Cl_4O_2P$

2,5-Dichlorophenyl dichlorophosphate, CA 53676-18-9: 383.15 (1).

$C_6H_3Cl_5Si$

(Dichlorophenyl)trichlorosilane, CA 27137-85-5: 414.26 (7).

$C_6H_3FN_2O_4$

2,4-Dinitrofluorobenzene, CA 70-34-8: Above 385.93 (1).

$C_6H_3F_2NO_2$

2,4-Difluoronitrobenzene, CA 446-35-5: 363.71 (1).

2,5-Difluoronitrobenzene, CA 364-74-9: 363.15 (1).

3,4-Difluoronitrobenzene, CA 369-34-6: 353.71 (1).

$C_6H_3F_3$

1,2,4-Trifluorobenzene, CA 367-23-7: 268.15* (4); 277.59 (1).

1,3,5-Trifluorobenzene, CA 372-38-3: 265.93 (1).

$C_6H_3F_4N$

2,3,5,6-Tetrafluoroaniline, CA 700-17-4: 335.93 (1).

$C_6H_3N_3O_7$

Picric acid, CA 88-89-1: 423.15 (7).

C_6H_4BrCl

2-Bromochlorobenzene, CA 694-80-4: 352.59 (1).

3-Bromochlorobenzene, CA 108-37-2: 353.71 (1).

C_6H_4BrClO

4-Bromo-2-chlorophenol, CA 3964-56-5: Above 385.93 (1).

C_6H_4BrF

1-Bromo-2-fluorobenzene, CA 1072-85-1: 316.48 (1); 321.15 (3).

1-Bromo-3-fluorobenzene, CA 1073-06-9: 312.04 (1); 319.15 (3).

1-Bromo-4-fluorobenzene, CA 460-00-4: 326.15 (3); 333.15 (1).

C_6H_4BrFO

2-Bromo-4-fluorophenol, CA 496-69-5: 358.15 (1).

$C_6H_4BrF_2N$

2-Bromo-4,6-difluoroaniline, CA 444-14-4: 353.71 (1).

C_6H_4BrI

1-Bromo-2-iodobenzene, CA 583-55-1: Above 385.93 (1).

1-Bromo-3-iodobenzene, CA 591-18-4: 383.15 (1).

$C_6H_4BrNO_2$

1-Bromo-2-nitrobenzene, CA 577-19-5: Above 385.93 (1).

$C_6H_4Br_2$

1,2-Dibromobenzene, CA 583-53-9: 364.82 (1).

1,3-Dibromobenzene, CA 108-36-1: 320.15 (3); 367.04 (1).

$C_6H_4Br_2O$

2,4-Dibromophenol, CA 615-58-7: 383.15 (1).

C_6H_4ClF

1-Chloro-2-fluorobenzene, CA 348-51-6: 304.26 (1).

1-Chloro-3-fluorobenzene, CA 625-98-9: 293.15 (1).

1-Chloro-4-fluorobenzene, CA 352-33-0: 302.59 (1).

78

C_6H_4ClFO

3-Chloro-4-fluorophenol, CA 2613-23-2: 382.04 (1).

$C_6H_4ClFO_2S$

4-Fluorobenzenesulfonyl chloride, CA 349-88-2: Above 385.93 (1).

$C_6H_4ClFO_4S_2$

3-Fluorosulfonylbenzenesulfonyl chloride, CA 2489-52-3: Above 385.93 (1).

C_6H_4ClI

1-Chloro-2-iodobenzene, CA 615-41-8: 384.82 (1).
1-Chloro-3-iodobenzene, CA 625-99-0: 374.82 (1).
1-Chloro-4-iodobenzene, CA 637-87-6: 381.48 (1).

$C_6H_4ClNO_2$

1-Chloro-2-nitrobenzene, CA 88-73-3: 397.04 (1); 400.15 (4,7).
1-Chloro-3-nitrobenzene, CA 121-73-3: 376.48 (1); 400.15 (4).
1-Chloro-4-nitrobenzene, CA 100-00-5: 383.15 (1); 400.15 (5,6).

$C_6H_4ClNO_2S$

4-Nitrobenzenesulfenyl chloride, technical, CA 937-32-6: Above 385.93 (1).

$C_6H_4ClO_2P$

1,2-Phenylene phosphorochloridite, CA 1641-40-3: Above 385.93 (1).

$C_6H_4Cl_2$

1,2-Dichlorobenzene, CA 95-50-1: 338.71 (1,3,4,5,6); 341.15 (2); 344.15 (4); 349.82* (2).
1,3-Dichlorobenzene, CA 541-73-1: 336.48 (1); 340.15 (3).
1,4-Dichlorobenzene, CA 106-46-7: 338.71 (1,3,5,6,7); 340.15 (4).

$C_6H_4Cl_2NO_4P$

4-Nitrophenyl phosphorodichloridate, CA 777-52-6: Above 385.93 (1).

$C_6H_4Cl_2O$

2,4-Dichlorophenol, CA 120-83-2: 387.04 (1,7); 387.04* (6).

$C_6H_4Cl_2O_2S$

4-Chlorobenzenesulfonyl chloride, CA 98-60-2: 380.93 (1).

$C_6H_4Cl_2S$

2,5-Dichlorobenzenethiol, CA 5858-18-4:   Above 385.93 (1).

2,6-Dichlorobenzenethiol, CA 24966-39-0:   Above 385.93 (1).

3,4-Dichlorobenzenethiol, CA 5858-17-3:   Above 385.93 (1).

$C_6H_4Cl_3OP$

2-Chlorophenyl dichlorophosphite, CA 56225-92-4:   320.93 (1).

$C_6H_4Cl_3O_2P$

2-Chlorophenyl dichlorophosphate, CA 15074-54-1:   378.15 (1).

4-Chlorophenyl dichlorophosphate, CA 772-79-2:   379.26 (1).

$C_6H_4Cl_4Si$

Chlorophenyltrichlorosilane, CA 26571-79-9:   397.04* (7).

C_6H_4FI

1-Fluoro-2-iodobenzene, CA 348-52-7:   344.26 (1).

1-Fluoro-3-iodobenzene, CA 1121-86-4:   340.37 (1).

1-Fluoro-4-iodobenzene, CA 352-34-1:   341.48 (1).

$C_6H_4FNO_2$

1-Fluoro-2-nitrobenzene, CA 1493-27-2:   367.59 (1,3).

1-Fluoro-3-nitrobenzene, CA 402-67-5:   349.82 (1); 360.15 (3).

1-Fluoro-4-nitrobenzene, CA 350-46-9:   356.48 (1); 363.15 (3).

$C_6H_4FNO_3$

5-Fluoro-2-nitrophenol, CA 446-36-6:   364.82 (1).

$C_6H_4FNO_4S$

2-Nitrobenzenesulfonyl fluoride, CA 433-98-7:   Above 385.93 (1).

$C_6H_4F_2$

1,2-Difluorobenzene, CA 367-11-3:   273.15 (3); 275.37 (1); 280.35* (4).

1,3-Difluorobenzene, CA 372-18-9:   262.05* (4); 273.15 (3); 275.37 (1).

1,4-Difluorobenzene, CA 540-36-3:   261.45* (4); 268.15 (7); 273.15 (3); 275.37 (1).

$C_6H_4F_2O$

2,3-Difluorophenol, CA 6418-38-8:   329.82 (1).

2,4-Difluorophenol, CA 367-27-1:   329.82 (1).

2,6-Difluorophenol, CA 28177-48-2:   332.04 (1).

$C_6H_4F_3N$

2,4,6-Trifluoroaniline, CA 363-81-5: 330.93 (1).

$C_6H_4INO_2$

1-Iodo-2-nitrobenzene, CA 609-73-4: Above 385.93 (1).

1-Iodo-3-nitrobenzene, CA 645-00-1: 344.82 (1).

$C_6H_4I_2$

1,2-Diiodobenzene, CA 615-42-9: Above 385.93 (1).

$C_6H_4N_2$

2-Cyanopyridine, CA 100-70-9: 362.59 (1).

3-Cyanopyridine, CA 100-54-9: 357.59 (1).

$C_6H_4N_2O_4$

1,2-Dinitrobenzene, CA 528-29-0: 423.15 (5,6,7).

$C_6H_4N_2S$

2,1,3-Benzothiadiazole, CA 273-13-2: 368.15 (1).

$C_6H_4O_2$

1,4-Benzoquinone, CA 106-51-4: 310.93 to 366.48 (6).

$C_6H_5BO_2$

Catecholborane, CA 274-07-7: 275.37 (1).

C_6H_5Br

Bromobenzene, CA 108-86-1: 324.26 (1,2,5,6,7); 338.15 (3).

C_6H_5BrO

2-Bromophenol, CA 95-56-7: 315.37 (1).

3-Bromophenol, CA 591-20-8: Above 385.93 (1).

C_6H_5BrS

2-Bromothiophenol, CA 6320-02-1: Above 385.93 (1).

3-Bromothiophenol, CA 6320-01-0: 378.15 (1).

$C_6H_5Br_2N$

2,5-Dibromoaniline, CA 3638-73-1: Above 385.93 (1).

C_6H_5Cl

Chlorobenzene, CA 108-90-7: 297.04 (1); 300.15 (2); 301.15 (4,5,6);
302.04* (2); 302.15 (3,7); 303.15 (4).

C_6H_5ClFN

3-Chloro-4-fluoroaniline, CA 367-21-5: Above 385.93 (1).

$C_6H_5ClN_2O_2$

2-Chloro-4-methyl-5-nitropyridine, CA 23056-33-9: Above 385.93 (1).

C_6H_5ClO

2-Chlorophenol, CA 95-57-8: 337.04 (1,6,7); 358.15 (5).

3-Chlorophenol, CA 108-43-0: Above 385.93 (1).

4-Chlorophenol, CA 106-48-9: 388.71 (1); 394.15 (3,5,6,7).

C_6H_5ClOS

2-Acetyl-5-chlorothiophene, CA 6310-09-4: 381.48 (1).

2-Thiopheneacetyl chloride, CA 39098-97-0: 358.15 (3); 374.82 (1).

$C_6H_5ClO_2S$

Benzenesulfonyl chloride, CA 98-09-9: Above 385.93 (1).

C_6H_5ClS

2-Chlorothiophenol, CA 6320-03-2: 361.48 (1).

3-Chlorothiophenol, CA 2037-31-2: 363.15 (1).

4-Chlorothiophenol, CA 106-54-7: Above 385.93 (1).

$C_6H_5ClS_2$

4-Chloro-1,3-benzenedithiol, CA 58593-78-5: Above 385.93 (1).

$C_6H_5Cl_2N$

2,3-Dichloroaniline, CA 608-27-5: Above 385.93 (1).

2,5-Dichloroaniline, CA 95-82-9: Above 385.93 (1).

2,6-Dichloroaniline, CA 608-31-1: Above 385.93 (1).

3,4-Dichloroaniline, CA 95-76-1: 439.26* (6).

3,5-Dichloroaniline, CA 626-43-7: Above 385.93 (1).

$C_6H_5Cl_2OP$

Phenylphosphonic dichloride, CA 824-72-6: Above 385.93 (1).

$C_6H_5Cl_2O_2P$

   Phenyl dichlorophosphate, CA 770-12-7:   Above 385.93 (1).

$C_6H_5Cl_2P$

   Dichlorophenylphosphine, CA 644-97-3:   Above 385.93 (1).

$C_6H_5Cl_2PS$

   Phenylphosphonothioic dichloride, CA 3497-00-5:   Above 385.93 (1).

$C_6H_5Cl_3Ge$

   Phenyltrichlorogermane, CA 1074-29-9:   Above 385.93 (1).

$C_6H_5Cl_3Si$

   Phenyltrichlorosilane, CA 98-13-5:   364.26 (1,3,5); 364.26* (6).

$C_6H_5Cl_3Sn$

   Phenyltin trichloride, CA 1124-19-2:   Above 385.93 (1).

C_6H_5F

   Fluorobenzene, CA 462-06-6:   258.15 (3,5,6,7); 260.15* (4); 260.37 (1).

$C_6H_5FN_2O_2$

   2-Fluoro-5-nitroaniline, CA 369-36-8:   364.26 (1).
   4-Fluoro-2-nitroaniline, CA 364-78-3:   362.59 (1).
   4-Fluoro-3-nitroaniline, CA 364-76-1:   364.26 (1).

C_6H_5FO

   2-Fluorophenol, CA 367-12-4:   319.82 (1).
   3-Fluorophenol, CA 372-20-3:   344.26 (1).
   4-Fluorophenol, CA 371-41-5:   341.48 (1).

$C_6H_5FO_2S$

   Benzenesulfonyl fluoride, CA 368-43-4:   360.37 (1); 364.26 (7).

C_6H_5FS

   4-Fluorothiophenol, CA 371-42-6:   327.59 (1).

$C_6H_5F_2N$

   2,3-Difluoroaniline, CA 4519-40-8:   344.26 (1).
   2,4-Difluoroaniline, CA 367-25-9:   335.93 (1).   *(continues)*

$C_6H_5F_2N$ *(continued)*

2,5-Difluoroaniline, CA 367-30-6: 342.04 (1).

2,6-Difluoroaniline, CA 5509-65-9: 316.48 (1).

3,4-Difluoroaniline, CA 3863-11-4: 358.15 (1).

3,5-Difluoroaniline, CA 372-39-4: 348.15 (1).

$C_6H_5F_7O_2$

Ethyl heptafluorobutyrate, CA 356-27-4: 287.59 (1).

C_6H_5I

Iodobenzene, CA 591-50-4: 347.59 (1); 350.15 (3).

C_6H_5IO

3-Iodophenol, CA 626-02-8: 383.15 (1).

C_6H_5NO

2-Pyridinecarboxaldehyde, CA 1121-60-4: 327.59 (1); 347.15 (3).

3-Pyridinecarboxaldehyde, CA 500-22-1: 333.15 (1); 354.15 (3).

4-Pyridinecarboxaldehyde, CA 872-85-5: 327.59 (1); 355.15 (3).

C_6H_5NOS

N-Thionylaniline, CA 1122-83-4: 357.59 (1).

$C_6H_5NO_2$

Nitrobenzene, CA 98-95-3: 360.93 (1,3,4,5,6,7).

C_6H_5NS

2-Thiopheneacetonitrile, CA 20893-30-5: 374.82 (1).

3-Thiopheneacetonitrile, CA 13781-53-8: Above 385.93 (1); 392.15 (3).

$C_6H_5N_3O_4$

2,4-Dinitroaniline, CA 97-02-9: 497.04 (6,7); 497.15* (5).

$C_6H_5N_3O_5$

4,6-Dinitro-2-aminophenol, CA 96-91-3: 483.15 (7).

C_6H_6

Benzene, CA 71-43-2: 260.93 (2); 262.04 (1,3,4,5,6,7).

1,5-Hexadien-3-yne, CA 821-08-9: Below 253.15 (5,6,7).

C_6H_6BrN

2-Bromoaniline, CA 615-36-1: Above 385.93 (1).

3-Bromoaniline, CA 591-19-5: Above 385.93 (1).

2-Bromo-5-methylpyridine, CA 3510-66-5: 376.48 (1).

C_6H_6ClN

2-Chloroaniline, CA 95-51-2: 370.93 (1,3).

3-Chloroaniline, CA 108-42-9: 397.04 (1).

6-Chloro-2-picoline, CA 18368-63-3: 347.04 (1).

C_6H_6ClNO

2-Chloro-6-methoxypyridine, CA 17228-64-7: 349.26 (1).

$C_6H_6Cl_2O_2$

1,2-Cyclobutanedicarboxylic acid chloride, *trans*, CA 3668-43-7: 366.48 (1).

C_6H_6FN

2-Fluoroaniline, CA 348-54-9: 333.15 (1).

3-Fluoroaniline, CA 372-19-0: 350.37 (1).

4-Fluoroaniline, CA 371-40-4: 347.04 (1).

$C_6H_6F_9O_3P$

Tris(2,2,2-trifluoroethyl) phosphite, CA 370-69-4: Above 385.93 (1).

C_6H_6IN

2-Iodoaniline, CA 615-43-0: Above 385.93 (1).

3-Iodoaniline, CA 626-01-7: Above 385.93 (1).

$C_6H_6N_2$

1,2-Dicyanocyclobutane, mixed isomers, CA 3396-17-6: Above 385.93 (1).

2-Methyleneglutaronitrile, CA 1572-52-7: Above 385.93 (1).

$C_6H_6N_2O_2$

4-Nitroaniline, CA 100-01-6: 438.15 (1); 472.04 (6,7); 472.15* (5).

C_6H_6O

Phenol, CA 108-95-2: 352.59 (1,3,4,6,7); 352.59* (2); 355.15 (5); 358.15* (4).

2-Vinylfuran, CA 1487-18-9: 275.37 (1).

C_6H_6OS

2-Acetylthiophene, CA 88-15-3: 364.26 (1).

4-Hydroxythiophenol, technical, CA 637-89-8: Above 385.93 (1).

3-Methyl-2-thiophenecarboxaldehyde, technical, CA 5834-16-2: 355.37 (1).

5-Methyl-2-thiophenecarboxaldehyde, CA 13679-70-4: 351.15 (3); 360.93 (1).

$C_6H_6O_2$

2-Acetylfuran, CA 1192-62-7: 344.26 (1).

Catechol, CA 120-80-9: 400.15 (4,5,6,7); 410.37 (1).

2-(Epoxyethyl)furan ±, CA 2745-17-7: 328.71 (1).

Hydroquinone, CA 123-31-9: 438.15 (4,6,7); 438.15* (5).

5-Methylfurfural, CA 620-02-0: 345.93 (1); 345.15 (3).

Resorcinol, CA 108-46-3: 400.15 (4,5,6,7); 444.26 (1).

$C_6H_6O_3$

5-(Hydroxymethyl)furfural, CA 67-47-0: 352.59 (1).

Methyl furoate, CA 611-13-2: 333.15 (3); 346.48 (1).

$C_6H_6O_4$

Dimethyl acetylenedicarboxylate, CA 762-42-5: 359.26 (1).

C_6H_6S

Thiophenol, CA 108-98-5: 323.71 (1); 328.15 (3).

$C_6H_6S_2$

1,2-Benzenedithiol, CA 17534-15-5: 377.04 (1).

1,3-Benzenedithiol, CA 626-04-0: Above 385.93 (1).

C_6H_6Se

Benzeneselenol, CA 645-96-5: 343.15 (1).

$C_6H_7ClN_2$

3-Chloro-2,5-dimethylpyrazine, CA 95-89-6: 356.48 (1).

$C_6H_7F_3O_2$

Ethyl 4,4,4-trifluorocrotanate, CA 25597-16-4: 298.71 (1).

$C_6H_7F_3O_3$

Ethyl 4,4,4-trifluoroacetoacetate, CA 372-31-6: 302.04 (1); 313.15 (3).

C_6H_7N

Aniline, CA 62-53-3: 343.15 (1,2,6,7); 349.15 (3,4,5).

Aniline hydrochloride, CA 142-04-1: 466.15* (5,6); 466.48 (1,7).

Dipropargylamine, CA 6921-28-4: 316.48 (1).

5-Hexynenitrile, CA 14918-21-9: 315.37 (1).

2-Methylpyridine, CA 109-06-8: 299.26 (1); 300.15 (5); 302.15 (3); 312.04* (6,7).

3-Methylpyridine, CA 108-99-6: 309.26 (1); 310.15 (3).

4-Methylpyridine, CA 108-89-4: 312.15 (3); 329.82 (1); 329.82* (6,7).

C_6H_7NO

2-Methoxypyridine, CA 1628-89-3: 305.37 (1).

1-Methyl-2-pyridone, CA 694-85-9: Above 385.93 (1).

1-Methyl-2-pyrrolecarboxaldehyde, CA 1192-58-1: 345.37 (1).

2-Picoline-N-oxide, CA 931-19-1: Above 385.93 (1).

3-Picoline-N-oxide, CA 1003-73-2: Above 385.93 (1).

2-Pyridylcarbinol, CA 586-98-1: Above 385.93 (1).

3-Pyridylcarbinol, CA 100-55-0: Above 385.93 (1).

4-Pyridylcarbinol, CA 586-95-8: Above 385.93 (1).

$C_6H_7NO_2$

2-Acetoxy-3-butenenitrile, CA 22581-05-1: 344.15 (3); 345.93 (1).

Acrylic acid, 2-cyanoethyl ester: 397.04* (6).

N-Ethylmaleimide, CA 128-53-0: 346.48 (1).

C_6H_7NS

2-Aminothiophenol, CA 137-07-5: 352.15 (3); 352.59 (1,7).

3-Aminothiophenol, technical, CA 22948-02-3: 352.59 (1).

4-Aminothiophenol, technical, CA 1193-02-8: Above 385.93 (1).

4-Methyl-5-vinylthiazole, CA 1759-28-0: 343.71 (1).

C_6H_8

1,3-Cyclohexadiene, CA 592-57-4: 262.15 (7); 265.15 (3); 299.82 (1).

1,4-Cyclohexadiene, CA 628-41-1: 263.15 (3); 266.48 (1).

1,3,5-Hexatriene, CA 2235-12-3: 310.93 (1).

Methyl cyclopentadiene, CA 26519-91-5: 322.04 (6).

C_6H_8ClNO

4-(Chloromethyl)-3,5-dimethylisoxazole, CA 19788-37-5: 368.71 (1).

$C_6H_8ClNO_3S$

Methyl 3-chlorocarbonyl-*levo*-thiazolidine-4-carboxylate: Above 385.93 (1).

$C_6H_8Cl_2$

1,3-Dichloro-2,4-hexadiene, CA 37248-93-4: 349.15 (5,6).

$C_6H_8Cl_2O_2$

Adipoyl chloride, CA 111-50-2: 345.37 (6); above 385.93 (1).

Methyl 2,2-dichloro-1-methylcyclopropanecarboxylate, CA 1447-13-8: 347.59 (1); 363.15 (3).

$C_6H_8Cl_2O_5$

Diethylene glycol, bis(chloroformate), CA 106-75-2: 419.26* (6); 433.15* (4); 455.35 (4).

$C_6H_8N_2$

Adiponitrile, CA 111-69-3: 366.48* (6,7); above 385.93 (1); 432.15 (4); 438.15* (5).

2-(Aminomethyl)pyridine, CA 3731-51-9: 363.15 (1).

3-(Aminomethyl)pyridine, CA 3731-52-0: 373.71 (1).

4-(Aminomethyl)pyridine, CA 3731-53-1: 381.48 (1).

2-Amino-3-picoline, CA 1603-40-3: 384.82 (1).

2-Amino-6-picoline, CA 1824-81-3: 376.48 (1).

2,3-Dimethylpyrazine, CA 5910-89-4: 327.59 (1).

2,5-Dimethylpyrazine, CA 123-32-0: 337.04 (1,5); 337.04* (6,7).

2,6-Dimethylpyrazine, CA 108-50-9: 325.93 (1).

4,6-Dimethylpyrimidine, technical, CA 1558-17-4: 319.26 (1).

Ethylpyrazine, CA 13925-00-3: 296.15 (3); 315.93 (1).

2-(Methylamino)pyridine, CA 4597-87-9: 360.93 (1).

2-Methylglutaronitrile ±, technical, CA 4553-62-2: 399.26 (1).

1,2-Phenylenediamine, CA 95-54-5: 429.26 (6).

1,4-Phenylenediamine, CA 106-50-3: 342.04 (1); 429.15 (5,7).

Phenylhydrazine, CA 100-63-0: 360.93 (6); 362.04 (1,3,5,7).

$C_6H_8N_2O$

5-Amino-2-methoxypyridine, CA 6628-77-9: Above 385.93 (1).

2-Cyanoethyl ether, CA 1656-48-0: Above 385.93 (1).

2-Methoxy-3-methylpyrazine, CA 2847-30-5: 328.71 (1).

C_6H_8O

2-Cyclohexen-1-one, CA 930-68-7: 307.15 (5,6,7); 329.26 (1); 331.15 (3).

(continues)

88

C_6H_8O *(continued)*

2,5-Dimethylfuran, CA 625-86-5: 271.48 (1); 280.15 (3); 280.37* (6); 289.15 (7).

2-Ethylfuran, CA 3208-16-0: 270.93 (1).

2,4-Hexadienal, CA 142-83-6: 340.93 (1); 340.93* (6).

3-Methyl-2-cyclopenten-1-one, CA 2758-18-1: 338.15 (1,3).

3-Methyl-1-penten-4-yn-3-ol, CA 3230-69-1: 301.15 (3).

3-Methyl-2-penten-4-yn-1-ol, *cis*, CA 6153-05-5: 332.15 (3).

3-Methyl-2-penten-4-yn-1-ol, *trans*, CA 6153-06-6: 349.15 (3).

C_6H_8OS

2-(2-Thienyl)ethanol, CA 5402-55-1: 374.26 (1).

2-(3-Thienyl)ethanol, CA 13781-67-4: 365.93 (1).

$C_6H_8O_2$

1-Acetoxy-1,3-butadiene, mixed isomers, CA 1515-76-0: 306.48 (1).

Acrolein dimer, CA 100-73-2: 314.15 (3); 320.93* (6,7).

Allyl acrylate, CA 999-55-3: 281.45* (4).

1,2-Cyclohexanedione, CA 765-87-7: 357.59 (1).

3,4-Dihydro-6-methyl-2*H*-pyran-2-one: 352.04 (1).

Methyl 1,3-butadiene-1-carboxylate, CA 1515-75-9: 310.15 (3).

Sorbic acid, CA 110-44-1: 399.15 to 403.15 (4); 399.82* (7).

Vinyl crotonate, CA 14861-06-4: 298.71* (6,7); 300.37 (1).

$C_6H_8O_3$

2-Acetylbutyrolactone, CA 517-23-7: Above 385.93 (1).

Glycidyl acrylate, CA 106-90-1: 333.71* (6,7); 349.26 (1).

3-Methylglutaric anhydride, CA 4166-53-4: Above 385.93 (1).

$C_6H_8O_4$

Dimethyl maleate, CA 624-48-6: 364.26 (1); 368.15 (3); 385.93* (6,7); 386.15 (5).

C_6H_8S

2,5-Dimethylthiophene, CA 638-02-8: 297.04 (1); 302.15 (3).

2-Ethylthiophene, CA 872-55-9: 294.26 (1).

$C_6H_8BrO_2$

Ethyl 4-bromocrotonate, technical, CA 37746-78-4: 370.37 (1).

C$_6$H$_9$ClO

  2-Chlorocyclohexanone, CA 822-87-7: 355.37 (1).

C$_6$H$_9$ClO$_3$

  Ethyl 2-chloroacetoacetate, CA 609-15-4: 323.15 (1); 327.15 (3).
  Ethyl 4-chloroacetoacetate, CA 638-07-3: 327.15 (3); 369.82 (1).
  Ethyl succinyl chloride, CA 14794-31-1: 357.59 (1).
  Methyl 4-(chloroformyl)butyrate, CA 1501-26-4: 355.37 (1).

C$_6$H$_9$Cl$_3$Si

  Cyclohexenyltrichlorosilane, CA 10137-69-6: 366.48* (7).

C$_6$H$_9$N

  Cyclopentyl cyanide, CA 4254-02-8: 329.15 (3).
  2,5-Dimethylpyrrole, CA 625-84-3: 327.59 (1).

C$_6$H$_9$NO

  4-Methyl-2-oxopentanenitrile, technical, CA 66582-16-9: 305.37 (1).
  5-Oxohexanenitrile, CA 10412-98-3: 380.93 (1).
  2,4,5-Trimethyloxazole, CA 20662-84-4: 306.48 (1).
  1-Vinyl-2-pyrrolidinone, CA 88-12-0: 367.04 (1); 371.15 (3); 371.48* (6,7);
  372.15 (5).

C$_6$H$_9$NOS

  4-Methyl-5-thiazoleethanol, CA 137-00-8: Above 385.93 (1).

C$_6$H$_9$NO$_2$

  1-Nitro-1-cyclohexene, CA 2562-37-0: 352.59 (1).

C$_6$H$_9$NO$_3$

  2-Nitrocyclohexanone, CA 4883-67-4: 383.15 (1).

C$_6$H$_9$NS

  2,4,5-Trimethylthiazole, CA 13623-11-5: 329.26 (1).

C$_6$H$_{10}$

  Cyclohexene, CA 110-83-8: 243.15 (3); 249.82 (2); below 253.15 (5);
  260.93 (1).
  2,3-Dimethyl-1,3-butadiene, CA 513-81-5: 250.93 (1); 272.15 (3).
  3,3-Dimethyl-1-butyne, CA 917-92-0: Below 238.71 (1); 248.15 (3).
  1,3-Hexadiene, mostly *trans*, CA 592-48-3: 269.26 (1).   *(continues)*

C_6H_{10} *(continued)*

1,4-Hexadiene, mixed isomers, CA 592-45-0: 247.59 (1); 252.15 (6,7).

1,4-Hexadiene, *cis*, CA 7318-67-4: 227.15 (3).

1,5-Hexadiene, CA 592-42-7: 227.15 (3,7); 245.93 (1).

2,4-Hexadiene, mixed isomers, CA 592-46-1: 265.37 (1).

2,4-Hexadiene, *cis*-2, *trans*-4, CA 5194-50-3: 227.15 (3); 265.37 (1).

2,4-Hexadiene, *trans*, *trans*, CA 5194-51-4: 227.15 (3); 265.37 (1).

1-Hexyne, CA 693-02-7: 352.04 (1); below 253.15 (5); 283.15 (3).

2-Hexyne, CA 764-35-2: 262.04 (1); below 263.15 (5,6).

1-Methyl-1-cyclopentene, CA 693-89-0: 255.93 (1); 303.15 (3).

Methylenecyclopentane, CA 1528-30-9: 253.71 (1); 291.15 (3).

2-Methyl-1,3-pentadiene, mixed isomers, CA 1118-58-7: Below 253.15 (5,6,7); 260.93 (1); 264.15 (3).

2-Methyl-1,3-pentadiene, *trans*, CA 926-54-5: 260.93 (1); 265.15 (3).

3-Methyl-1,3-pentadiene, mixed isomers, CA 4549-74-0: 244.26 (1).

3-Methyl-1,4-pentadiene, CA 1115-08-8: Below 238.71 (1).

4-Methyl-1,3-pentadiene, CA 926-56-7: 239.15 (6,7); 255.15 (3).

Methylpentadiene, mixed isomers, CA 54363-49-4: Below 253.15 (7).

$C_6H_{10}BrClO$

6-Bromohexanoyl chloride, CA 22809-37-6: Above 385.93 (1).

$C_6H_{10}Br_2$

1,2-Dibromocyclohexane, (±)-*trans*, CA 7429-37-0: Above 385.93 (1).

$C_6H_{10}ClNO_2$

N,N-Dimethyl-2-chloroacetoacetamide, CA 5810-11-7: 383.15 (3).

$C_6H_{10}Cl_2$

1,2-Dichlorocyclohexane, *trans*, CA 822-86-6: 339.26 (1).

$C_6H_{10}N_2$

1-(Dimethylamino)pyrrole, CA 78307-76-3: 309.82 (1).

2-Ethyl-4-methylimidazole, CA 931-36-2: 410.93 (1).

1-Piperidinecarbonitrile, CA 1530-87-6: 370.37 (1).

$C_6H_{10}N_2O_4$

Diethyl azodicarboxylate, CA 1972-28-7: 354.15 (3); above 385.93 (1).

$C_6H_{10}O$

Allyl ether, CA 557-40-4: 266.48 (1,7); 266.48* (6).   *(continues)*

$C_6H_{10}O$ (continued)

Cyclohexanone, CA 108-94-1: 316.15 (4,5); 317.15 (3,4,6,7); 319.15* (4); 319.82 (1); 327.15* (2,4).

Cyclohexene oxide, CA 286-20-4: 300.37 (1,3,7).

2-Cyclohexen-1-ol, CA 822-67-3: 331.48 (1).

1,2-Epoxy-5-hexene, CA 10353-53-4: 288.71 (1).

1,5-Hexadien-3-ol, CA 924-41-4: 302.59 (1).

2,4-Hexadien-1-ol, *trans, trans*, CA 17102-64-6: 345.37 (1).

2-Hexenal, *trans*, CA 6728-26-3: 303.15 (3); 308.15 (1).

4-Hexen-3-one, mostly *trans*, CA 25659-22-7: 307.59 (1).

5-Hexen-2-one, CA 109-49-9: 297.04 (1,3).

3-Hexen-1-ol, CA 1002-28-4: 336.48 (1).

2-Isopropylacrolein, CA 4417-80-5: 284.82 (1).

Isopropyl ether, CA 108-20-3: 245.15 (4).

Mesityl oxide, CA 141-79-7: 301.48 (2); 302.15* (4); 303.71 (1,3,4,5,6,7).

3-Methylcyclopentanone ±, CA 1757-42-2: 307.15 (3); 309.82 (1).

3-Methylcyclopentanone, (*R*)-(+), CA 6672-30-6: 309.82 (1).

Methyl 1-methylcyclopropyl ketone, CA 1567-75-5: 297.04 (1).

3-Methyl-1-pentyl-3-ol, CA 77-75-8: 299.82 (1); 301.15 (3); 311.48* (4,6,7).

7-Oxabicyclo[2,2,1]heptane, CA 279-49-2: 285.93 (1).

2-Propylacrolein, CA 1070-13-9: 290.93 (1).

$C_6H_{10}O_2$

Allyl glycidyl ether, CA 106-92-3: 330.37 (1); 330.37* (7).

epsilon-Caprolactone, CA 502-44-3: 382.59 (1).

Cyclopentanecarboxylic acid, CA 3400-45-1: 367.04 (1).

3,4-Dihydro-2-methoxy-2*H*-pyran, CA 4454-05-1: 289.82 (1); 299.15 (3).

5,6-Dihydro-4-methoxy-2*H*-pyran, CA 17327-22-9: 319.82 (1).

3,4-Dihydro-2*H*-pyran-2-methanol, technical, CA 3749-36-8: 362.04 (1).

2,3-Dimethoxy-1,3-butadiene, CA 3588-31-6: 305.93 (1).

3-Ethoxymethacrolein, CA 42588-57-8: 308.71 (1); 350.15 (3).

Ethyl crotonate, CA 623-70-1: 275.37 (3,6,7).

Ethyl cyclopropanecarboxylate, CA 4606-07-9: 291.48 (1); 303.15 (3).

Ethyl methacrylate, CA 97-63-2: 288.71 (1); 292.15 (3); 293.15* (6,7); 294.15* (2,4); 300.15 (5); 308.15* (4).

2,3-Hexanedione, technical, CA 3848-24-6: 301.48 (1).

2,5-Hexanedione, CA 110-13-4: 343.15 (2); 352.04 (1,5,6,7).

2-Hexenoic acid, *trans*, CA 13419-69-7: Above 385.93 (1).

3-Hexenoic acid, *trans*, CA 1577-18-0: Above 385.93 (1).

3-Hexyne-2,5-diol, CA 3031-66-1: Above 385.93 (1).

(continues)

92

$C_6H_{10}O_2$ *(continued)*

2-Hydroxycyclohexanone dimer, CA 30282-14-5: 352.59 (1).

3-Methyl-2,4-pentanedione, technical, CA 815-57-6: 329.82 (1).

2-Methyl-2-pentenoic acid, *trans*, CA 3142-72-1: 380.93 (1).

Vinyl butyrate, CA 123-20-6: 293.15* (6,7).

$C_6H_{10}O_3$

4-Acetoxy-2-butanone, technical, CA 10150-87-5: 349.26 (1).

4-Acetylbutyric acid, CA 3128-06-1: Above 385.93 (1).

2,5-Dimethoxy-2,5-dihydrofuran, mixed isomers, CA 332-77-4: 320.37 (1,3).

Ethyl acetoacetate, CA 141-97-9: 330.37 (6); 338.15 (5); 357.59 (1,3,4).

2-Hydroxyethyl methacrylate, CA 868-77-9: 370.37 (1); 381.15* (4).

2-Methoxyethyl acrylate, CA 3121-61-7: 339.25* (4); 355.15 (5); 355.37* (6,7).

3-Methyl-2-oxopentanoic acid ±, CA 39748-49-7: 355.37 (1).

Methyl propionylacetate, CA 30414-53-0: 344.26 (1).

Mevalonic lactone ±, CA 674-26-0: Above 385.93 (1).

Propionic anhydride, CA 123-62-6: 335.93 (6); 347.04 (1,3,5); 347.04* (7).

Propylene glycol, monoacrylate, CA 999-61-1: 338.15 (6).

$C_6H_{10}O_4$

Adipic acid, CA 124-04-9: 469.15* (5); 469.26 (1,6,7); 483.15* (4).

Diethyl oxalate, CA 95-92-1: 348.71 (1,3,5,6); 348.71* (2,7).

Dimethyl succinate, CA 106-65-0: 358.15 (1); 363.15 (3).

Ethyl acetylglycolate, CA 623-86-9: 355.15 (5,6).

Ethylene glycol diacetate, CA 111-55-7: 355.93 (1); 358.15 (4); 359.15 (3); 361.48 (2,6); 369.26* (7); 372.15* (2,4); 374.85* (4); 377.15 (4,6).

Methyl 4-methoxyacetoacetate, CA 41051-15-4: 360.15 (3); 362.59 (1).

Monomethyl glutarate, CA 1501-27-5: Above 385.93 (1).

$C_6H_{10}O_4S_2$

Ethylene glycol, bis(thioglycolate), CA 123-81-9: Above 385.93 (1); 475.37 (6).

$C_6H_{10}O_5$

Diethyl pyrocarbonate, CA 1609-47-8: 342.59 (1).

Dimethyl methoxymalonate, CA 5018-30-4: 380.93 (1).

$C_6H_{10}O_6$

Dimethyl tartrate, *levo*, CA 608-68-4: Above 385.93 (1).

Dimethyl tartrate, *dextro*, CA 5057-96-5: Above 385.93 (1).

$C_6H_{10}S$

Diallyl sulfide, CA 592-88-1: 296.15 (3); 319.26 (1).

$C_6H_{10}S_2$

Diallyl disulfide, CA 2179-57-9: 324.15 (3).

$C_6H_{11}Br$

6-Bromo-1-hexene, CA 2695-47-8: 325.37 (1); 327.15 (3).
5-Bromo-2-methyl-2-pentene, CA 2270-59-9: 295.93 (1).
Cyclohexyl bromide, CA 108-85-0: 335.93 (1,2,3).

$C_6H_{11}BrO$

2-(Bromomethyl)tetrahydro-2H-pyran, CA 34723-82-5: 338.15 (1).
1-Bromopinacolone, CA 5496-26-1: 318.15 (3); as 90% solution, 339.82 (1).

$C_6H_{11}BrO_2$

4-Bromobutyl acetate, CA 4753-59-7: 382.59 (1).
2-(2-Bromoethyl)-1,3-dioxane, CA 33884-43-4: 369.82 (1); 371.15 (3).
2-Bromohexanoic acid ±, CA 616-05-7: 336.15 (3); above 385.93 (1).
6-Bromohexanoic acid, CA 4224-70-8: 340.93 (1).
tert-Butyl bromoacetate, CA 5292-43-3: 322.59 (1).
Ethyl 2-bromobutyrate, CA 533-68-6: 329.15 (3); 331.48 (1).
Ethyl 4-bromobutyrate, CA 2969-81-5: 331.15 (3); 363.71 (1).
Ethyl 2-bromoisobutyrate, CA 600-00-0: 330.15 (3); 333.15 (1).
Methyl 5-bromovalerate, CA 5454-83-1: 372.59 (1).

$C_6H_{11}Cl$

Cyclohexyl chloride, CA 542-18-7: 302.04 (1); 305.15 (5,6); 320.15 (3).

$C_6H_{11}ClO$

tert-Butylacetyl chloride, CA 7065-46-5: 294.26 (1); 309.15 (3).
2-Chlorocyclohexanol ±, technical, CA 1561-86-0: 343.15 (1).
Hexanoyl chloride, CA 142-61-0: 323.15 (1); 355.15 (3).

$C_6H_{11}ClO_2$

tert-Butyl chloroacetate, CA 107-59-5: 319.82 (1); 333.15 (3).
4-Chlorobutyl acetate, CA 6962-92-1: 343.15 (1).
Chloromethyl pivalate, CA 18997-19-8: 313.15 (1,3).
2-(3-Chloropropyl)-1,3-dioxolane, CA 16686-11-6: 352.15 (3).
Ethyl 4-chlorobutyrate, CA 3153-36-4: 324.82 (1); 344.15 (3).

$C_6H_{11}Cl_3Si$

Cyclohexyltrichlorosilane, CA 98-12-4: 364.15 (5); 364.26* (6).

$C_6H_{11}I$

Cyclohexyl iodide, CA 626-62-0: 343.15 (3); 344.82 (1).

$C_6H_{11}IO_2$

4-Iodobutyl acetate, technical, CA 40596-44-9: 363.71 (1).

$C_6H_{11}N$

Diallylamine, CA 124-02-7: 280.15 (3); 288.71 (1); 294.15 (7).

2,5-Dimethyl-3-pyrroline, mixed isomers, CA 59480-92-1: 278.15 (1).

Hexanenitrile, CA 628-73-9: 316.48 (1).

$C_6H_{11}NO$

epsilon-Caprolactam, CA 105-60-2: 398.15 (4).

1,5-Dimethyl-2-pyrrolidinone, CA 5075-92-3: 362.59 (1).

5,5-Dimethyl-1-pyrroline *N*-oxide, CA 3317-61-1: 368.71 (1).

1-Ethyl-2-pyrrolidinone, CA 2687-91-4: 349.26 (1).

1-Formylpiperidine, CA 2591-86-8: 364.82 (1).

4-Hydroxy-4-methylpentanenitrile, CA 6789-52-2: 341.48 (6).

1-Methyl-2-piperidone, CA 931-20-4: 364.26 (1).

1-Methyl-4-piperidone, CA 1445-73-4: 333.15 (1,3).

2,4,4-Trimethyl-2-oxazoline, CA 1772-43-6: 285.93 (1).

$C_6H_{11}NO_2$

4-Acetylmorpholine, CA 1696-20-4: 385.93 (6); 385.93* (7); above 385.93 (1).

N,N-Dimethylacetoacetamide, CA 2044-64-6: 395.37* (7).

Ethyl 3-aminocrotonate, CA 7318-00-5: 370.37 (1).

2-Methoxy-1-pyrrolidinecarboxaldehyde, mixed isomers, CA 61020-06-2: 377.04 (1).

Nitrocyclohexane, CA 1122-60-7: 347.59 (1); 361.15 (5).

$C_6H_{11}NO_3$

Ethyl acetamidoacetate, CA 1906-82-7: Above 385.93 (1).

C_6H_{12}

Cyclohexane, CA 110-82-7: 253.15 (6); 254.82 (1,3,5); 256.15 (4,7); 260.93 (2).

2,3-Dimethyl-1-butene, CA 563-78-0: 254.82 (1); below 253.15 (5,6); 265.15 (3).

(continues)

C_6H_{12} *(continued)*

2,3-Dimethyl-2-butene, CA 563-79-1: Below 253.15 (5,6); 256.48 (1); 265.15 (3).

3,3-Dimethyl-1-butene, CA 558-37-2: 244.26 (1); 265.15 (3); 280.37 (2).

2-Ethyl-1-butene, CA 760-21-4: 247.04 (1); below 253.15 (5,6,7); 263.15 (3).

Ethylcyclobutane, CA 4806-61-5: Below 253.15 (5); below 257.59 (6,7).

1-Hexene, CA 592-41-6: 247.04 (1,2,7); 248.15 (3); below 253.15 (5); as 97% solution, 244.82 (1).

2-Hexene, mixed isomers, CA 592-43-8: 252.59 (1,2,7).

2-Hexene, *cis*, CA 7688-21-3: 248.15 (3); below 253.15 (5,6).

2-Hexene, *trans*, CA 4050-45-7: 248.15 (3); 252.59 (1).

3-Hexene, *trans*, CA 13269-52-8: 248.15 (3); 260.93 (1).

Methylcyclopentane, CA 96-37-7: 244.15 (3); 245.93 (2); 249.26 (1).

2-Methyl-1-pentene, CA 763-29-1: 245.15 (7); 246.15 (3); 247.04 (1,2).

2-Methyl-2-pentene, CA 625-27-4: 246.15 (3); 249.82 (1,2).

3-Methyl-1-pentene, CA 760-20-3: 244.26 (1); 246.15 (3).

3-Methyl-2-pentene, mixed isomers, CA 922-61-2: 266.48 (1).

3-Methyl-2-pentene, *cis*, CA 922-62-3: 246.15 (3).

3-Methyl-2-pentene, *trans*, CA 616-12-6: 246.15 (3).

4-Methyl-1-pentene, CA 691-37-2: 241.48 (1,2); below 253.15 (5); 266.15 (3,7).

4-Methyl-2-pentene, mixed isomers, CA 4461-48-7: Below 266.48 (6,7).

4-Methyl-2-pentene, *cis*, CA 691-38-3: 239.82 (1); 241.48 (2).

4-Methyl-2-pentene, *trans*, CA 674-76-0: 244.26 (2,7).

$C_6H_{12}BrCl$

1-Bromo-6-chlorohexane, CA 6294-17-3: 374.26 (1).

$C_6H_{12}Br_2$

1,6-Dibromohexane, CA 629-03-8: 305.37 (1).

$C_6H_{12}ClNO$

2-Chloro-*N,N*-diethylacetamide, CA 2315-36-8: Above 385.93 (1).

$C_6H_{12}ClNSi$

Chloro(3-cyanopropyl)dimethylsilane, CA 18156-15-5: 381.48 (1).

$C_6H_{12}Cl_2$

1,1-Dichloro-3,3-dimethylbutane, CA 6130-96-7: 309.82 (1).

1,6-Dichlorohexane, CA 2163-00-0: 347.04 (1); 350.15 (3).

96

$C_6H_{12}Cl_2O$

  Bis(2-chloro-1-methylethyl) ether, CA 108-60-1: 358.15 (5); 358.15*
  (4,6,7).

$C_6H_{12}Cl_2O_2$

  1,2-Bis(2-chloroethoxy)ethane, CA 112-26-5: 394.26 (1,5); 394.26* (2,6,7).
  Dichloroacetaldehyde diethyl acetal, CA 619-33-0: 333.15 (1).

$C_6H_{12}Cl_3O_3P$

  Tris(2-chloroethyl)phosphite, CA 140-08-9: 463.71 (1); 463.71* (2).

$C_6H_{12}Cl_3O_4P$

  Tris(2-chloroethyl)phosphate, CA 115-96-8: 489.26* (7); 505.37 (1).

$C_6H_{12}F_3NOSi$

  N-Methyl-N-(trimethylsilyl)trifluoroacetamide, CA 24589-78-4: 298.15 (1);
  299.15 (3).

$C_6H_{12}I_2$

  1,6-Diiodohexane, CA 629-09-4: Above 385.93 (1).

$C_6H_{12}NO_3P$

  Diethyl cyanomethylphosphonate, CA 2537-48-6: 373.15 (3); above 385.93 (1).

$C_6H_{12}N_2$

  6-Aminocapronitrile, CA 2432-74-8: Above 385.93 (1).
  (Butylamino)acetonitrile, CA 3010-04-6: Above 385.93 (1).
  2-(Diethylamino)acetonitrile, CA 3010-02-4: 327.04 (1).
  Triethylenediamine, CA 280-57-9: 323.15 (3); above 323.15 (4).

$C_6H_{12}N_2O$

  1,3-Dimethyl-3,4,5,6-tetrahydro-2(1H)-pyrimidinone, CA 7226-23-5: 373.15
  (3); above 385.93 (1).

$C_6H_{12}N_2Si$

  1-(Trimethylsilyl)imidazole, CA 18156-74-6: 278.71 (1); 327.15 (3).

$C_6H_{12}N_4$

  Hexamethylenetetramine, CA 100-97-0: 523.15 (1,7).

$C_6H_{12}O$

Allyl propyl ether, CA 1471-03-0: 268.15 (1).

Butyl vinyl ether, CA 111-34-2: 254.82 (7); 261.15 (3); 263.71 (1); 263.71* (2,6,7); 272.15 (4).

Cyclohexanol, CA 108-93-0: 335.93 (2); 340.93 (1,3,4,5,6,7); 340.93* (2,4).

Cyclopentanemethanol, CA 3637-61-4: 335.37 (1).

3,3-Dimethyl-1,2-epoxybutane, CA 2245-30-9: 269.26 (1).

2,5-Dimethyltetrahydrofuran, mixed isomers, CA 1003-38-9: 261.15 (3); 299.82 (1).

1,2-Epoxyhexane, CA 1436-34-6: 288.15 (3); 288.71 (1).

2-Ethylbutyraldehyde, CA 97-96-1: 284.15 (5); 294.26 (1,3); 294.26* (6,7).

Hexanal, CA 66-25-1: 298.15 (3); 305.37 (1,5); 305.37* (6,7).

2-Hexanone, CA 591-78-6: 296.15 (3,5); 298.15 (2,6); 301.15* (2,4); 308.15 (1); 308.15* (4,7).

3-Hexanone, CA 589-38-8: 287.15 (3); 287.15* (7); 308.15 (1,4,5); 308.15* (6).

1-Hexen-3-ol, CA 4798-44-1: 308.15 (1).

2-Hexen-1-ol, *cis*, CA 928-94-9: 334.82 (1).

2-Hexen-1-ol, *trans*, CA 928-95-0: 327.59 (1); 337.15 (3).

3-Hexen-1-ol, *cis*, CA 928-96-1: 317.59 (1); 327.59 (6).

3-Hexen-1-ol, *trans*, CA 928-97-2: 332.04 (1).

4-Hexen-1-ol, mostly *trans*, CA 928-92-7: 334.26 (1).

4-Hexen-3-ol, mixed isomers, CA 4798-58-7: 319.15 (3).

5-Hexen-1-ol, CA 821-41-0: 320.37 (1); 321.15 (3).

Isobutyl vinyl ether, CA 109-53-5: 259.82 (1); 263.71* (2); 264.15 (5,6,7); 266.15 (3,4); 266.48* (2).

1-Methylcyclopentanol, CA 1462-03-9: 313.71 (1).

2-Methylcyclopentanol, *trans*, CA 25144-05-2: 319.26 (1).

3-Methylcyclopentanol ±, CA 18729-48-1: 328.15 (1).

2-Methyl-3-pentanone, CA 565-69-5: 284.15 (3); 287.04 (1).

3-Methyl-2-pentanone, CA 565-61-7: 285.37 (1).

4-Methyl-2-pentanone, CA 108-10-1: 286.48 (1); 287.15 (3,5); 288.71 (2); 290.55 (6,7); 293.15* (2); 296.15 (4); 296.15* (4).

3-Methyl-1-penten-3-ol ±, CA 918-85-4: 298.71 (1).

3-Methyltetrahydropyran ±, CA 26093-63-0: 275.15 (3); 279.82 (1).

2-Methylvaleraldehyde, CA 123-15-9: 289.82 (1,6); 293.15* (6); 295.37* (2).

Oxepane, CA 592-90-5: 283.15 (1); 289.15 (3).

Pinacolone, CA 75-97-8: 276.15 (3); 285.15 (7); 297.04 (1).

$C_6H_{12}OS$

S-tert-Butyl thioacetate, CA 999-90-6: 307.15 (3).

$C_6H_{12}OS_2$

1,5-Dithiacyclooctan-3-ol, CA 86944-00-5: Above 385.93 (1).

$C_6H_{12}OSi$

(Propargyloxy)trimethylsilane, CA 5582-62-7: 354.15 (3).

$C_6H_{12}O_2$

Butyl acetate, CA 123-86-4: 295.37 (1,4,6,7); 297.15 (3); 300.15 (5); 301.15* (2); 302.15* (4); 306.15* (4).

sec-Butyl acetate ±, CA 105-46-4: 265.37 (7); 289.26 (1); 292.04* (2); 292.15 (5); 304.15* (4,6).

tert-Butyl acetate, CA 540-88-5: 271.15 (3); 288.71 (1).

tert-Butylacetic acid, CA 1070-83-3: 362.04 (1).

1,3-Cyclohexanediol, mixed isomers, CA 504-01-8: Above 385.93 (1).

1,4-Cyclohexanediol, mixed isomers, CA 556-48-9: 338.71 (1).

2,6-Dimethyl-1,4-dioxane, CA 10138-17-7: 297.04 (7); 297.04* (6).

2-Ethoxytetrahydrofuran ±, CA 13436-46-9: 289.26 (1).

Ethyl butyrate, CA 105-54-4: 292.59 (1); 296.15 (2); 297.04 (6); 298.15 (4,5); 299.15 (3).

2-Ethylbutyric acid, CA 88-09-5: 360.93 (1); 372.04* (6); 372.15 (2,3,5).

Ethyl isobutyrate, CA 97-62-1: 287.04 (1); 291.15 (7); 293.15 (3); 294.15 (5,6).

Glycidyl isopropyl ether ±, CA 4016-14-2: 306.48 (1).

Hexanoic acid, CA 142-62-1: 374.82* (4,6,7); 375.15 (5); 377.59 (1).

5-Hexene-1,2-diol, technical, CA 36842-44-1: Above 385.93 (1).

4-Hydroxy-4-methyl-2-pentanone, CA 123-42-2: 285.93 (1); 320.15 (4); 329.15 (3); 331.15 (5,6); 334.15* (4); 335.37* (2); 337.15 (4,7); 341.48* (2).

Isobutyl acetate, CA 110-19-0: 290.93 (1,3,4,5,6,7); 293.71 (2); 295.15 (4); 297.15* (2,4).

Isopentyl formate, CA 110-45-2: 295.15 (5); 299.15 (3).

2-Methyl-2-ethyl-1,3-dioxolane, CA 126-39-6: 296.48* (6).

Methyl isovalerate, CA 556-24-1: 289.15 (3).

Methyl 2-methylbutyrate ±, CA 53955-81-0: 291.15 (3); 305.37 (1).

5-Methyltetrahydrofuran-2-methanol, mixed isomers, CA 6126-49-4: 313.71 (1).

Methyl tetrahydrofurfuryl ether, CA 19354-27-9: 306.48 (1).

Methyl trimethylacetate, CA 598-98-1: 280.15 (1,3).

Methyl valerate, CA 624-24-8: 295.37 (1); 300.15 (3).

2-Methylvaleric acid ±, CA 97-61-0: 364.26 (1); 380.37* (6,7).

3-Methylvaleric acid, CA 105-43-1: 358.15 (1).

4-Methylvaleric acid, CA 646-07-1: 370.37 (1).

Pentyl formate, CA 638-49-3: 299.82 (2,6,7).   *(continues)*

$C_6H_{12}O_2$ *(continued)*

Propyl propionate, CA 106-36-5: 292.59 (1); 294.15 (5); 295.15 (3); 352.59* (6,7).

Tetrahydropyran-2-methanol, CA 100-72-1: 349.15 (3); 366.48 (1,7); 366.48* (2,6).

$C_6H_{12}O_2S$

2,4-Dimethylsulfolane, CA 1003-78-7: 416.48* (7); 418.71 (1).

$C_6H_{12}O_2S_2$

2,5-Dihydroxy-2,5-dimethyl-1,4-dithiane, technical, CA 55704-78-4: 315.93 (1).

$C_6H_{12}O_3$

Acetylacetaldehyde dimethyl acetal, technical, CA 5436-21-5: 322.59 (1); 334.15 (3).

3-Allyloxy-1,2-propanediol, CA 123-34-2: Above 385.93 (1).

Butyl glycolate, CA 7397-62-8: 334.26 (6); 341.15 (5); 347.15 (3).

2,5-Dimethoxytetrahydrofuran, mixed isomers, CA 696-59-3: Below 283.15 (7); 308.15 (1); 311.15 (3).

2,2-Dimethyl-1,3-dioxolane-4-methanol, (S)-(+), CA 22323-82-6: 353.15 (1); 363.15 (3).

2,2-Dimethyl-1,3-dioxolane-4-methanol, (R)-(-), CA 14347-78-5: 352.15 (3); 353.15 (1).

2-Ethoxyethyl acetate, CA 111-15-9: 320.37 (6,7); 322.15 (4); 324.15 (4,5,6); 327.59 (2); 330.15* (4); 330.37 (1); 332.15* (2,4).

Ethyl 3-Hydroxybutyrate, CA 5405-41-4: 337.59 (1); 350.15 (3).

Ethyl 2-hydroxyisobutyrate, CA 80-55-7: 314.15 (3); 317.59 (1).

2-Hydroxycaproic acid ±, CA 6064-63-7: Above 385.93 (1).

Isopropyl lactate (S), CA 63697-00-7: 327.59* (6); 330.37 (1).

Methyl 2,2-dimethyl-3-hydroxypropionate, CA 14002-80-3: 349.26 (1).

Paraldehyde, CA 123-63-7: 290.15 (7); 297.15 (3); 300.15 (5); 308.71* (2,6).

Propylene glycol methyl ether acetate, mixed isomers, CA 84540-57-8: 315.37 (6); 316.48 (1); 318.15 (2).

Solketal, CA 100-79-8: 353.15 (1).

$C_6H_{12}O_4$

Ethyl 2-ethoxy-2-hydroxyacetate, technical, CA 49653-17-0: 322.59 (1).

Methyl 3,3-dimethoxypropionate, CA 7424-91-1: 338.71 (1).

$C_6H_{12}O_5S$

Ethyl *levo*-2-[(methylsulfonyl)oxy]propionate, CA 58742-64-6: Above 385.93 (1).

$C_6H_{12}S$

    Cyclohexyl mercaptan, CA 1569-69-3:   314.15 (3); 316.48 (1,6); 322.04* (2).

$C_6H_{12}Si$

    1-(Trimethylsilyl)-1-propyne, CA 6224-91-5:   269.26 (1); 277.15 (3).

$C_6H_{13}Br$

    1-Bromo-2-ethylbutane, CA 3814-34-4:   308.71 (1).

    1-Bromohexane, CA 111-25-1:   309.15 (3); 330.37 (1).

$C_6H_{13}BrO$

    6-Bromo-1-hexanol, CA 4286-55-9:   Above 385.93 (1).

$C_6H_{13}BrO_2$

    2-Bromo-1,1-diethoxyethane, CA 2032-35-1:   324.82 (1); 330.15 (3).

$C_6H_{13}BrSi$

    2-Bromo-3-trimethylsilylpropane, CA 81790-10-5:   306.15 (3).

$C_6H_{13}Cl$

    1-Chlorohexane, CA 544-10-5:   299.82 (1); 302.15 (3); 308.15 (2,6).

    3-Chloro-3-methylpentane, CA 918-84-3:   282.59 (1).

$C_6H_{13}ClO$

    6-Chloro-1-hexanol, CA 2009-83-8:   372.04 (1).

$C_6H_{13}ClO_2$

    Chloroacetaldehyde diethyl acetal, CA 621-62-5:   302.59 (1); 322.15 (3).

    4-Chlorobutyraldehyde dimethyl acetal, technical, CA 29882-07-3:
329.26 (1).

$C_6H_{13}ClO_3$

    2-[2-(2-Chloroethoxy)ethoxy]ethanol, CA 5197-62-6:   380.37 (1).

$C_6H_{13}ClSi$

    Allyl(chloromethyl)dimethylsilane, CA 75422-66-1:   300.93 (1).

    (3-Chloroallyl)trimethylsilane, CA 18187-39-8:   290.15 (3).

$C_6H_{13}F$

    1-Fluorohexane, CA 373-14-8:   300.37 (1).

$C_6H_{13}I$

1-Iodohexane, CA 638-45-9:   334.26 (1); 340.15 (3).

$C_6H_{13}N$

Cyclohexylamine, CA 108-91-8:   300.15 (3); 304.26 (2,6); 305.37 (1,5); 305.37* (4).

2,5-Dimethylpyrrolidine, mixed isomers, technical, CA 3378-71-0: 280.37 (1).

N-Ethyl-2-methylallylamine, CA 18328-90-0:   280.37 (1).

Hexamethyleneimine, CA 111-49-9:   291.48 (1); 295.15 (3,7).

1-Methylpiperidine, CA 626-67-5:   276.48 (1,3); 296.15 (7).

2-Methylpiperidine, CA 109-05-7:   281.48 (1); 283.15 (3,7).

3-Methylpiperidine, CA 626-56-2:   275.59 (7); 290.37 (1); 294.15 (3).

4-Methylpiperidine, CA 626-58-4:   280.37 (1); 282.15 (7); 286.15 (3).

$C_6H_{13}NO$

1-Amino-1-cyclopentanemethanol, CA 10316-79-7:   368.71 (1).

N-Butylacetamide, CA 1119-49-9:   388.71 (6).

N,N-Diethylacetamide, CA 658-91-6:   343.71 (1); 349.15 (3); 349.82 (7).

2,6-Dimethylmorpholine, mixed isomers, CA 141-91-3:   317.59* (6,7); 322.04 (1).

4-Ethylmorpholine, CA 100-74-3:   300.93 (1); 302.15 (5); 305.15* (6,7); 308.75* (4); 316.15 (3).

1-Ethyl-3-pyrrolidinol, CA 30727-14-1:   347.04 (1).

1-(2-Hydroxyethyl)pyrrolidine, CA 2955-88-6:   329.26 (1).

3-Hydroxy-1-methylpiperidine, CA 3554-74-3:   343.15 (1).

4-Hydroxy-1-methylpiperidine, CA 106-52-5:   Above 385.93 (1).

2-(Methoxymethyl)pyrrolidine, (S)-(+), CA 63126-47-6:   318.71 (1).

1-Methyl-2-pyrrolidinemethanol, (S)-(-), CA 34381-71-0:   336.48 (1,3).

3-Piperidinemethanol, CA 4606-65-9:   Above 385.93 (1).

$C_6H_{13}NO_2$

2-Dimethylamino-1,3-dioxane, CA 19449-32-2:   328.15 (3).

N,N-Dimethylglycine, ethyl ester, CA 33229-89-9:   305.15 (3); 317.59 (1).

Ethyl 3-aminobutyrate, CA 5303-65-1:   315.37 (1).

4-(2-Hydroxyethyl)morpholine, CA 622-40-2:   372.59 (1,3,5); 372.59* (6,7).

Methyl 3-dimethylaminopropionate, CA 3853-06-3:   324.26 (1).

1-Nitrohexane, CA 646-14-0:   345.15 (3); 346.48 (1).

$C_6H_{13}NO_2Si$

3-Trimethylsilyl-2-oxazolidinone, CA 43112-38-5:   349.82 (1); 353.15 (3).

$C_6H_{13}NS_2$

Thialdine, CA 638-17-5: 366.15 (5); 366.48* (6).

$C_6H_{13}O_3P$

Diethyl vinylphosphonate, CA 682-30-4: Above 385.93 (1).

$C_6H_{13}O_5P$

Ethyl dimethylphosphonoacetate, CA 311-46-6: 343.15 (3).

C_6H_{14}

2,2-Dimethylbutane, CA 75-83-2: 225.37 (4,6); below 238.71 (1); 241.48 (2); 244.15 (3).

2,3-Dimethylbutane, CA 79-29-8: 239.82 (1); 244.15 (3,6); below 253.15 (5); 254.26 (7).

Hexane, CA 110-54-3: 247.15 (3); 249.82 (1,2,7); 251.15 (4,6); below 253.15 (5).

Isohexane, mixed isomers, CA 73513-42-5: Below 244.26 (6).

2-Methylpentane, CA 107-83-5: 249.82 (1,2); below 253.15 (5); 266.48 (3,7).

3-Methylpentane, CA 96-14-0: Below 253.15 (5); 266.48 (1,3,7); below 266.48 (6).

$C_6H_{14}BrO_3P$

Diethyl 2-bromoethylphosphonate, CA 5324-30-1: Above 385.93 (1).

$C_6H_{14}N_2$

1-(2-Aminoethyl)pyrrolidine, CA 7154-73-6: 320.93 (1).

1-Aminohomopiperidine, CA 5906-35-4: 329.26 (1).

4-(Aminomethyl)piperidine, CA 7144-05-0: 352.04 (1).

1,2-Diaminocyclohexane, mixed isomers, CA 694-83-7: 343.15 (3); 348.15 (1).

1,2-Diaminocyclohexane, ± trans, CA 1121-22-8: 342.04 (1).

1,4-Dimethylpiperazine, CA 106-58-1: 291.48 (1).

2,5-Dimethylpiperazine, mixed isomers, CA 106-55-8: 331.48 (1).

2,5-Dimethylpiperazine, cis, CA 6284-84-0: 341.15 (5).

2,6-Dimethylpiperazine, CA 108-49-6: 318.15 (1).

$C_6H_{14}N_2O$

4-(2-Aminoethyl)morpholine, CA 2038-03-1: 353.15 (5); 448.15 (1); 448.15* (7).

1-Amino-2-(methoxymethyl)pyrrolidine, (S)-(-), CA 59983-39-0: 345.37 (1).

1-Amino-2-(methoxymethyl)pyrrolidine, (R)-(+), CA 72748-99-3: 345.37 (1).

1-(2-Hydroxyethyl)piperazine, CA 103-76-4: Above 385.93 (1); 397.04* (6,7).

$C_6H_{14}O$

tert-Amyl methyl ether, CA 994-05-8: 261.48 (1); 266.15 (3).

Butyl ethyl ether, CA 628-81-9: 267.59 (1); 272.15 (3); 277.59 (6,7).

tert-Butyl ethyl ether, CA 637-92-3: 253.71 (1); 272.15 (3).

2,3-Dimethyl-2-butanol, CA 594-60-5: 302.59 (1).

3,3-Dimethyl-1-butanol, CA 624-95-3: 320.15 (3); 320.93 (1).

3,3-Dimethyl-2-butanol ±, CA 464-07-3: 299.15 (3); 302.04 (1).

2-Ethyl-1-butanol, CA 97-95-0: 330.37* (6,7); 331.15* (2); 331.48 (1,3,5); 336.15 (4).

1-Hexanol, CA 111-27-3: 333.15 (1); 336.15 (4,5,6,7); 347.15 (3); 347.15* (2).

2-Hexanol, CA 626-93-7: 314.82 (1); 331.15 (3,5,6).

3-Hexanol, CA 623-37-0: 314.82 (1); 318.15 (3).

Isopropyl ether, CA 108-20-3: 244.15 (3); 245.37 (2,6,7); below 253.15 (5); 260.37 (1); 263.71* (2).

2-Methyl-1-pentanol, CA 105-30-6: 319.15 (5,7); 323.71 (1); 326.15 (3); 327.04 (6).

2-Methyl-2-pentanol, CA 590-36-3: 294.26 (1); 303.15 (3).

2-Methyl-3-pentanol, CA 565-67-3: 309.15 (3); 319.26 (1).

3-Methyl-1-pentanol, CA 589-35-5: 332.04 (1); 333.15 (3).

3-Methyl-2-pentanol, CA 565-60-6: 313.71 (1).

3-Methyl-3-pentanol, CA 77-74-7: 297.15 (3); 319.26 (1,5,6).

4-Methyl-1-pentanol, CA 626-89-1: 324.82 (1); 330.15 (3).

4-Methyl-2-pentanol, CA 108-11-2: 314.26 (1,2,5,6,7); 319.25 (4); 327.15 (3); 328.15* (2).

Propyl ether, CA 111-43-3: 245.15 (3); 277.59 (1); below 294.15 (5); 294.26 (6,7).

$C_6H_{14}OSi$

(Allyloxy)trimethylsilane, CA 18146-00-4: 272.59 (1); 273.15 (3).

Dimethylethoxyvinyl silane, CA 5356-83-2: 269.82 (1).

$C_6H_{14}O_2$

Acetal, CA 105-57-7: 252.04 (1,6,7); 277.15 (3); 286.15 (5).

2-Butoxyethanol, CA 111-76-2: 333.15 (1,2,3,4); 334.15 (4,5,6); 338.71 (2,6); 342.55* (4); 344.26* (7); 347.04* (2).

3,3-Dimethyl-1,2-butanediol, technical, CA 59562-82-2: 372.04 (1).

Ethylene glycol, diethyl ether, CA 629-14-4: 295.15 (3); 299.82* (2); 308.15 (5,6); 308.15* (2,4,6,7); 311.48 (1).

2-Ethyl-2-methyl-1,3-propanediol, CA 77-84-9: Above 385.93 (1).

1,2-Hexanediol ±, CA 6920-22-5: 374.82* (6); above 385.93 (1).

1,5-Hexanediol, CA 928-40-5: Above 385.93 (1).

1,6-Hexanediol, CA 629-11-8: 374.82 (1); 402.58* (2); 403.15 (7).

(continues)

$C_6H_{14}O_2$ *(continued)*

2,5-Hexanediol, CA 2935-44-6: 374.82 (1); 377.15 (5); 377.59* (2); 383.15 (6,7).

2-Isobutoxyethanol, CA 4439-24-1: 330.93 (6).

2-Methyl-1,3-pentanediol, CA 149-31-5: 383.15 (5,6).

2-Methyl-2,4-pentanediol, CA 107-41-5: 363.15 (3); 367.04 (1); 369.15 (5); 369.26* (6,7); 372.04* (2).

2-Methyl-2,4-pentanediol, (R)-(-), CA 99210-90-9: 374.82 (1).

Pinacol, CA 76-09-5: 350.37 (1).

Propylene glycol, isopropyl ether, CA 3944-36-3: 322.04* (2).

Propylene glycol, propyl ether, CA 1569-01-3: 331.15 (4); 336.15* (4).

$C_6H_{14}O_2S$

Propyl sulfone, CA 598-03-8: 399.82 (1).

3,3'-Thiodipropanol, CA 10595-09-2: Above 385.93 (1).

$C_6H_{14}O_2Si$

Methyl (trimethylsilyl)acetate, CA 2916-76-9: 283.15 (3); 305.93 (1).

Trimethylsilylmethyl acetate, CA 2917-65-9: 293.15 (1,3).

$C_6H_{14}O_3$

Dipropylene glycol, mixed isomers, CA 110-98-5: 391.15 (4,5); 394.26* (6); 399.82 (2); 410.93 (1); 410.93* (6,7).

2-(2-Ethoxyethoxy)ethanol, CA 111-90-0: 356.45 (4); 363.15 (3); 364.15 (2); 367.15 (5,6); 367.15* (7); 369.26 (1); 369.26* (2,4,6); 372.04* (2).

2-Methoxyethyl ether, CA 111-96-6: 326.15 (3); 340.37 (6); 343.15 (1); 343.15* (2,4); 348.71* (2).

1,2,6-Trihydroxyhexane, CA 106-69-4: 352.59 (1); 463.71* (2,6,7).

1,1,3-Trimethoxypropane, CA 14315-97-0: 313.71 (1).

Trimethylolpropane, CA 77-99-6: 452.59* (2).

$C_6H_{14}O_3S$

2-Methylbutyl methanesulfonate, (S)-(+): 383.15 (1).

$C_6H_{14}O_4$

Triethylene glycol, CA 112-27-6: 438.71 (1); 445.15 (4); 449.82 (2,7); 449.82* (5,6).

$C_6H_{14}S$

1-Hexanethiol, CA 111-31-9: 293.71 (1); 303.15 (3); 324.82* (2).

Isopropyl sulfide, CA 625-80-9: 280.37 (1).

Propyl sulfide, CA 111-47-7: 301.48 (1); 305.15 (3).

$C_6H_{14}S_2$

1,6-Hexanedithiol, CA 1191-43-1:   363.71 (1).

Isopropyl disulfide, CA 4253-89-8:   291.48 (1).

Propyl disulfide, CA 629-19-6:   337.15 (3); 339.26 (1).

$C_6H_{14}Si$

Allyltrimethylsilane, CA 762-72-1:   280.37 (1); 289.15 (3).

$C_6H_{15}Al$

Triethylaluminum, CA 97-93-8:   Below 220.37 (7); ignites spontaneously in air (6).

$C_6H_{15}Al_2Cl_3$

Ethylaluminum sesquichloride, CA 12075-68-2:   Ignites spontaneously in air (6).

$C_6H_{15}BO_3$

Triethyl borate, CA 150-46-9:   284.26 (1,3,5,6,7).

$C_6H_{15}BrSi$

Bromotriethylsilane, CA 1112-48-7:   325.93 (1).

$C_6H_{15}BrSn$

Triethyltin bromide, CA 2767-54-6:   372.59 (1).

$C_6H_{15}ClOSi$

Chloromethyldimethylisopropoxysilane, CA 18171-11-4:   301.48 (1).

2-(Trimethylsilyl)ethoxymethyl chloride, CA 76513-69-4:   273.15 (3); 319.82 (1).

$C_6H_{15}ClO_3Si$

(3-Chloropropyl)trimethoxysilane, CA 2530-87-2:   348.15 (3); 351.48 (1).

$C_6H_{15}ClSi$

tert-Butyldimethylsilyl chloride, CA 18162-48-6:   295.93 (1,3).

Chlorotriethylsilane, CA 994-30-9:   295.15 (3); 302.59 (1).

$C_6H_{15}N$

2-Amino-3,3-dimethylbutane, CA 3850-30-4:   274.26 (1).

2-Aminohexane ±, CA 68107-05-1:   285.93* (2).

2-Amino-4-methylpentane, CA 108-09-8:   285.93 (1,5); 285.93* (6,7).

(continues)

$C_6H_{15}N$ *(continued)*

Diisopropylamine, CA 108-18-9: 262.04 (2); 264.26 (2); 266.15 (3,5,7); 267.04 (1,4); 266.48* (2); 272.04* (6).

3,3-Dimethylbutylamine, CA 15673-00-4: 278.71 (1).

Dipropylamine, CA 142-84-7: 277.04 (1); 280.15 (3,5); 280.37* (2); 284.82 (2); 290.37* (6,7).

N-Ethylbutylamine, CA 13360-63-9: 285.15 (3); 286.48 (2); 290.93* (6); 291.48 (1).

2-Ethylbutylamine, CA 617-79-8: 287.04 (1); 290.93* (7); 294.26* (2).

Hexylamine, CA 111-26-2: 282.04 (1); 302.15 (3,5); 302.59* (6,7).

Triethylamine, CA 121-44-8: 258.15 (3); 260.93 (2); 264.26* (6) 266.48 (1,4); 266.48* (2,7); 267.59 (2).

$C_6H_{15}NO$

2-Amino-1-hexanol ±, CA 5665-74-7: 373.15 (1).

2-(Butylamino)ethanol, CA 111-75-1: 349.82* (2,6,7); 350.15 (5); 360.15 (3).

2-(*tert*-Butylamino)ethanol, CA 4620-70-6: 342.04 (1); 360.15 (3).

2-(Diethylamino)ethanol, CA 100-37-8: 322.04 (1); 323.15 (3); 324.82 (2); 333.15 (5); 333.15* (6,7).

2-Dimethylamino-2-methyl-1-propanol, CA 7005-47-2: 326.15 (3); 338.71* (2).

Isoleucinol, *levo*, CA 24629-25-2: 373.71 (1).

3-Isopropoxypropylamine, CA 2906-12-9: 312.59 (1).

Leucinol, (S)-(+), CA 7533-40-6: 363.71 (1).

Leucinol, (R)-(-), CA 53448-09-2: 363.71 (1).

$C_6H_{15}NOSi$

N-Methyl-N-trimethylsilylacetamide, CA 7449-74-3: 321.15 (3).

$C_6H_{15}NO_2$

Aminoacetaldehyde diethyl acetal, CA 645-36-3: 318.71 (1).

Diisopropanolamine, CA 110-97-4: 266.48* (2); 272.15* (4); 399.82 (1,5); 399.82* (6,7).

Di(2-methoxyethyl)amine, CA 111-95-5: 341.48 (2).

N,N-Dimethylacetamide dimethyl acetal, CA 18871-66-4: 281.48 (1); 297.15 (3).

Dimethylaminoacetaldehyde dimethyl acetal,, CA 38711-20-5: 305.37 (1).

N-Ethyldiethanolamine, CA 139-87-7: 397.04 (1,5); 410.93* (6).

$C_6H_{15}NO_2Si$

Trimethylsilyl N,N-dimethylcarbamate, CA 32115-55-2: 300.15 (3).

$C_6H_{15}NO_3$

Triethanolamine, CA 102-71-6: 452.15* (2,4,5); 452.59 (7); 458.15 (1); 467.59* (2); 469.26 (6).

$C_6H_{15}NS$

2-(Butylamino)ethanethiol, CA 5842-00-2: 345.93 (1).

$C_6H_{15}N_3$

1-(2-Aminoethyl)piperazine, CA 140-31-8: 356.15 (3); 366.48 (1); 366.48* (4,6,7); 374.82 (2).

1,3,5-Trimethylhexahydro-1,3,5-triazine, CA 108-74-7: 322.59 (1).

$C_6H_{15}N_3O$

Acetaldehyde ammonia trimer, CA 75-39-8: 328.71 (1).

$C_6H_{15}O_3P$

Diisopropyl phosphite, CA 1809-20-7: Above 385.93 (1).

Dipropyl phosphite, CA 1809-21-8: 369.26 (1).

Triethyl phosphite, CA 122-52-1: 299.15 (3); 327.59 (1); 327.59* (2).

$C_6H_{15}O_3PS$

Triethyl phosphorothioate, CA 126-68-1: 380.37* (7).

$C_6H_{15}O_4P$

Triethyl phosphate, CA 78-40-0: 388.71 (1); 388.71* (6,7); 389.15 (5).

$C_6H_{15}P$

Triethylphosphine, CA 554-70-1: 255.93 (1).

$C_6H_{16}BFO_4$

Dimethoxyfluoroborane diethyl ether complex, CA 367-46-4: 283.15 (3).

$C_6H_{16}BNO$

Borane-4-ethylmorpholine complex: 325.37 (1).

$C_6H_{16}Cl_2OSi_2$

1,3-Bis(chloromethyl)tetramethyldisiloxane, CA 2362-10-9: 347.04 (1).

$C_6H_{16}Cl_2Si_2$

1,2-Bis(chlorodimethylsilyl)ethane, CA 13528-93-3: 338.15 (3); as 80% solution, 339.82 (1).

$C_6H_{16}N_2$

N,N-Diethylethylenediamine, CA 100-36-7:   303.71 (1); 314.15 (3); 319.26*
(6,7).

N,N'-Diethylethylenediamine, CA 111-74-0:   306.48 (1).

N,N-Dimethyl-N'-ethylethylenediamine, CA 123-83-1:   297.04 (1).

1,6-Hexanediamine, CA 124-09-4:   354.26 (1); 358.15* (4).

N,N,N',N'-Tetramethylethylenediamine, CA 110-18-9:   283.15 (1); 290.15*
(4); 294.15 (3).

$C_6H_{16}OSi$

tert-Butyldimethylsilanol, CA 18173-64-3:   318.15 (1).

Triethylsilanol, CA 597-52-4:   321.15 (3).

3-(Trimethylsilyl)-1-propanol, CA 2917-47-7:   325.93 (1).

$C_6H_{16}O_2Si$

Diethoxydimethylsilane, CA 78-62-6:   279.15 (3); 284.82 (1).

$C_6H_{16}O_3SSi$

(3-Mercaptopropyl)trimethoxysilane, CA 4420-74-0:   322.04 (1); 358.15 (3).

$C_6H_{16}O_3Si$

Triethoxysilane, CA 998-30-1:   299.15 (3); 299.82 (1).

$C_6H_{16}P_2$

1,2-Bis(dimethylphosphino)ethane, CA 23936-60-9:   256.48 (1).

$C_6H_{16}Si$

tert-Butyldimethylsilane, CA 29681-57-0:   250.37 (1).

Triethylsilane, CA 617-86-7:   253.15 (3); 269.26 (1).

$C_6H_{17}Cl_2NSi_2$

1,3-Bis(chloromethyl)-1,1,3,3-tetramethyldisilazane, CA 14579-91-0: 362.04
(1); 366.15 (3).

$C_6H_{17}NO_3Si$

3-Aminopropyltrimethoxysilane, CA 13822-56-5:   356.48 (1); 365.15 (3).

$C_6H_{17}N_3$

3,3'-Iminobispropylamine, CA 56-18-8:   391.48 (1); 394.15 (4).

$C_6H_{17}O_3PSi$

Dimethyl trimethylsilylmethylphosphonate, CA 13433-42-6: 349.15 (3).

$C_6H_{18}BN$

Borane-diisopropylamine complex, CA 55124-35-1: 260.93 (1).

Borane-triethylamine complex, CA 1722-26-5: 266.48 (1); 348.15 (3).

$C_6H_{18}BN_3$

Tris(dimethylamino)borane, CA 4375-83-1: 252.59 (1).

$C_6H_{18}N_2Si$

Bis(dimethylamino)dimethylsilane, CA 3768-58-9: 265.37 (1); 279.15 (3).

$C_6H_{18}N_3OP$

Hexamethylphosphoramide, CA 680-31-9: 378.71 (1).

$C_6H_{18}N_3P$

Tris(dimethylamino)phosphine, CA 1608-26-0: 310.15 (3); as 85% solution, 299.82 (1).

$C_6H_{18}N_4$

Triethylenetetramine, technical, CA 112-24-3: 391.15 (4); 408.15 (5,6,7); 416.48 (1); 416.48* (2).

Triethylenetetramine hydrate: 383.15 (1).

Tris(2-aminoethyl)amine, CA 4097-89-6: Above 385.93 (1).

$C_6H_{18}OSi_2$

Hexamethyldisiloxane, CA 107-46-0: 270.37 (1); 272.15 (3,7).

$C_6H_{18}O_3Si_3$

Hexamethylcyclotrisiloxane, CA 541-05-9: 305.15 (1,3).

$C_6H_{18}O_4SSi_2$

Bis(trimethylsilyl)sulfate, CA 18306-29-1: 270.37 (1).

$C_6H_{18}SSi_2$

1,1,1,3,3,3-Hexamethyldisilathiane, CA 3385-94-2: 296.15 (3); 299.26 (1).

$C_6H_{18}Si_2$

Hexamethyldisilane, CA 1450-14-2: 271.48 (1); 284.15 (3).

$C_6H_{18}Sn_2$

Hexamethylditin, CA 661-69-8: 334.26 (1).

$C_6H_{19}NOSi_2$

N,O-Bis(trimethylsilyl)hydroxylamine, CA 22737-37-7: 302.04 (1,3).

$C_6H_{19}NSi_2$

1,1,1,3,3,3-Hexamethyldisilazane, CA 999-97-3: 282.04 (1); 287.15 (3,7).

$C_6H_{21}N_3Si_3$

2,2,4,4,6,6-Hexamethylcyclotrisilazane, CA 1009-93-4: 338.15 (3).

C_7ClF_5O

Pentafluorobenzoyl chloride, CA 2251-50-5: Nonflammable (1).

C_7F_5N

Pentafluorobenzonitrile, CA 773-82-0: 302.59 (1).

C_7F_8

Octafluorotoluene, CA 434-64-0: 293.71 (1).

C_7F_{14}

Perfluoro(methylcyclohexane), CA 355-02-2: Nonflammable (1).

C_7F_{16}

Perfluoroheptane, CA 335-57-9: 315.15 (3).

C_7HF_4N

2,3,5,6-Tetrafluorobenzonitrile, CA 5216-17-1: 343.15 (1).

C_7HF_5O

Pentafluorobenzaldehyde, CA 653-37-2: 350.93 (1).

$C_7H_2BrF_5$

alpha-Bromo-2,3,4,5,6-pentafluorotoluene, CA 1765-40-8: 355.93 (1).

$C_7H_2Cl_2F_3NO_2$

2,4-Dichloro-5-nitrobenzotrifluoride, CA 400-70-4: Above 385.93 (1).

$C_7H_2Cl_6$

1,2,3,4,7,7-Hexachlorobicyclo(2,2,1)-2,5-heptadiene, CA 3389-71-7: 362.04 (1).

$C_7H_3BrClF_3$

5-Bromo-2-chlorobenzotrifluoride, CA 445-01-2: 354.26 (1).

$C_7H_3ClF_2O$

2,4-Difluorobenzoyl chloride, CA 72482-64-5: 355.37 (1).

2,5-Difluorobenzoyl chloride, CA 35730-09-7: 332.59 (1).

2,6-Difluorobenzoyl chloride, CA 18063-02-0: 322.59 (1).

3,4-Difluorobenzoyl chloride, CA 76903-88-3: 351.48 (1).

3,5-Difluorobenzoyl chloride: 342.59 (1).

$C_7H_3ClF_3NO_2$

2-Chloro-5-nitrobenzotrifluoride, CA 777-37-7: 372.04 (1); 408.15 (5,6).

4-Chloro-3-nitrobenzotrifluoride, CA 121-17-5: 374.82 (1).

5-Chloro-2-nitrobenzotrifluoride, CA 118-83-2: 375.93 (1).

$C_7H_3Cl_2F_3$

2,4-Dichlorobenzotrifluoride, CA 320-60-5: 345.37 (1).

2,5-Dichlorobenzotrifluoride, CA 320-50-3: 349.26 (1).

3,4-Dichlorobenzotrifluoride, CA 328-84-7: 338.71 (1).

$C_7H_3Cl_2NO$

2,3-Dichlorophenyl isocyanate, CA 41195-90-8: Above 385.93 (1).

2,5-Dichlorophenyl isocyanate, CA 5392-82-5: Above 385.93 (1).

2,6-Dichlorophenyl isocyanate, CA 39920-37-1: 349.82 (1).

3,4-Dichlorophenyl isocyanate, CA 102-36-3: Above 385.93 (1); 415.15* (4); as 90% solution, 383.15 (1).

$C_7H_3Cl_3O$

2,4-Dichlorobenzoyl chloride, CA 89-75-8: 410.93 (1).

2,6-Dichlorobenzoyl chloride, CA 4659-45-4: Above 385.93 (1).

3,4-Dichlorobenzoyl chloride, CA 3024-72-4: 415.37 (1).

3,5-Dichlorobenzoyl chloride, CA 2905-62-6: 383.15 (1).

$C_7H_3Cl_4F$

1-(Trichloromethyl)-2-chloro-6-fluorobenzene, CA 84473-83-6: Above 385.93 (1).

$C_7H_3FN_2O_3$

    4-Fluoro-3-nitrophenyl isocyanate, CA 65303-82-4: 383.15 (1).

$C_7H_3F_2N$

    2,3-Difluorobenzonitrile, CA 21524-39-0: 348.15 (1).

    2,4-Difluorobenzonitrile, CA 3939-09-1: Above 385.93 (1).

    2,5-Difluorobenzonitrile, CA 64248-64-2: 350.93 (1).

    2,6-Difluorobenzonitrile, CA 1897-52-5: 353.15 (1).

    3,4-Difluorobenzonitrile, CA 64248-62-0: 342.59 (1).

$C_7H_3F_2NO$

    2,4-Difluorophenyl isocyanate, CA 59025-55-7: 323.71 (1).

$C_7H_3F_3N_2O_4$

    3,5-Dinitrobenzotrifluoride, CA 401-99-0: Above 385.93 (1).

$C_7H_3F_4NO_2$

    4-Fluoro-3-nitrobenzotrifluoride, CA 367-86-2: 306.48 (1).

$C_7H_3F_5$

    2,3,4,5,6-Pentafluorotoluene, CA 771-56-2: 307.59 (1).

$C_7H_3F_5O$

    2,3,4,5,6-Pentafluoroanisole, CA 389-40-2: 305.37 (1).

    2,3,4,5,6-Pentafluorobenzyl alcohol, CA 440-60-8: 360.93 (1).

C_7H_4BrClO

    2-Bromobenzoyl chloride, CA 7154-66-7: Above 385.93 (1).

    3-Bromobenzoyl chloride, CA 1711-09-7: 380.37 (1).

    4-Bromobenzoyl chloride, CA 586-75-4: Above 385.93 (1).

$C_7H_4BrF_3$

    2-Bromobenzotrifluoride, CA 392-83-6: 324.82 (1).

    3-Bromobenzotrifluoride, CA 401-78-5: 316.48 (1).

    4-Bromobenzotrifluoride, CA 402-43-7: 322.04 (1).

C_7H_4BrN

    2-Bromobenzonitrile, CA 2042-37-7: Above 385.93 (1).

    3-Bromobenzonitrile, CA 6952-59-6: Above 385.93 (1).

C_7H_4BrNO

2-Bromophenyl isocyanate, CA 1592-00-3: 380.93 (1).

3-Bromophenyl isocyanate, CA 23138-55-8: 381.48 (1).

4-Bromophenyl isocyanate, CA 2493-02-9: 382.59 (1).

C_7H_4BrNS

2-Bromophenyl isothiocyanate, CA 13037-60-0: Above 385.93 (1).

3-Bromophenyl isothiocyanate, CA 2131-59-1: Above 385.93 (1).

C_7H_4ClFO

2-Chloro-6-fluorobenzaldehyde, CA 387-45-1: 374.82 (1).

2-Fluorobenzoyl chloride, CA 393-52-2: 355.37 (1).

3-Fluorobenzoyl chloride, CA 1711-07-5: 355.37 (1).

4-Fluorobenzoyl chloride, CA 403-43-0: 355.37 (1).

$C_7H_4ClFO_3S$

3-(Fluorosulfonyl)benzoyl chloride, CA 454-93-3: Above 385.93 (1).

$C_7H_4ClF_3$

2-Chlorobenzotrifluoride, CA 88-16-4: 317.15 (3); 332.04 (1,5,6).

3-Chlorobenzotrifluoride, CA 98-15-7: 311.48 (1); as 80% solution, 309.26 (1).

4-Chlorobenzotrifluoride, CA 98-56-6: 319.15 (3); 320.37 (1,5), 327.15* (4).

Chlorobenzotrifluoride, mixed isomers, CA 52181-51-8: 320.37 (6).

C_7H_4ClIO

2-Iodobenzoyl chloride, CA 609-67-6: Above 385.93 (1).

C_7H_4ClN

2-Chlorobenzonitrile, CA 873-32-5: 381.48 (1).

3-Chlorobenzonitrile, CA 766-84-7: 370.37 (1).

C_7H_4ClNO

2-Chlorobenzoxazole, CA 615-18-9: 358.15 (1).

2-Chlorophenyl isocyanate, CA 3320-83-0: 362.04 (1).

3-Chlorophenyl isocyanate, CA 2909-38-8: 359.82 (1); 374.82* (7).

4-Chlorophenyl isocyanate, CA 104-12-1: 383.15 (1,3); 383.15* (4).

$C_7H_4ClNO_3$

2-Nitrobenzoyl chloride, CA 610-14-0: Above 385.93 (1).   *(continues)*

$C_7H_4ClNO_3$ *(continued)*

3-Nitrobenzoyl chloride, CA 121-90-4: Above 385.93 (1).

$C_7H_4ClNO_3S$

4-Chlorobenzenesulfonyl isocyanate, technical, CA 5769-15-3: 322.59 (1).

C_7H_4ClNS

2-Chlorobenzothiazole, CA 615-20-3: Above 385.93 (1).

2-Chlorophenyl isothiocyanate, CA 2740-81-0: Above 385.93 (1).

3-Chlorophenyl isothiocyanate, CA 2392-68-9: Above 385.93 (1).

4-Chlorophenyl isothiocyanate, CA 2131-55-7: Above 385.93 (1).

$C_7H_4Cl_2O$

2-Chlorobenzoyl chloride, CA 609-65-4: 383.15 (1).

3-Chlorobenzoyl chloride, CA 618-46-2: 383.15 (1).

4-Chlorobenzoyl chloride, CA 122-01-0: 378.15 (1).

3,4-Dichlorobenzaldehyde, CA 6287-38-3: Above 385.93 (1).

$C_7H_4Cl_2OS$

4-Chlorophenyl chlorothioformate, CA 13606-10-5: 416.48 (2).

$C_7H_4Cl_3F$

1-(Dichloromethyl)-2-chloro-6-fluorobenzene, CA 62476-62-4: Above 385.93 (1).

2-Fluorobenzotrichloride, CA 488-98-2: Above 385.93 (1).

$C_7H_4Cl_4$

2-Chlorobenzotrichloride, CA 2136-89-2: 371.48 (1).

4-Chlorobenzotrichloride, CA 5216-25-1: Above 385.93 (1).

C_7H_4FN

2-Fluorobenzonitrile, CA 394-47-8: 347.04 (1,3).

3-Fluorobenzonitrile, CA 403-54-3: 340.93 (1).

4-Fluorobenzonitrile, CA 1194-02-1: 338.71 (1).

C_7H_4FNO

2-Fluorophenyl isocyanate, CA 16744-98-2: 323.71 (1).

3-Fluorophenyl isocyanate, CA 404-71-7: 314.82 (1); 335.15 (3).

4-Fluorophenyl isocyanate, CA 1195-45-5: 325.93 (1).

$C_7H_4FNO_3S$

3-Fluorosulfonylphenyl isocyanate, CA 402-36-8:   Above 385.93 (1).

C_7H_4FNS

2-Fluorophenyl isothiocyanate, CA 38985-64-7:   370.37 (1).

3-Fluorophenyl isothiocyanate, CA 404-72-8:   358.15 (1).

4-Fluorophenyl isothiocyanate, CA 1544-68-9:   362.59 (1).

$C_7H_4F_2O$

2,3-Difluorobenzaldehyde, CA 2646-91-5:   331.48 (1).

2,4-Difluorobenzaldehyde, CA 1550-35-2:   328.15 (1).

2,5-Difluorobenzaldehyde, CA 2646-90-4:   332.04 (1).

2,6-Difluorobenzaldehyde, CA 437-81-0:   346.48 (1).

3,4-Difluorobenzaldehyde, CA 34036-07-2:   338.71 (1).

3,5-Difluorobenzaldehyde, CA 32085-88-4:   329.26 (1).

$C_7H_4F_3I$

2-Iodobenzotrifluoride, CA 444-29-1:   354.26 (1).

3-Iodobenzotrifluoride, CA 401-81-0:   343.15 (1).

$C_7H_4F_3NO_2$

2-Nitrobenzotrifluoride, CA 384-22-5:   368.71 (1).

3-Nitrobenzotrifluoride, CA 98-46-4:   360.93 (1); 375.93* (6,7); 376.15 (5).

4-Nitrobenzotrifluoride, CA 402-54-0:   361.48 (1).

$C_7H_4F_3NO_3$

2-Nitro-4-(trifluoromethyl)phenol, CA 400-99-7:   368.15 (1).

$C_7H_4F_4$

2-Fluorobenzotrifluoride, CA 392-85-8:   290.93 (1).

3-Fluorobenzotrifluoride, CA 401-80-9:   280.37 (1).

4-Fluorobenzotrifluoride, CA 402-44-8:   283.71 (1).

2,3,5,6-Tetrafluorotoluene, CA 5230-78-4:   296.48 (1).

$C_7H_4F_4O$

2,3,5,6-Tetrafluoroanisole, CA 2324-98-3:   314.82 (1).

$C_7H_4N_2O_3$

2-Nitrophenyl isocyanate, CA 3320-86-3:   Above 385.93 (1).

3-Nitrophenyl isocyanate, CA 3320-87-4:   Above 385.93 (1).

$C_7H_5BrCl_2$

   1-(Bromomethyl)-2,6-dichlorobenzene, CA 20443-98-5: Above 385.93 (1).

$C_7H_5BrF_2$

   1-(Bromomethyl)-2,4-difluorobenzene, CA 23915-07-3: 313.15 (1).

   1-(Bromomethyl)-2,5-difluorobenzene: 288.71 (1).

   1-(Bromomethyl)-2,6-difluorobenzene: Above 385.93 (1).

   1-(Bromomethyl)-3,4-difluorobenzene: 352.59 (1).

   1-(Bromomethyl)-3,5-difluorobenzene: 354.82 (1).

$C_7H_5BrF_3N$

   3-Amino-4-bromo-1-trifluoromethylbenzene, CA 454-79-5: Above 385.93 (1).

C_7H_5BrO

   Benzoyl bromide, CA 618-32-6: 363.71 (1).

   2-Bromobenzaldehyde, CA 6630-33-7: 368.15 (1).

   3-Bromobenzaldehyde, CA 3132-99-8: 369.26 (1).

   4-Bromobenzaldehyde, CA 1122-91-4: 382.04 (1).

$C_7H_5BrO_2$

   4-Bromo-1,2-(methylenedioxy)benzene, CA 2635-13-4: Above 385.93 (1).

$C_7H_5Br_2F$

   3-Fluorobenzal bromide, CA 455-34-5: Above 385.93 (1).

$C_7H_5ClF_3N$

   2-Amino-5-chlorobenzotrifluoride, CA 445-03-4: 368.15 (1).

   3-Amino-4-chlorobenzotrifluoride, CA 121-50-6: Nonflammable (1).

   5-Amino-2-chlorobenzotrifluoride, CA 320-51-4: Above 385.93 (1).

$C_7H_5ClN_2$

   2-Amino-5-chlorobenzonitrile, CA 5922-60-1: Above 385.93 (1).

C_7H_5ClO

   Benzoyl chloride, CA 98-88-4: 342.04 (1); 345.15 (4,5,6,7); 375.15 (5).

   2-Chlorobenzaldehyde, CA 89-98-5: 360.93 (1).

   3-Chlorobenzaldehyde, CA 587-04-2: 361.48 (1).

   4-Chlorobenzaldehyde, CA 104-88-1: 360.93 (1,6).

C_7H_5ClOS

Phenyl chlorothioformate, CA 13464-19-2: 388.71 (2).

Phenyl chlorothionocarbonate, CA 1005-56-7: 354.82 (1).

$C_7H_5ClO_2$

5-Chloro-1,3-benzodioxole, CA 7228-38-8: 367.04 (1).

Phenyl chloroformate, CA 1885-14-9: 348.71 (1); 350.15 (4).

$C_7H_5Cl_2F$

2-Chloro-6-fluorobenzyl chloride, CA 55117-15-2: 366.48 (1).

4-Fluorobenzal chloride, CA 456-19-9: 363.71 (1).

$C_7H_5Cl_2N$

Phenyl isocyanide dichloride, CA 622-44-6: 352.59 (1).

$C_7H_5Cl_2NO_2$

4-Chloro-2-nitrobenzyl chloride, CA 938-71-6: Above 385.93 (1).

$C_7H_5Cl_3$

1-Chloromethyl-2,4-dichlorobenzene, CA 94-99-5: Above 385.93 (1).

1-Chloromethyl-2,6-dichlorobenzene, CA 2014-83-7: Above 385.93 (1).

1-Chloromethyl-3,4-dichlorobenzene, CA 102-47-6: 383.15 (1).

(Trichloromethyl)benzene, CA 98-07-7: 370.37 (1); above 372.15 (4);
399.82 (6); 400.15* (4).

$C_7H_5Cl_3O$

2,3,6-Trichloroanisole, CA 50375-10-5: Above 385.93 (1).

$C_7H_5Cl_3S$

Methyl 2,4,5-trichlorophenyl sulfide, CA 4163-78-4: Above 385.93 (1).

$C_7H_5CoO_2$

Cyclopentadienylcobalt dicarbonyl, technical, CA 12078-25-0: 299.82 (1).

C_7H_5FO

Benzoyl fluoride, CA 455-32-3: 322.04 (1).

2-Fluorobenzaldehyde, CA 446-52-6: 328.15 (1).

3-Fluorobenzaldehyde, CA 456-48-4: 329.82 (1).

4-Fluorobenzaldehyde, CA 459-57-4: 329.82 (1).

118

$C_7H_5F_3$

(Trifluoromethyl)benzene, CA 98-08-8:   285.37 (1,3,5,6,7); 288.75* (4).

$C_7H_5F_4N$

1-Amino-2-trifluoromethyl-4-fluorobenzene, CA 393-39-5:   345.37 (1).

1-Amino-3-trifluoromethyl-4-fluorobenzene, CA 2357-47-3:   364.82 (1).

1-Amino-5-trifluoromethyl-2-fluorobenzene, CA 535-52-4:   343.15 (1).

C_7H_5N

Benzonitrile, CA 100-47-0:   343.15 (3); 344.82 (1); 352.15* (4).

C_7H_5NO

Anthranil, CA 271-58-9:   367.04 (1).

1,2-Benzisoxazole, CA 271-95-4:   359.26 (1).

Benzoxasole, CA 273-53-0:   331.48 (1).

2-Furanacrylonitrile, mixed isomers, CA 7187-01-1:   360.37 (1).

Phenyl isocyanate, CA 103-71-9:   324.15 (3); 328.71 (1,7); 333.15* (4).

$C_7H_5NO_3$

2-Nitrobenzaldehyde, CA 552-89-6:   Above 385.93 (1).

$C_7H_5NO_3S$

Benzenesulfonyl isocyanate, CA 2845-62-7:   Above 385.93 (1).

$C_7H_5NO_4$

1,2-(Methylenedioxy)-4-nitrobenzene, CA 2620-44-2:   415.93 (1).

C_7H_5NS

Benzothiazole, CA 95-16-9:   Above 385.93 (1).

Phenyl isothiocyanate, CA 103-72-0:   360.93 (1,3).

$C_7H_5N_3O_6$

2,4,6-Trinitrotoluene, CA 118-96-7:   Explodes (6).

C_7H_6BrCl

2-Bromo-5-chlorotoluene, CA 14495-51-3:   368.71 (1).

2-Chlorobenzyl bromide, CA 611-17-6:   382.04 (1).

3-Chlorobenzyl bromide, CA 766-80-3:   Above 385.93 (1).

C_7H_6BrClO

4-Bromo-6-chloro-*ortho*-cresol, CA 7530-27-0: Above 385.93 (1).

C_7H_6BrF

2-Fluorobenzyl bromide, CA 446-48-0: 355.93 (1).

3-Fluorobenzyl bromide, CA 456-41-7: 334.82 (1).

4-Fluorobenzyl bromide, CA 459-46-1: Above 385.93 (1).

$C_7H_6BrNO_2$

2-Bromo-6-nitrotoluene, CA 55289-35-5: Above 385.93 (1).

2-Nitrobenzyl bromide, CA 3958-60-9: Above 385.93 (1).

$C_7H_6Br_2$

Benzal bromide, CA 618-31-5: 383.15 (1).

3-Bromobenzyl bromide, CA 823-78-9: Above 385.93 (1).

2,5-Dibromotoluene, CA 615-59-8: Above 385.93 (1).

Dibromotoluene, mixed isomers, CA 40088-48-0: Nonflammable (2).

$C_7H_6Br_2O$

2,6-Dibromo-4-methylphenol, CA 2432-14-6: Above 385.93 (1).

C_7H_6ClF

2-Chloro-4-fluorotoluene, CA 452-73-3: 323.15 (1).

2-Chloro-6-fluorotoluene, CA 443-83-4: 319.26 (1); 322.15 (3).

4-Chloro-2-fluorotoluene, CA 33406-96-1: 324.26 (1).

2-Fluorobenzyl chloride, CA 345-35-7: 330.37 (1).

3-Fluorobenzyl chloride, CA 456-42-8: 332.04 (1).

4-Fluorobenzyl chloride, CA 352-11-4: 333.71 (1).

C_7H_6ClFO

2-Chloro-6-fluorobenzyl alcohol, CA 56456-50-9: Above 385.93 (1).

C_7H_6ClI

2-Iodobenzyl chloride, CA 59473-45-9: Above 385.93 (1).

$C_7H_6ClNO_2$

2-Chloro-6-nitrotoluene, CA 83-42-1: 398.15 (1).

4-Chloro-2-nitrotoluene, CA 89-59-8: Above 385.93 (1).

4-Chloro-3-nitrotoluene, technical, CA 89-60-1: Above 385.93 (1).

(continues)

120

$C_7H_6ClNO_2$ *(continued)*

2-Nitrobenzyl chloride, CA 612-23-7: Above 385.93 (1).

3-Nitrobenzyl chloride, CA 619-23-8: Above 385.93 (1).

$C_7H_6Cl_2$

2-Chlorobenzyl chloride, CA 611-19-8: 355.37 (1).

3-Chlorobenzyl chloride, CA 620-20-2: 372.04 (1).

4-Chlorobenzyl chloride, CA 104-83-6: 370.93 (1).

(Dichloromethyl)benzene, CA 98-87-3: 365.37 (1).

2,3-Dichlorotoluene, CA 32768-54-0: 356.48 (1).

2,4-Dichlorotoluene, CA 95-73-8: 352.59 (1,3); 365.93* (2).

2,5-Dichlorotoluene, CA 19398-61-9: 352.59 (1).

2,6-Dichlorotoluene, CA 118-69-4: 352.15 (3); 355.93 (1).

3,4-Dichlorotoluene, CA 95-75-0: 358.71 (1); 360.93* (2); 363.15 (3).

$C_7H_6Cl_2O$

2,3-Dichloroanisole, CA 1984-59-4: Above 385.93 (1).

2,6-Dichloroanisole, CA 1984-65-2: 364.26 (1).

3,5-Dichloroanisole, CA 33719-74-3: 379.26 (1).

alpha,4-Dichloroanisole, CA 21151-56-4: Above 385.93 (1).

2,4-Dichlorobenzyl alcohol, CA 1777-82-8: Above 385.93 (1).

3,4-Dichlorobenzyl alcohol, CA 1805-32-9: Above 385.93 (1).

$C_7H_6Cl_2S$

Chloromethyl 4-chlorophenyl sulfide, CA 7205-90-5: Above 385.93 (1).

$C_7H_6Cl_4O_2$

5,5-Dimethoxy-1,2,3,4-tetrachlorocyclopentadiene, CA 2207-27-4: 339.82 (1).

C_7H_6FI

2-Fluoro-4-iodotoluene, CA 39998-81-7: 352.59 (1).

$C_7H_6FNO_2$

2-Fluoro-4-nitrotoluene, CA 1427-07-2: 347.04 (1).

2-Fluoro-5-nitrotoluene, CA 455-88-9: 378.15 (1).

2-Fluoro-6-nitrotoluene, CA 769-10-8: 362.04 (1).

3-Fluoro-4-nitrotoluene, CA 446-34-4: Above 385.93 (1).

4-Fluoro-2-nitrotoluene, CA 446-10-6: 372.04 (1).

4-Fluoro-3-nitrotoluene, CA 446-11-7: Above 385.93 (1).

5-Fluoro-2-nitrotoluene, CA 446-33-3: 360.93 (1).

$C_7H_6F_2$

 2,4-Difluorotoluene, CA 452-76-6: 288.15 (1).

$C_7H_6F_2O$

 2,4-Difluoroanisole, CA 452-10-8: 324.26 (1).
 2,3-Difluorobenzyl alcohol, CA 75853-18-8: 367.04 (1).
 2,4-Difluorobenzyl alcohol, CA 56456-47-4: 363.15 (1).
 2,5-Difluorobenzyl alcohol, CA 75853-20-2: 367.04 (1).
 2,6-Difluorobenzyl alcohol, CA 19064-18-7: 362.04 (1).
 3,4-Difluorobenzyl alcohol: 370.93 (1).
 3,5-Difluorobenzyl alcohol, CA 79538-20-8: 367.59 (1).

$C_7H_6F_3N$

 2-Aminobenzotrifluoride, CA 88-17-5: 328.15 (1).
 3-Aminobenzotrifluoride, CA 98-16-8: 358.15 (1,4).
 4-Aminobenzotrifluoride, CA 455-14-1: 359.82 (1).

$C_7H_6N_2$

 3-Aminobenzonitrile, CA 2237-30-1: Above 385.93 (1).
 Anthranilonitrile, CA 1885-29-6: Above 385.93 (1).
 Imidazo[1,2,a]pyridine, CA 274-76-0: Above 385.93 (1).
 2-Pyridylacetonitrile, CA 2739-97-1: 367.04 (1); 373.15 (3).
 3-Pyridylacetonitrile, CA 6443-85-2: 371.48 (1).

$C_7H_6N_2O_4$

 2,4-Dinitrotoluene, CA 121-14-2: 479.82 (6,7); 485.15* (5).
 3,4-Dinitrotoluene, CA 610-39-9: Above 385.93 (1).

C_7H_6O

 Benzaldehyde, CA 100-52-7: 335.93 (1,6); 337.15 (3,4,5,7); 347.05* (4).

C_7H_6OS

 Thiobenzoic acid, technical, CA 98-91-9: Above 385.93 (1).

$C_7H_6O_2$

 1,3-Benzodioxole, CA 274-09-9: 328.15 (1); 345.15 (3).
 Benzoic acid, CA 65-85-0: 394.26 (1,5,6).
 3-(2-Furyl)acrolein, CA 623-30-3: 372.59 (1).
 Phenyl formate, CA 1864-94-4: 353.15 (3).

(continues)

$C_7H_6O_2$ *(continued)*

Salicylaldehyde, CA 90-02-8:   334.15 (3); 349.82 (1); 350.93 (6).

Tropolone, CA 533-75-5:   Above 385.93 (1).

$C_7H_6O_3$

Salicylic acid, CA 69-72-7:   430.15 (4,6).

C_7H_7Br

Benzyl bromide, CA 100-39-0:   359.82 (1).

2-Bromotoluene, CA 95-46-5:   352.04 (1,2,6); 353.15 (3).

3-Bromotoluene, CA 591-17-3:   333.15 (1); 354.15 (3).

4-Bromotoluene, CA 106-38-7:   358.15 (1,3,6).

C_7H_7BrO

2-Bromoanisole, CA 578-57-4:   369.82 (1); 371.15 (3).

3-Bromoanisole, CA 2398-37-0:   366.48 (1); 371.15 (3).

4-Bromoanisole, CA 104-92-7:   367.59 (1); 371.15 (3).

3-Bromobenzyl alcohol, CA 15852-73-0:   Above 385.93 (1).

2-Bromo-4-methylphenol, CA 6627-55-0:   Above 385.93 (1).

C_7H_7BrS

2-Bromothioanisole, CA 19614-16-5:   Above 385.93 (1).

4-Bromothioanisole, CA 104-95-0:   Above 385.93 (1).

C_7H_7Cl

Benzyl chloride, CA 100-44-7:   333.15 (3,5); 340.15 (4,6,7); 347.04 (1); 347.15* (4).

2-Chlorotoluene, CA 95-49-8:   315.15 (3); 319.15 (2); 320.37 (1).

3-Chlorotoluene, CA 108-41-8:   323.71 (1); 325.15 (3).

4-Chlorotoluene, CA 106-43-4:   322.59 (1); 325.15 (4); 326.15 (3); 333.15* (2).

Chlorotoluene, mixed isomers, CA 25168-05-2:   325.15 (5); 325.37* (6).

C_7H_7ClO

2-Chloroanisole, CA 766-51-8:   349.26 (1).

3-Chloroanisole, CA 2845-89-8:   346.48 (1).

4-Chloroanisole, CA 623-12-1:   347.15 (3); 351.48 (1).

3-Chlorobenzyl alcohol, CA 873-63-2:   Above 385.93 (1).

2-Chloro-5-methylphenol, CA 615-74-7:   354.26 (1).

4-Chloro-2-methylphenol, CA 1570-64-5:   Above 385.93 (1).

$C_7H_7ClO_2S$

4-Methylbenzenesulfonyl chloride, CA 98-59-9: 383.15* (7).

$C_7H_7ClO_3S$

4-Methoxybenzenesulfonyl chloride, CA 98-68-0: Above 385.93 (1).

C_7H_7ClS

2-Chlorobenzyl mercaptan, CA 17733-22-1: 342.59 (1).

4-Chlorobenzyl mercaptan, CA 6258-66-8: 349.82 (1).

Chloromethyl phenyl sulfide, CA 7205-91-6: 374.26 (1).

4-Tolylsulfenyl chloride, CA 933-00-6: Nonflammable (1).

$C_7H_7Cl_2N$

2,4-Dichlorobenzylamine, CA 95-00-1: Above 385.93 (1).

3,4-Dichlorobenzylamine, CA 102-49-8: Above 385.93 (1).

2,6-Dichloro-3-methylaniline, CA 64063-37-2: Above 385.93 (1).

C_7H_7F

2-Fluorotoluene, CA 95-52-3: 281.15 (3); 285.93 (1).

3-Fluorotoluene, CA 352-70-5: 282.59 (1); 285.15 (3).

4-Fluorotoluene, CA 352-32-9: 283.15 (3,7); 313.71 (1).

C_7H_7FO

2-Fluoroanisole, CA 321-28-8: 326.15 (3); 333.15 (1).

3-Fluoroanisole, CA 456-49-5: 317.04 (1).

4-Fluoroanisole, CA 459-60-9: 316.48 (1).

2-Fluorobenzyl alcohol, CA 446-51-5: 363.71 (1).

3-Fluorobenzyl alcohol, CA 456-47-3: 363.71 (1).

4-Fluorobenzyl alcohol, CA 459-56-3: 363.15 (1).

$C_7H_7FO_2S$

4-Toluenesulfonyl fluoride, CA 455-16-3: 378.71 (1).

$C_7H_7F_2N$

2,5-Difluorobenzylamine, CA 85118-06-5: 348.71 (1).

3,4-Difluorobenzylamine, CA 72235-53-1: 352.59 (1).

C_7H_7I

2-Iodotoluene, CA 615-37-2: 363.15 (1).

3-Iodotoluene, CA 625-95-6: 355.37 (1).

4-Iodotoluene, CA 624-31-7: 363.15 (1).

C_7H_7IO

2-Iodoanisole, CA 529-28-2: Above 385.93 (1).

3-Iodoanisole, CA 766-85-8: Above 385.93 (1).

4-Iodoanisole, CA 696-62-8: Above 385.93 (1).

3-Iodobenzyl alcohol, CA 57455-06-8: Above 385.93 (1).

C_7H_7N

2-Vinylpyridine, CA 100-69-6: 319.82 (1); 321.15 (3).

4-Vinylpyridine, CA 100-43-6: 321.15 (3); 324.82 (1).

C_7H_7NO

2-Acetylpyridine, CA 1122-62-9: Above 385.93 (1).

3-Acetylpyridine, CA 350-03-8: 423.15 (1).

4-Acetylpyridine, CA 1122-54-9: Above 385.93 (1).

2-Aminobenzaldehyde, CA 529-23-7: Above 385.93 (1).

syn-Benzaldehyde oxime, CA 622-31-1: 382.04 (1).

Formanilide, CA 103-70-8: Above 385.93 (1).

6-Methyl-2-pyridinecarboxaldehyde, CA 1122-72-1: 341.48 (1).

$C_7H_7NO_2$

3-Acetoxypyridine, CA 11747-43-2: 373.71 (1).

3,4-(Methylenedioxy)aniline, CA 14268-66-7: 415.93 (1).

Methyl isonicotinate, CA 2459-09-8: 355.37 (1).

Methyl nicotinate, CA 93-60-7: 368.71 (1).

2-Nitrotoluene, CA 88-72-2: 368.15 (3); 379.26 (1,4,5,6,7).

3-Nitrotoluene, CA 99-08-1: 374.82 (1); 379.15 (4,5,6); 385.15 (4,7).

4-Nitrotoluene, CA 99-99-0: 363.15 (3); 379.26 (1,4,5,6,7).

$C_7H_7NO_3$

3-Methyl-2-nitrophenol, CA 4920-77-8: 380.37 (1).

4-Methyl-2-nitrophenol, CA 119-33-5: 381.48 (1).

5-Methyl-2-nitrophenol, CA 700-38-9: 382.59 (1).

2-Nitroanisole, CA 91-23-6: Above 385.93 (1).

3-Nitroanisole, CA 555-03-3: 383.15 (1).

3-Nitrobenzyl alcohol, CA 619-25-0: Above 385.93 (1).

$C_7H_7N_3S$

Azidomethyl phenyl sulfide, CA 77422-70-9: 380.93 (1).

C_7H_8

Bicyclo(2,2,1)hepta-2,5-diene, CA 121-46-0: 243.71 (7); 252.04 (6); 262.04 (1,3).

Cycloheptatriene, CA 544-25-2: 276.15 (3); 277.15 (7); as 90% solution, 299.82 (1).

Quadricyclane, CA 278-06-8: 284.26 (1).

Toluene, CA 108-88-3: 277.59 (1,2,3,4,6,7); 279.15 (5); 280.37 (1).

C_7H_8BrN

2-Bromo-4-methylaniline, CA 583-68-6: Above 385.93 (1).

4-Bromo-2-methylaniline, CA 583-75-5: Above 385.93 (1).

C_7H_8ClN

2-Chlorobenzylamine, CA 89-97-4: 362.04 (1).

4-Chlorobenzylamine, CA 104-86-9: 363.71 (1).

2-Chloro-4-methylaniline, CA 615-65-6: 372.59 (1).

2-Chloro-5-methylaniline, CA 95-81-8: 380.37 (1).

2-Chloro-6-methylaniline, CA 87-63-8: 372.04 (1).

3-Chloro-2-methylaniline, CA 87-60-5: Above 385.93 (1).

3-Chloro-4-methylaniline, CA 95-74-9: 373.15 (1).

4-Chloro-N-methylaniline, CA 932-96-7: 324.82 (1).

4-Chloro-2-methylaniline, CA 95-69-2: 372.59 (1).

5-Chloro-2-methylaniline, CA 95-79-4: 433.15 (1).

C_7H_8ClNO

3-Chloro-para-anisidine, technical, CA 5345-54-0: Above 385.93 (1).

C_7H_8ClOP

Methylphenylphosphinic chloride, CA 5761-97-7: Above 385.93 (1).

$C_7H_8Cl_2Si$

Dichloromethylphenylsilane, CA 149-74-6: 340.15 (3); 355.93 (1).

C_7H_8FN

2-Fluorobenzylamine, CA 89-99-6: 340.37 (1).

3-Fluorobenzylamine, CA 100-82-3: 344.26 (1).

4-Fluorobenzylamine, CA 140-75-0: 339.82 (1); 346.15 (3).

2-Fluoro-5-methylaniline, CA 452-84-6: 354.82 (1).

3-Fluoro-2-methylaniline, CA 443-86-7: 359.26 (1).

3-Fluoro-4-methylaniline, CA 452-77-7: 362.04 (1).

5-Fluoro-2-methylaniline, CA 367-29-3: 363.15 (1).

126

$C_7H_8N_2$

1-(2-Cyanoethyl)pyrrole, CA 43036-06-2: Above 385.93 (1).

1,5-Dimethyl-2-pyrrolecarbonitrile, CA 56341-36-7: 377.59 (1).

1-Methyl-2-pyrroleacetonitrile, CA 24437-41-0: Above 385.93 (1).

$C_7H_8N_2O$

N-Methyl-N-(2-pyridyl)formamide, CA 67242-59-5: 382.59 (1).

$C_7H_8N_2O_2$

N-Methyl-2-nitroaniline, CA 612-28-2: Above 385.93 (1).

4-Methyl-2-nitroaniline, CA 89-62-3: 430.15 (5,6).

4-Methyl-3-nitroaniline, CA 119-32-4: 430.37 (7).

$C_7H_8N_2O_3$

3-Ethoxy-2-nitropyridine, CA 74037-50-6: 311.48 (1).

C_7H_8O

Anisole, CA 100-66-3: 316.15 (3,5); 324.82 (1); 324.82* (2,6,7).

Benzyl alcohol, CA 100-51-6: 366.48 (6); 369.15 (3); 373.71 (1,4,5,7); 377.55* (4).

2-Hydroxytoluene, CA 95-48-7: 354.26 (1,5,6,7); 354.26* (2).

3-Hydroxytoluene, CA 108-39-4: 359.26 (1,5,6); 367.15 (4,7).

4-Hydroxytoluene, CA 106-44-5: 359.15 (5,6); 362.59 (1); 367.59 (7).

C_7H_8OS

2-Acetyl-3-methylthiophene, CA 13679-72-6: 365.93 (1).

2-Acetyl-4-methylthiophene, CA 13679-73-7: 380.37 (1).

2-Methoxybenzenethiol, CA 7217-59-6: 358.15 (3); 368.71 (1).

3-Methoxybenzenethiol, CA 15570-12-4: 358.15 (3); 369.26 (1).

4-Methoxybenzenethiol, CA 696-63-9: 358.15 (3); 369.26 (1).

2-(Methylmercapto)phenol, CA 1073-29-6: 368.71 (1).

Methyl phenyl sulfoxide, CA 1193-82-4: 359.26 (1).

1-(2-Thienyl)-1-propanone, CA 13679-75-9: 373.71 (1).

$C_7H_8O_2$

2-Acetyl-5-methylfuran, CA 1193-79-9: 353.15 (1).

2,5-Dihydroxytoluene, CA 95-71-6: 445.37* (6).

Guaicol, CA 90-05-1: 355.37 (1); 355.37* (7).

3-Methoxyphenol, technical, CA 150-19-6: Above 385.93 (1).

4-Methoxyphenol, CA 150-76-5: Above 385.93 (1); 405.37* (6).

(continues)

$C_7H_8O_2$ *(continued)*

  3,3a,6,6a-Tetrahydro-2H-cyclopenta[b]furan-2-one, *cis*-(+), CA 54483-22-6:
  Above 385.93 (1).

$C_7H_8O_2S$

  Ethyl 2-thiophenecarboxylate, CA 2810-04-0: 362.04 (1).

  S-Furfuryl thioacetate, CA 13678-68-7: 365.93 (1).

$C_7H_8O_3$

  Ethyl 2-furoate, CA 614-99-3: 343.15 (1); 352.15 (3).

  Ethyl 3-furoate, CA 614-98-2: 332.59 (1).

  Furfuryl acetate, CA 623-17-6: 338.71 (1); 358.15 (3).

  3-Methoxycatechol, CA 934-00-9: Above 385.93 (1).

  Methyl 2-methyl-3-furancarboxylate, CA 6141-58-8: 337.04 (1).

$C_7H_8O_3S$

  4-Toluenesulfonic acid, CA 104-15-4: 457.04 (6).

C_7H_8S

  Benzyl mercaptan, CA 100-53-8: 343.15 (1,3,5,6,7).

  Thioanisole, CA 100-68-5: 330.37 (1); 345.15 (3).

  2-Thiocresol, CA 137-06-4: 337.04 (1).

  3-Thiocresol, CA 108-40-7: 345.93 (1).

  4-Thiocresol, CA 106-45-6: 341.48 (1).

$C_7H_8S_2$

  3,4-Dimercaptotoluene, technical, CA 496-74-2: Above 385.93 (1).

C_7H_9ClO

  3-Chloro-2-norbornanone, CA 30860-22-1: 367.59 (1).

$C_7H_9F_6O_5P$

  Bis(2,2,2-trifluoroethyl) (methoxycarbonylmethyl)phosphonate: Above
  385.93 (1).

C_7H_9N

  Benzylamine, CA 100-46-9: 333.15 (1); 338.15 (3).

  1-Cyclopentaneacetonitrile, CA 22734-04-9: 342.59 (1).

  2,3-Dimethylpyridine, CA 583-61-9: 323.15 (1).

  2,4-Dimethylpyridine, CA 108-47-4: 310.37 (1); 320.15 (3).

  2,5-Dimethylpyridine, CA 589-93-5: 320.93 (1).   *(continues)*

128

C_7H_9N *(continued)*

2,6-Dimethylpyridine, CA 108-48-5: 306.48 (1); 313.15 (3).

3,4-Dimethylpyridine, CA 583-58-4: 320.15 (3); 327.04 (1).

3,5-Dimethylpyridine, CA 591-22-0: 320.15 (3); 326.48 (1).

2-Ethylpyridine, CA 100-71-0: 302.59 (1); 312.15 (3).

3-Ethylpyridine, CA 536-78-7: 322.04 (1); 324.15 (3).

4-Ethylpyridine, CA 536-75-4: 320.93 (1); 323.15 (3).

N-Methylaniline, CA 100-61-8: 352.04 (1); 358.15 (3).

2-Methylaniline, CA 95-53-4: 358.15 (1,3,5,6,7).

3-Methylaniline, CA 108-44-1: 358.71 (1); 363.15 (3).

4-Methylaniline, CA 106-49-0: 360.15 (3,5,6,7); 362.04 (1).

C_7H_9NO

2-Acetyl-1-methylpyrrole, CA 932-16-1: 341.48 (1).

2-Methoxyaniline, CA 90-04-0: 372.04 (1); 390.93* (6).

3-Methoxyaniline, CA 536-90-3: Above 385.93 (1).

2-(2-Hydroxyethyl)pyridine, CA 103-74-2: 365.93 (1).

C_7H_9NOS

5-Acetyl-2,4-dimethylthiazole, CA 38205-60-6: 377.59 (1).

$C_7H_9NO_2$

2,6-Dimethoxypyridine, CA 6231-18-1: 334.82 (1).

$C_7H_9NO_3$

Ethyl 4-methyloxazole-5-carboxylate, CA 20485-39-6: 369.26 (1).

C_7H_9NS

2-(Methylmercapto)aniline, CA 2987-53-3: Above 385.93 (1).

3-(Methylmercapto)aniline, CA 1783-81-9: Above 385.93 (1).

4-(Methylmercapto)aniline, CA 104-96-1: 283.15 (1).

C_7H_{10}

1,3-Cycloheptadiene, CA 4054-38-0: 284.26 (1).

1-Methyl-1,4-cyclohexadiene, CA 4313-57-9: 283.71 (1).

Norbornylene, CA 498-66-8: 258.15 (1).

$C_7H_{10}ClNO_3$

1-Chlorocarbonylproline, *levo*, methyl ester, CA 85665-59-4: Above 385.93 (1).

$C_7H_{10}Cl_2O_2$

Diethylmalonyl dichloride, CA 54505-72-5: 347.04 (1).

3-Methyladipoyl chloride, CA 44987-62-4: Above 385.93 (1).

Pimeloyl chloride, CA 142-79-0: Above 385.93 (1).

$C_7H_{10}N_2$

2-(2-Aminoethyl)pyridine, CA 2706-56-1: 373.71 (1).

1,5-Dicyanopentane, CA 646-20-8: Above 385.93 (1).

2-Dimethylaminopyridine, CA 5683-33-0: 348.15 (1).

2-Ethyl-3-methylpyrazine, CA 15707-23-0: 330.37 (1).

1-Methyl-1-phenylhydrazine, CA 618-40-6: 365.15 (3); 369.26 (1).

2,3,5-Trimethylpyrazine, CA 14667-55-1: 327.59 (1).

$C_7H_{10}N_2O$

2-Ethyl-3-methoxypyrazine, CA 25680-58-4: 342.04 (1).

2-Furaldehyde dimethylhydrazone, CA 14064-21-2: 363.71 (1).

$C_7H_{10}O$

2-Cyclohepten-1-one, technical, CA 1121-66-0: 341.48 (1).

Dicyclopropyl ketone, CA 1121-37-5: 312.59 (1).

exo-2,3-Epoxynorbornane, CA 3146-39-2: 283.15 (1).

1-Ethynylcyclopentanol, CA 17356-19-3: 322.04 (1).

2,4-Heptadienal, trans,trans, technical, CA 4313-03-5: 338.71 (1).

1-Methoxy-1,3-cyclohexadiene, technical, CA 2161-90-2: 299.82 (1).

1-Methoxy-1,4-cyclohexadiene, technical, CA 2886-59-1: 309.15 (3); 309.82 (1).

3-Methyl-2-cyclohexen-1-one, CA 1193-18-6: 341.48 (1).

3-Methyl-3-cyclohexen-1-one, CA 31883-98-4: 348.15 (3).

5-Norbornen-2-ol, mixed isomers, CA 13080-90-5: 323.71 (1).

Norcamphor, CA 497-38-1: 307.04 (1).

1,2,3,6-Tetrahydrobenzaldehyde, CA 100-50-5: 320.15 (3); 330.37 (1); 330.37* (2,6).

$C_7H_{10}O_2$

2-Acetylcyclopentanone, CA 1670-46-8: 345.93 (1); 348.15 (3).

Allyl methacrylate, CA 96-05-9: 307.04 (1); 311.15 (3).

2-Cyclopentene-1-acetic acid, CA 13668-61-6: Above 385.93 (1).

2-Cyclopenten-1-one ethylene ketal, CA 695-56-7: 317.59 (1).

3,4-Dihydro-4,4-dimethyl-2H-pyran-2-one, CA 76897-39-7: 340.37 (1).

$C_7H_{10}O_3$

alpha-Acetyl-alpha-methyl-gamma-butyrolactone, CA 1123-19-9: 384.82 (1).

Allyl acetoacetate, CA 1118-84-9: 348.71 (1).

Diallyl carbonate, CA 15022-08-9: 332.04 (1).

2,2-Dimethylglutaric anhydride, CA 2938-48-9: Above 385.93 (1).

Ethyl 2-formyl-1-cyclopropanecarboxylate, mostly trans, CA 20417-61-2: 357.59 (1).

Glycidyl methacrylate, CA 106-91-2: 349.26 (1); 350.15 (3); 357.15* (4).

Methyl 2-oxocyclopentanecarboxylate, CA 10472-24-9: 352.15 (3); above 385.93 (1).

Triacetylmethane, CA 815-68-9: 360.93 (1).

2,2,6-Trimethyl-1,3-dioxen-4-one, CA 5394-63-8: 331.15 (3); 359.82 (1); as 85% solution, 292.59 (1).

$C_7H_{10}O_4$

1,1-Diacetoxy-2-propene, CA 869-29-4: 351.48 (1); 355.38* (6,7).

Dimethyl ethylidenemalonate, CA 17041-60-0: 370.37 (1).

Dimethyl itaconate, CA 617-52-7: 373.71 (1).

Ethyl 2,4-dioxovalerate, CA 615-79-2: Above 385.93 (1).

Methyl 2-acetylacetoacetate, CA 4619-66-3: 356.15 (3).

$C_7H_{10}O_5$

Diethyl ketomalonate, CA 609-09-6: Above 385.93 (1).

Dimethyl 1,3-acetonedicarboxylate, CA 1830-54-2: 336.15 (3); above 385.93 (1).

$C_7H_{11}Br$

exo-2-Bromonorbornane, CA 2534-77-2: 333.15 (1).

$C_7H_{11}BrO_2$

Ethyl 1-bromocyclobutanecarboxylate, CA 35120-18-4: 352.59 (1).

$C_7H_{11}BrO_4$

Diethyl bromomalonate, CA 685-87-0: Above 385.93 (1).

$C_7H_{11}Cl$

exo-2-Chloronorbornane ±, CA 67844-27-3: 317.59 (1).

$C_7H_{11}ClO$

Cyclohexanecarboxylic acid chloride, CA 2719-27-9: 339.82 (1).

$C_7H_{11}ClO_4$

Diethyl chloromalonate, CA 14064-10-9: 376.48 (1).

$C_7H_{11}FO_4$

Diethyl fluoromalonate, CA 685-88-1: 335.37 (1).

$C_7H_{11}N$

Cyclohexyl isocyanide, CA 931-53-3: 349.82 (1).
1,2,5-Trimethylpyrrole, CA 930-87-0: 325.37 (1).

$C_7H_{11}NO$

Cyclohexanone cyanohydrin, CA 931-97-5: 333.15 (1,4).
Cyclohexyl isocyanate, CA 3173-53-3: 322.04 (1); 327.15* (4).

$C_7H_{11}NO_2$

1-Acetyl-4-piperidone, CA 32161-06-1: Above 385.93 (1).
tert-Butyl cyanoacetate, CA 1116-98-9: 364.82 (1).
tert-Butyl isocyanoacetate, CA 2769-72-4: 355.15 (3).

$C_7H_{11}NO_3$

Methyl 2-oxo-1-pyrrolidineacetate, CA 59776-88-4: 383.15 (1).

$C_7H_{11}NO_6$

Diethyl nitromalonate, CA 603-67-8: Above 385.93 (1).

$C_7H_{11}NS$

Cyclohexyl isothiocyanate, CA 1122-82-3: 368.15 (3); 368.71 (1).
2-(Dimethylaminomethyl)thiophene, CA 26019-17-0: 332.04 (1).
2-Isobutylthiazole, CA 18640-74-9: 330.93 (1).

C_7H_{12}

Cycloheptene, CA 628-92-2: 266.48 (1); below 296.15 (7); 318.15 (3).
2,4-Dimethyl-1,3-pentadiene, CA 1000-86-8: 283.15 (1).
1,6-Heptadiene, CA 3070-53-9: 263.15 (1).
1-Heptyne, CA 628-71-7: 263.15 (2); 270.93 (1).
1-Methyl-1-cyclohexene, CA 591-49-1: 169.26 (1); 272.15 (3).
3-Methyl-1-cyclohexene, CA 591-48-0: 269.26 (1); 272.15 (3).
4-Methyl-1-cyclohexene, CA 591-47-9: 266.15 (3); 272.04 (1,7).
Methylenecyclohexane, CA 1192-37-6: 266.48 (1); 271.15 (3).
2-Methyl-1,5-hexadiene, CA 27477-37-8: 260.93 (1).

(continues)

C_7H_{12}

  3-Methyl-1,5-hexadiene, CA 1541-33-9: 254.26 (1).

  Vinylcyclopentane, CA 3742-34-5: 283.15 (1).

$C_7H_{12}BrN$

  7-Bromoheptanenitrile, technical, CA 20965-27-9: Above 385.93 (1).

$C_7H_{12}N_2$

  2-(2-Aminoethyl)-1-methylpyrrole, CA 83732-75-6: 372.04 (1).

  1-Butylimidazole, CA 4316-42-1: 371.15 (3).

  1,5-Diazabicyclo(4,3,0)non-5-ene, CA 3001-72-7: 367.59 (1,3).

$C_7H_{12}N_2O$

  4-(2-Cyanoethyl)morpholine, CA 4542-47-6: 388.15 (3).

  2-Morpholinoethyl isocyanide, CA 78375-48-1: 383.15 (3).

$C_7H_{12}O$

  2-Butylacrolein, CA 1070-66-2: 306.48 (1).

  Cycloheptanone, CA 502-42-1: 328.71 (1); 330.15 (3).

  Cyclohexanecarboxaldehyde, CA 2043-61-0: 313.71 (1).

  3-Cyclohexene-1-methanol ±, CA 72581-32-9: 349.26 (1).

  2,4-Dimethylcyclopentanone ±, CA 1121-33-1: 314.82 (1).

  2,2-Dimethyl-4-pentenal, CA 5497-67-6: 291.48 (1).

  3-Ethyl-1-pentyn-3-ol, CA 6285-06-9: 310.15 (3).

  1,6-Heptadien-4-ol, CA 2883-45-6: 313.15 (1).

  2-Heptenal, *trans*, technical, CA 18829-55-5: 326.48 (1).

  2-Methylcyclohexanone, CA 583-60-8: 319.82 (1); 321.15 (3,5).

  3-Methylcyclohexanone ±, CA 591-24-2: 321.15 (3); 324.82 (1).

  3-Methylcyclohexanone, (*R*)-(+), CA 13368-65-5: 324.82 (1).

  4-Methylcyclohexanone, CA 589-92-4: 272.04* (6); 313.71 (1); 321.15 (3).

  Methylcyclohexanone, mixed isomers, CA 1331-22-2: 320.93 (3,6,7); 326.15 (2).

  3-Methyl-2-cyclohexen-1-ol, CA 21378-21-2: 344.82 (1).

  5-Methyl-3-hexen-2-one, technical, CA 5166-53-0: 320.93 (1).

  1,2,3,6-Tetrahydrobenzyl alcohol, CA 1679-51-2: 346.15 (3).

$C_7H_{12}O_2$

  Allyl butyrate, CA 2051-78-7: 314.82 (1).

  Butyl acrylate, CA 141-32-2: 309.15 (3); 310.15 (5); 312.15* (4); 312.59 (1); 320.93* (6); 322.04* (2,4,7).

(continues)

$C_7H_{12}O_2$ *(continued)*

tert-Butyl acrylate, CA 1663-39-4: 292.15* (4); 309.15 (3).

Cyclohexanecarboxylic acid, CA 98-89-5: Above 385.93 (1).

Cyclohexyl formate, CA 4351-54-6: 324.15 (5,6).

Cyclopentylacetic acid, CA 1123-00-8: 382.59 (1).

3,3-Diethoxy-1-propyne, CA 10160-87-9: 304.15 (3); 305.37 (1).

3,4-Dihydro-2-ethoxy-2*H*-pyran, CA 103-75-3: 297.59 (1); 317.04* (6,7).

2,2-Dimethyl-4-pentenoic acid, technical, CA 16386-93-9: 361.48 (1).

Ethyl cyclobutanecarboxylate, CA 14924-53-9: 314.82 (1).

Ethyl 3,3-dimethylacrylate, CA 638-10-8: 307.04 (1).

Ethyl 1-methylcyclopropanecarboxylate, CA 71441-76-4: 303.71 (1).

Ethyl 2-methylcyclopropanecarboxylate, CA 20913-25-1: 315.37 (1).

3-Ethyl-2,4-pentanedione, CA 1540-34-7: 339.82 (1).

2-Ethyl-*delta*-valerolactone, CA 32821-68-4: 382.04 (1).

Isobutyl acrylate, CA 106-63-8: 303.15* (6); 305.15 (3); 315.15* (4).

2-Methoxycyclohexanone, CA 7429-44-9: 342.59 (1).

4-Penten-1-yl acetate, CA 1576-85-8: 318.71 (1).

Pivaloylacetaldehyde, CA 23459-13-4: 311.15 (3).

$C_7H_{12}O_2S_2$

Ethyl 1,3-dithiane-2-carboxylate, CA 20462-00-4: 327.59 (1); 328.15 (3).

$C_7H_{12}O_3$

beta,beta-Dimethyl-*gamma*-(hydroxymethyl)-*gamma*-butyrolactone ±, CA 52398-48-8: Above 385.93 (1)

Ethyl 3-ethoxyacrylate, CA 1001-26-9: 349.26 (1).

Ethyl levulinate, CA 539-88-6: 363.71 (1).

Ethyl 2-methylacetoacetate, CA 609-14-3: 330.15 (3); 335.93 (1).

Ethyl 3-methyl-2-oxobutyrate, CA 20201-24-5: 316.48 (1).

Ethyl propionylacetate, CA 4949-44-4: 350.93 (1).

4-Hydroxybutyl acrylate, CA 2478-10-6: 383.15 (1).

2-Hydroxypropyl methacrylate, mixed isomers, CA 923-26-2: 368.15 (3); 369.82 (1); 394.15* (4).

$C_7H_{12}O_4$

Adipic acid, monomethyl ester, CA 627-91-8: Above 385.93 (1).

Cyclohexanecarboxaldehyde, CA 2043-61-0: 314.15 (3).

Diethyl malonate, CA 105-53-3: 348.15 (4); 353.15 (3); 366.15 (5); 366.48* (6,7); 373.15 (1).

2,5-Dimethoxy-3-tetrahydrofurancarboxaldehyde, technical, CA 50634-05-4: 354.26 (1).

(continues)

134

$C_7H_{12}O_4$ *(continued)*

Dimethyl ethylmalonate, CA 26717-67-9: 365.15 (3).

Dimethyl glutarate, CA 1119-40-0: 370.15 (3); 376.48 (1).

Dimethyl methylsuccinate, CA 1604-11-1: 356.48 (1).

2-Methoxyethyl acetoacetate, CA 22502-03-0: 376.48 (1).

Methyl 2,2-dimethyl-1,3-dioxolane-4-carboxylate, CA 60456-21-5: 348.15 (3); 351.48 (1).

Methyl 2,3-O-isopropylideneglycerate, *dextro*, CA 52373-72-5: 348.15 (3).

2,4,8,10-Tetraoxaspiro[5,5]undecane, CA 126-54-5: 381.48 (1).

$C_7H_{12}O_5$

Glycerol diacetate, CA 25395-31-7: 419.26* (2).

$C_7H_{12}O_6Si$

Methyltriacetoxysilane, CA 4253-34-3: 358.15 (1)

$C_7H_{13}Br$

Cyclohexylmethyl bromide, CA 2550-36-9: 330.37 (1).

Cycloheptyl bromide, CA 2404-35-5: 341.48 (1).

$C_7H_{13}BrO_2$

Ethyl 2-bromovalerate, CA 615-83-8: 350.37 (1).

Ethyl 5-bromovalerate, CA 14660-52-7: 377.04 (1).

$C_7H_{13}ClO$

Heptanoyl chloride, CA 2528-61-2: 331.48 (1).

$C_7H_{13}ClO_2$

2-(4-Chlorobutyl)-1,3-dioxolane: 358.15 (3).

2-(3-Chloropropyl)-2-methyl-1,3-dioxolane, CA 5978-08-5: 345.15 (3); 345.93 (1).

Hexyl chloroformate, CA 6092-54-2: 334.82 (1).

$C_7H_{13}Cl_3O_3$

Tris(2-chloroethyl) orthoformate, CA 18719-58-9: 337.04 (1).

$C_7H_{13}N$

exo-2-Aminonorbornane, CA 7242-92-4: 308.15 (1).

3-Diethylamino-1-propyne, CA 4079-68-9: 293.15 (3).

1,1-Diethylpropargylamine, CA 3234-64-8: 294.26 (1).

$C_7H_{13}NO$

2-Azacyclooctanone, CA 673-66-5: Above 385.93 (1).

1-Aza-2-methoxy-1-cycloheptene, CA 2525-16-8: 324.26 (1).

N-Cyclohexylformamide, CA 766-93-8: Above 385.93 (1).

2,4,4-Trimethyl-1-pyrroline, *N*-oxide, CA 6931-11-9: 379.82 (1).

$C_7H_{13}NO_2$

1,4-Dioxa-8-azaspiro(4,5)decane, CA 177-11-7: 354.82 (1).

1-(3-Hydroxypropyl)-2-pyrrolidinone, CA 62012-15-1: Above 385.93 (1).

2-(Methoxymethyl)-1-pyrrolidinecarboxaldehyde, (*S*)-(-), CA 63126-45-4: Above 385.93 (1).

2-Methoxy-1-piperidinecarboxaldehyde, mixed isomers, CA 61020-07-3: 372.04 (1).

$C_7H_{13}NO_4$

2-(2-Nitroethoxy)tetrahydropyran, CA 75233-61-3: 320.93 (1).

C_7H_{14}

Cycloheptane, CA 291-64-5: 279.26 (1); 288.15 (3,7).

1,1-Dimethylcyclopentane, CA 1638-26-2: Below 294.26 (2).

1,2-Dimethylcyclopentane, CA 2452-99-5: Below 294.26 (2).

1,3-Dimethylcyclopentane, CA 2453-00-1: Below 294.26 (2).

2,3-Dimethyl-1-pentene, CA 3404-72-6: 256.15 (3).

4,4-Dimethyl-1-pentene, CA 762-62-9: 261.15 (3).

Ethylcyclopentane, CA 1640-89-7: 288.71 (1); below 294.15 (5,6,7).

3-Ethyl-1-pentene, CA 4038-04-4: 253.15 (3).

1-Heptene, CA 592-76-7: 264.26 (1); 269.26 (2); 272.15 (3); below 273.15 (5,6,7).

2-Heptene, *cis*, CA 6443-92-1: 255.37 (2); 267.04 (1).

2-Heptene, *trans*, CA 14686-13-6: 272.04 (1,3); below 273.15 (5,6).

3-Heptene, *cis*, CA 7642-10-6: 265.93 (1).

3-Heptene, *trans*, CA 14686-14-7: 267.04 (1); 272.15 (3).

3-Heptene, mixed isomers, CA 592-78-9: 266.15 (3); below 266.15 (7); 267.04 (2,6).

Methylcyclohexane, CA 108-87-2: 269.26 (1,3,5,6,7); 294.26 (2).

2-Methyl-1-hexene, CA 6094-02-6: 265.15 (3); 266.48 (1).

3-Methyl-1-hexene, CA 3404-61-3: 266.48 (1).

4-Methyl-1-hexene, CA 3769-23-1: 263.15 (3).

5-Methyl-1-hexene, CA 3524-73-0: 263.15 (3).

2,3,3-Trimethyl-1-butene, CA 594-56-9: 255.93 (1); below 273.15 (5,6).

136

$C_7H_{14}Br_2$

1,7-Dibromoheptane, CA 4549-31-9:   Above 385.93 (1).

$C_7H_{14}N_2$

1,3-Diisopropylcarbodiimide, CA 693-13-0:   304.15 (3); 307.04 (1).

Diisopropylcyanamide, CA 3085-76-5:   352.04 (1).

$C_7H_{14}N_2O$

1-(3-Aminopropyl)-2-pyrrolidinone, technical, CA 7663-77-6:   Above 385.93 (1).

$C_7H_{14}N_2O_2$

Ethyl 1-piperazinecarboxylate, CA 120-43-4:   Above 385.93 (1).

$C_7H_{14}O$

Allyl butyl ether, CA 3739-64-8:   290.15 (3).

Cycloheptanol, CA 502-41-0:   331.15 (3); 344.26 (1).

Cyclohexylmethanol, CA 100-49-2:   344.26 (1); 350.15 (3).

2,4-Dimethyl-3-pentanone, CA 565-80-0:   288.15 (3); 288.71 (1); 322.15 (7).

4,4-Dimethyl-2-pentanone, CA 590-50-1:   292.04 (1).

2,3-Dimethylvaleraldehyde, CA 32749-94-3:   307.59* (2,6); 331.48 (1).

Heptaldehyde, CA 111-71-7:   308.15 (1); 311.15 (3).

2-Heptanone, CA 110-43-0:   312.04 (2,6); 314.15 (3); 318.71* (2); 320.15* (4); 320.37 (1); 322.15 (4,5); 322.15* (4,7).

3-Heptanone, CA 106-35-4:   311.15 (3); 314.15* (4); 314.26 (1); 319.26* (6,7); 324.82 (2).

4-Heptanone, CA 123-19-3:   307.15 (5); 322.04 (1,3,6,7); 322.15* (2).

1-Hepten-3-ol, CA 4938-52-7:   327.15 (3).

1-Methylcyclohexanol, CA 590-67-0:   340.93 (1).

2-Methylcyclohexanol, ± cis, CA 7443-70-1:   332.04 (1).

2-Methylcyclohexanol, ± trans, CA 7443-52-9:   332.04 (1).

2-Methylcyclohexanol, mixed isomers, CA 583-59-5:   332.04 (1); 337.15 (3); 338.15 (6); 341.15 (5).

3-Methylcyclohexanol, CA 591-23-1:   335.93 (1); 343.15 (5,6,7).

4-Methylcyclohexanol, cis, CA 7731-28-4:   343.15 (1).

4-Methylcyclohexanol, trans, CA 7731-29-5:   343.15 (1).

4-Methylcyclohexanol, mixed isomers, CA 589-91-3:   312.15 (3); 343.15 (1,5,6).

Methylcyclohexanol, mixed isomers, CA 25639-42-3:   314.15 (3); 340.93 (7).

2-Methyl-3-hexanone, CA 7379-12-6:   297.04 (1).

5-Methyl-2-hexanone, CA 110-12-3:   308.71 (2,6); 313.15 (3); 314.26 (1); 314.26* (2,4); 316.15* (4,7).

(continues)

$C_7H_{14}O$ *(continued)*

5-Methyl-3-hexanone, CA 623-56-3: 299.82 (1).

$C_7H_{14}OSi$

1-(Trimethylsilyloxy)-1,3-butadiene, mixed isomers, CA 6651-43-0: 298.15 (1).

2-(Trimethylsilyoxy)-1,3-butadiene, CA 38053-91-7: 285.37 (1).

$C_7H_{14}O_2$

Acrolein diethyl acetal, CA 3054-95-3: 277.59 (1); 288.15 (3).

Amyl acetate, CA 628-63-7: 288.71 (6); 297.04 (1); 298.15 (3,4,7); 299.26* (2); 310.15 (5).

Amyl acetate, mixed isomers: 309.82 (1).

Butyl glycidyl ether, CA 2426-08-6: 295.15 (3); 328.71 (1).

tert-Butyl glycidyl ether, CA 7665-72-7: 314.15 (3); 316.48 (1).

Butyl propionate, CA 590-01-2: 290.37 (2); 308.37 (6,7).

tert-Butyl propionate, CA 20487-40-5: 289.15 (3); 289.82 (1).

Ethyl isovalerate, CA 108-64-5: 295.15 (3); 298.15 (7); 299.82 (1).

Ethyl 2-methylbutyrate, CA 7452-79-1: 299.15 (3).

Ethyl trimethylacetate, CA 3938-95-2: 288.15 (3); 289.26 (1).

Ethyl valerate, CA 539-82-2: 307.15 (3); 312.04 (1).

Heptanoic acid, CA 111-14-8: Above 385.93 (1).

Isoamyl acetate, CA 123-92-2: 298.15 (1,3,4,5,6,7).

Isobutyl propionate, CA 540-42-1: Below 295.15 (7); 311.15 (5).

1-Methylbutyl acetate, CA 626-38-0: 296.15 (7); 305.15 (5,6).

Methyl hexanoate, CA 106-70-7: 316.15 (3).

2-Methylhexanoic acid, technical, CA 4536-23-6: 378.71 (1).

Propyl butyrate, CA 105-66-8: 310.15 (3,5,6); 312.04 (1).

$C_7H_{14}O_3$

Butyl ethyl carbonate, CA 30714-78-4: 323.15 (5,6).

Butyl lactate ±, CA 138-22-7: 334.15 (5); 342.59 (1); 344.15 (2); 344.26* (6,7); 348.15 (3).

Dipropyl carbonate, CA 623-96-1: 328.15 (1); 337.15 (4).

Ethyl 3-ethoxypropionate, CA 763-69-9: 355.37* (7).

Ethyl 2-hydroxypentanoate, CA 6938-26-7: 334.15 (3).

3-Methoxybutyl acetate, CA 4435-53-4: 333.15 (5); 349.82 (6,7).

$C_7H_{14}O_3S$

Methyl *tert*-butylsulfinylacetate, CA 56535-32-1: 313.15 (3).

138

$C_7H_{14}O_4$

Diethoxymethyl acetate, CA 14036-06-7: 300.93 (1).

Diethylene glycol, monomethyl ether acetate, CA 629-38-9: 355.15 (4,5); 355.37* (6,7).

(-)-2,3-O-Isopropylidene-*dextro*-threitol, CA 73346-74-4: Above 385.93 (1).

(+)-2,3-O-Isopropylidene-*levo*-threitol, CA 50622-09-8: Above 385.93 (1).

$C_7H_{15}Br$

1-Bromoheptane, CA 629-04-9: 333.71 (1); 338.15 (3).

2-Bromoheptane, CA 1974-04-5: 320.37 (1).

$C_7H_{15}Cl$

1-Chloroheptane, CA 629-06-1: 311.15 (3); 314.82 (1).

$C_7H_{15}ClO_2$

3-Chloro-1,1-diethoxypropane, technical, CA 355-93-4: 309.82 (1); 327.15 (3).

$C_7H_{15}Cl_2N_2O_2P$

Cyclophosphamide monohydrate, CA 6055-19-2: Above 385.93 (1).

$C_7H_{15}F_3O_3SSi$

tert-Butyldimethylsilyl trifluoromethanesulfonate, CA 69739-34-0: 309.82 (1); 326.15 (3).

Triethylsilyl trifluoromethanesulfonate, CA 79271-56-0: 339.15 (3); 345.37 (1).

$C_7H_{15}I$

1-Iodoheptane, CA 4282-40-0: 349.15 (3); 352.04 (1).

$C_7H_{15}N$

Cycloheptylamine, CA 5452-35-7: 315.37 (1); 320.15 (3).

Cyclohexanemethylamine, CA 3218-02-8: 316.48 (1); 321.15 (3).

2,6-Dimethylpiperidine, CA 504-03-0: 284.82 (1); 289.15 (3,7).

3,3-Dimethylpiperidine, CA 1193-12-0: 298.15 (3).

3,5-Dimethylpiperidine, mixed isomers, CA 35794-11-7: 305.93 (1).

1-Ethylpiperidine, CA 766-09-6: 290.15 (3); 292.04 (1,7).

2-Ethylpiperidine, CA 1484-80-6: 290.15 (3); 304.26 (1).

Heptamethyleneimine, CA 1121-92-2: 302.59 (1).

N-Methylcyclohexylamine, mixed isomers, CA 100-60-7: 302.59 (1); 312.15 (3).

2-Methylcyclohexylamine, mixed isomers, CA 7003-32-9: 294.82 (1); 312.15 (3).

(continues)

$C_7H_{15}N$ *(continued)*

3-Methylcyclohexylamine, mixed isomers, CA 6850-35-7: 295.37 (1).

4-Methylcyclohexylamine, mixed isomers, CA 6321-23-9: 299.82 (1).

$C_7H_{15}NO$

Diethylaminoacetone, CA 1620-14-0: 311.48 (1).

N,N-Diethylpropionamide, CA 1114-51-8: 345.93 (1).

N,N-Diisopropylformamide, CA 2700-30-3: 348.15 (1,3).

4-Dimethylamino-3-methyl-2-butanone, CA 22104-62-7: 311.48 (1).

1-Ethyl-3-hydroxypiperidine, CA 13444-24-1: 320.37 (1); 352.15 (3).

1-Ethyl-4-hydroxypiperidine, CA 3518-83-0: 368.71 (1).

1-Methyl-2-piperidinemethanol, CA 20845-34-5: 354.26 (1).

1-Methyl-3-piperidinemethanol, CA 7583-53-1: 367.59 (1).

1-Methyl-2-pyrrolidineethanol, CA 67004-64-2: 357.59 (1).

1-Piperidineethanol, CA 3040-44-6: 342.04 (1).

2-Piperidineethanol, CA 1484-84-0: 375.37 (1).

4-Piperidineethanol, CA 622-26-4: Above 385.93 (1).

$C_7H_{15}NO_2$

N-Butylurethane, CA 591-62-8: 364.82 (6).

Ethyl 3-dimethylaminopropionate, CA 20120-21-2: 331.15 (3).

3-Pyrrolidino-1,2-propanediol: Above 385.93 (1).

$C_7H_{15}NO_3$

3-Morpholino-1,2-propanediol, CA 6425-32-7: Above 385.93 (1).

$C_7H_{15}O_5P$

Ethyl diethoxyphosphinylformate, CA 1474-78-8: Above 385.93 (1).

Methyl diethylphosphinoacetate, CA 1067-74-9: Above 385.93 (1).

C_7H_{16}

2,2-Dimethylpentane, CA 590-35-2: 264.15 (3); 288.71 (1).

2,3-Dimethylpentane, CA 565-59-3: Below 249.82 (2); 266.48 (1,7); below 266.48 (6).

2,4-Dimethylpentane, CA 108-08-7: 249.82 (2); 260.93 (1,6,7); 266.15 (3).

3,3-Dimethylpentane, CA 562-49-2: 266.48 (1,3).

Heptane, CA 142-82-5: 269.15 (2,3,5,6,7); 272.04 (1); 277.15 (4).

Heptane, mixed isomers: Below 269.15 (5).

2-Methylhexane, CA 591-76-4: Below 255.37 (6,7); 269.26 (1); 272.15 (3).

(continues)

C_7H_{16} *(continued)*

3-Methylhexane, CA 589-34-4: 269.26 (1,2,6); 272.15 (3); below 273.15 (5).

2,2,3-Trimethylbutane, CA 464-06-2: 253.15 (3); 266.48 (1); below 273.15 (5,6).

$C_7H_{16}N_2$

1-Amino-2,6-dimethylpiperidine, technical, CA 39135-39-2: 315.37 (1).

2-(2-Aminoethyl)-1-methylpyrrolidine, CA 51387-90-7: 338.15 (1).

1-(2-Aminoethyl)piperidine, CA 27578-60-5: 330.93 (1,3).

3-Amino-1-ethylpiperidine, CA 6789-94-2: 333.71 (1).

2-(Aminomethyl)-1-ethylpyrrolidine, CA 26116-12-1: 333.71 (1).

N'-*tert*-Butyl-N,N-dimethylformamide, CA 23314-06-9: 307.59 (1).

1-Methyl-4-(methylamino)piperidine, CA 7149-42-0: 328.71 (1).

$C_7H_{16}N_2O$

4-(3-Aminopropyl)morpholine, CA 123-00-2: 372.04 (1); 377.60* (6,7).

$C_7H_{16}O$

2,2-Dimethyl-3-pentanol, CA 3970-62-5: 310.93 (1).

2,3-Dimethyl-3-pentanol, CA 595-41-5: 313.71 (1).

2,4-Dimethyl-3-pentanol, CA 600-36-2: 310.37 (1); 322.04 (6); 323.15 (5).

4,4-Dimethyl-2-pentanol, CA 6144-93-0: 309.82 (1).

3-Ethyl-3-pentanol, CA 597-49-9: 313.15 (1).

1-Heptanol, CA 111-70-6: 343.15 (3); 347.04 (1).

2-Heptanol ±, CA 543-49-7: 314.26 (1); 332.15 (3); 344.15 (5,6); 344.26* (2,7).

3-Heptanol ±, CA 589-82-2: 327.59 (1); 333.15 (2,3,5,6); 333.15* (7).

4-Heptanol, CA 589-55-9: 327.15 (3).

2-Methyl-2-hexanol, CA 625-23-0: 313.71 (1).

2-Methyl-3-hexanol, CA 617-29-8: 313.71 (1).

5-Methyl-2-hexanol ±, CA 627-59-8: 319.26 (1).

$C_7H_{16}OS$

3-(Methylthio)-1-hexanol, CA 51755-66-9: 380.93 (1).

$C_7H_{16}O_2$

1-*tert*-Butoxy-2-methoxyethane, CA 66728-50-5: 298.15 (1).

1,1-Diethoxypropane, CA 4744-08-5: 285.15 (3); 285.93 (1).

2,2-Diethoxypropane, CA 126-84-1: 280.93 (1).

2,2-Diethyl-1,3-propanediol, CA 115-76-4: 374.15 (5); 374.82* (6,7).

(continues)

$C_7H_{16}O_2$ *(continued)*

2,4-Dimethyl-2,4-pentanediol, CA 24892-49-7: 369.82 (1).

1,7-Heptanediol, CA 629-30-1: Above 385.93 (1).

2-Methyl-2-propyl-1,3-propanediol, CA 78-26-2: Above 385.93 (1).

Propylene glycol, butyl ether, CA 5131-66-8: 332.15 (4); 341.15* (4); 344.26* (2).

$C_7H_{16}O_2Si$

Diethoxymethylvinylsilane, CA 5507-44-8: 290.93 (1).

Ethyl (trimethylsilyl)acetate, CA 4071-88-9: 308.15 (1); 321.15 (3).

$C_7H_{16}O_3$

Diethylene glycol, methyl ethyl ether, CA 1002-67-1: 355.15* (2).

Dipropylene glycol, methyl ether, mixed isomers, CA 20324-32-7: 347.59 (1,2); 352.15 (2,4); 355.15* (4); 358.15 (5,6,7); 358.15* (2).

Methoxyacetaldehyde diethyl acetal, CA 4819-75-4: 308.15 (1).

3-Methoxybutyraldehyde dimethyl acetal, CA 10138-89-3: 320.37 (1).

Triethyl orthoformate, CA 122-51-0: 303.15 (1,3,5,6,7).

1,2,3-Trimethoxybutane, CA 6607-66-5: 318.71 (1).

Trimethyl orthobutyrate, CA 43083-12-1: 308.15 (1); 323.15 (3).

$C_7H_{16}O_4$

Ethylene glycol, monomethyl ether formal, CA 4431-82-7: 341.48* (6).

1,1,3,3-Tetramethoxypropane, CA 102-52-3: 327.59 (1); 339.15 (3); 349.82 (6).

Triethylene glycol, methyl ether, CA 112-35-6: 387.59 (2,4); 391.48* (4,6,7).

$C_7H_{16}S$

1-Heptanethiol, CA 1639-09-4: 314.15 (3); as 90% solution, 319.26 (1).

$C_7H_{16}S_2Si$

2-Trimethylsilyl-1,3-dithiane, CA 13411-42-2: 369.26 (1).

$C_7H_{17}BO_2$

Diisopropoxymethylborane, CA 86595-27-9: 280.37 (1).

$C_7H_{17}ClNOP$

N,N-Diisopropylmethylphosphonamidic chloride, CA 86030-43-5: 345.93 (1).

$C_7H_{17}N$

2-Aminoheptane, CA 123-82-0: 327.59 (1).   *(continues)*

$C_7H_{17}N$ *(continued)*

  3-Aminoheptane, technical, CA 28292-42-4: As 70% solution, 313.15 (1).

  2-Amino-5-methylhexane, CA 28292-43-5: 301.15 (3).

  N-tert-Butylisopropylamine, CA 7515-80-2: 298.15 (3).

  Heptylamine, CA 111-68-2: 308.15 (1); 317.15 (3); 327.59* (2,6).

  N-Hexylmethylamine, CA 35161-70-7: 304.15 (3).

$C_7H_{17}NO$

  1-Diethylamino-2-propanol, CA 4402-32-8: 306.48 (1,3).

  3-Diethylamino-1-propanol, CA 622-93-5: 338.71 (1).

$C_7H_{17}NOSi$

  4-(Trimethylsilyl)morpholine, CA 13368-42-8: 300.93 (1).

$C_7H_{17}NO_2$

  3-Diethylamino-1,2-propanediol, CA 621-56-7: 372.04* (2); 380.93 (1).

  N,N-Dimethylaminodiethoxymethane, CA 1188-33-6: 295.37 (1); 301.15 (3).

$C_7H_{17}NO_3$

  1-[*N,N*-Bis(2-hydroxyethyl)amino]-2-propanol, CA 6712-98-7: Above 385.93 (1).

$C_7H_{17}NSi$

  1-(Trimethylsilyl)pyrrolidine, CA 15097-49-1: 277.59 (1).

$C_7H_{17}N_3$

  4-Amino-1,2-diethylpyrazolidine, CA 70180-92-6: 347.04 (1).

$C_7H_{17}O_3PS$

  Diethyl (ethylthiomethyl)phosphonate, CA 54091-78-0: Above 385.93 (1).

$C_7H_{17}O_5PSi$

  Trimethylsilyl *P,P*-dimethylphosphonoacetate, CA 85169-29-5: 292.15 (3).

$C_7H_{18}N_2$

  1,7-Diaminoheptane, CA 646-19-5: 360.93 (1).

  3-Diethylaminopropylamine, CA 104-78-9: 325.15 (3); 332.04 (1); 332.04* (6,7).

  N,N'-Diethyl-1,3-propanediamine, CA 10061-68-4: 320.15 (3); 323.71 (1).

  1-(Isopropylamino)-2-methyl-1,2-propanediamine, CA 5448-29-3: 363.71 (1).

(continues)

$C_7H_{18}N_2$ *(continued)*

N,N,N',N'-Tetramethyl-1,3-propanediamine, CA 110-95-2: 304.82 (1).
N,N,2,2-Tetramethyl-1,3-propanediamine, CA 53369-71-4: 308.71 (1).

$C_7H_{18}N_2O$

1,3-Bis(dimethylamino)-2-propanol, CA 5966-51-8: Above 385.93 (1).
2-{[2-(Dimethylamino)ethyl]methylamino} ethanol, CA 2212-32-0: 359.82 (1).

$C_7H_{18}O_3Si$

Methyltriethoxysilane, CA 2031-67-6: 297.04 (1); 308.15 (4).

$C_7H_{19}NSi$

N,N-Diethyltrimethylsilylamine, CA 996-50-9: 283.15 (1); 295.15 (3).

$C_7H_{19}N_3$

3,3'-Diamino-*N*-methyldipropylamine, CA 105-83-9: 375.93 (1); 377.59 (7).
Spermidine, CA 124-20-9: Above 385.93 (1).
Tris(dimethylamino)methane, CA 5762-56-1: 257.59 (1); 284.15 (3).

$C_7H_{20}N_4$

N,N'-Bis(2-aminoethyl)-1,3-propanediamine, technical, CA 4741-99-5:
383.15 (1).

$C_7H_{20}Si_2$

Bis(trimethylsilyl)methane, CA 2117-28-4: 285.93 (1); 298.15 (3).

$C_7H_{21}NSi_2$

Heptamethyldisilazane, CA 920-68-3: 300.93 (1,3).

$C_8F_{14}O_3$

Heptafluorobutyric anhydride, CA 336-59-4: Nonflammable (1).

C_8F_{16}

Perfluoro-1,3-dimethylcyclohexane, technical, CA 335-27-3:
Nonflammable (1).

$C_8F_{17}I$

Perfluorooctyl iodide, CA 507-63-1: Nonflammable (1).

$C_8H_3BrF_6$

3,5-Bis(trifluoromethyl)bromobenzene, CA 328-70-1: Above 385.93 (1).

$C_8H_3F_5$

2,3,4,5,6-Pentafluorostyrene, CA 653-34-9: 307.59 (1).

$C_8H_3F_5O$

2',3',4',5',6',-Pentafluoroacetophenone, CA 652-29-9: 338.71 (1).

$C_8H_3F_5O_2$

Methyl pentafluorobenzoate, CA 36629-42-2: 351.48 (1).

$C_8H_3F_6NO_2$

3,5-Bis(trifluoromethyl)nitrobenzene, CA 328-75-6: 350.93 (1).

$C_8H_3F_7$

2,3,5,6-Tetrafluoro-4-(trifluoromethyl)toluene, CA 778-35-8: 314.82 (1).

$C_8H_3F_{15}O$

2,2,3,3,4,4,5,5,6,6,7,7,8,8,8-Pentadecafluoro-1-octanol, CA 307-30-2:
Above 385.93 (1).

$C_8H_4ClF_3O$

2-(Trifluoromethyl)benzoyl chloride, CA 312-94-7: 368.71 (1).

3-(Trifluoromethyl)benzoyl chloride, CA 2251-65-2: Above 385.93 (1).

4-(Trifluoromethyl)benzoyl chloride, CA 329-15-7: 351.48 (1).

C_8H_4ClNO

3-Cyanobenzoyl chloride, CA 1711-11-1: Above 385.93 (1).

$C_8H_4ClNO_2$

2-Isocyanatobenzoyl chloride, CA 5100-23-2: Above 385.93 (1).

$C_8H_4Cl_2O_2$

Isophthaloyl dichloride, CA 99-63-8: 453.15* (1,6,7).

Phthaloyl dichloride, CA 88-95-9: Above 385.93 (1).

Terephthaloyl dichloride, CA 100-20-9: 453.15 (1,6); 453.15* (7).

$C_8H_4F_3N$

2-(Trifluoromethyl)benzonitrile, CA 447-60-9: 362.59 (1).

3-(Trifluoromethyl)benzonitrile, CA 368-77-4: 345.37 (1).

4-(Trifluoromethyl)benzonitrile, CA 455-18-5: 344.82 (1).

$C_8H_4F_3NO$

2-(Trifluoromethyl)phenyl isocyanate, CA 2285-12-3: 332.04 (1).

3-(Trifluoromethyl)phenyl isocyanate, CA 329-01-1: 332.04 (1).

4-(Trifluoromethyl)phenyl isocyanate, CA 1548-13-6: 343.15 (1).

$C_8H_4F_3NO_4$

4-Nitrophenyl trifluoroacetate, CA 658-78-6: Above 385.93 (1).

$C_8H_4F_3NS$

3-(Trifluoromethyl)phenyl isothiocyanate, CA 1840-19-3: 368.15 (1).

$C_8H_4F_6$

1,3-Bis(trifluoromethyl)benzene, CA 402-31-3: 299.26 (1).

1,4-Bis(trifluoromethyl)benzene, CA 433-19-2: 294.82 (1).

$C_8H_4F_6O$

3,5-Bis(trifluoromethyl) phenol, CA 349-58-6: Above 385.93 (1).

$C_8H_4N_2O_2$

1,3-Phenylene diisocyanate, CA 123-61-5: 380.15* (4).

$C_8H_4O_3$

Phthalic anhydride, CA 85-44-9: 424.15 (4); 425.15 (5,6,7); 438.15* (4).

C_8H_5ClFN

2-Chloro-6-fluorophenylacetonitrile, CA 75279-55-9: Above 385.93 (1).

$C_8H_5Cl_2F_3O_2S$

4-Chlorophenyl 2-chloro-1,1,2-trifluoroethyl sulfone, CA 26574-59-4: Above 385.93 (1).

$C_8H_5Cl_2N$

3,4-Dichlorophenylacetonitrile, CA 3218-49-3: Above 385.93 (1).

$C_8H_5Cl_3O$

2,2',4'-Trichloroacetophenone, CA 4252-78-2: Above 385.93 (1).

$C_8H_5F_2N$

2,4-Difluorophenylacetonitrile: 366.48 (1).

2,5-Difluorophenylacetonitrile, CA 69584-87-8: 375.93 (1).

2,6-Difluorophenylacetonitrile: 326.48 (1).

$C_8H_5F_3O$

  2,2,2-Trifluoroacetophenone, CA 434-45-7: 314.82 (1).

  2-Trifluoromethylbenzaldehyde, CA 447-61-0: 334.26 (1).

  3-Trifluoromethylbenzaldehyde, CA 454-89-7: 341.48 (1).

  4-Trifluoromethylbenzaldehyde, CA 455-19-6: 338.71 (1)

$C_8H_5F_3O_2S$

  Thenoyltrifluoroacetone, CA 326-91-0: 384.82 (1).

$C_8H_5F_5O$

  1-(Pentafluorophenyl)ethanol ±, CA 75853-08-6: 360.37 (1).

$C_8H_5F_6N$

  3,5-Bis(trifluoromethyl)aniline, CA 328-74-5: 356.48 (1).

C_8H_5NO

  Benzoyl cyanide, CA 613-90-1: 357.59 (1).

C_8H_5NOS

  Benzoyl isothiocyanate, CA 532-55-8: 383.15 (1).

C_8H_6

  Phenylacetylene, CA 536-74-3: 300.15 (3); 304.26 (1).

$C_8H_6BrF_3$

  1-Bromomethyl-3-trifluoromethylbenzene, CA 402-23-3: 360.93 (1).

  1-Bromomethyl-4-trifluoromethylbenzene, CA 402-49-3: 362.04 (1).

C_8H_6BrN

  2-Bromophenylacetonitrile, CA 19472-74-3: Above 385.93 (1).

  3-Bromophenylacetonitrile, CA 31938-07-5: 383.15 (1).

  4-Bromophenylacetonitrile, CA 16532-79-9: Above 385.93 (1).

  Bromophenylacetonitrile, mixed isomers, CA 5798-79-8: Nonflammable (7).

C_8H_6ClFO

  2-Chloro-4'-fluoroacetophenone, CA 456-04-2: Above 385.93 (1).

$C_8H_6ClF_3$

  1-Chloromethyl-2-trifluoromethylbenzene, CA 21742-00-7: 343.15 (1).

  1-Chloromethyl-3-trifluoromethylbenzene, CA 705-29-3: 322.04 (1).

$C_8H_6ClF_5Si$

Chlorodimethylpentafluorophenylsilane, CA 20082-71-7: 293.71 (1); above 368.15 (3).

C_8H_6ClN

2-Chlorobenzyl cyanide, CA 2856-63-5: Above 385.93 (1).

3-Chlorobenzyl cyanide, CA 1529-41-5: Above 385.93 (1).

4-Chlorobenzyl cyanide, CA 140-53-4: Above 385.93 (1).

4-Chloroindole, CA 25235-85-2: Above 385.93 (1).

3-Chloro-4-methylbenzonitrile, CA 21423-81-4: 357.04 (1).

$C_8H_6ClO_4$

Methyl 4-chloro-2-nitrobenzoate, CA 42087-80-9: Above 385.93 (1).

4-Nitrobenzyl chloroformate, CA 4457-32-3: Above 385.93 (1).

$C_8H_6Cl_2$

alpha,beta-Dichlorostyrene, CA 6607-45-0: 380.15 (5); 380.37* (6,7).

2,6-Dichlorostyrene, CA 28469-92-3: 344.26 (1).

$C_8H_6Cl_2O$

3-(Chloromethyl)benzoyl chloride, CA 63024-77-1: Above 385.93 (1).

2-Chloro-2-phenylacetyl chloride ±, CA 2912-62-1: 332.15 (3); 377.59 (1).

2,2-Dichloroacetophenone, CA 2648-61-5: Above 385.93 (1).

2',4'-Dichloroacetophenone, CA 2234-16-4: Above 385.93 (1).

2',5'-Dichloroacetophenone, CA 2476-37-1: Above 385.93 (1).

$C_8H_6Cl_2O_2$

4-Chlorophenoxyacetyl chloride, CA 4122-68-3: Above 385.93 (1).

Methyl 2,5-dichlorobenzoate, CA 2905-69-3: 383.15 (1).

C_8H_6FN

5-Fluoroindole, CA 399-52-0: Above 385.93 (1).

2-Fluorophenylacetonitrile, CA 326-62-5: 316.48 (1).

3-Fluorophenylacetonitrile, CA 10036-43-8: Above 385.93 (1).

4-Fluorophenylacetonitrile, CA 459-22-3: 381.48 (1).

$C_8H_6F_2O$

2',4'-Difluoroacetophenone, CA 364-83-0: 339.82 (1).

2',6'-Difluoroacetophenone, CA 13670--99-0: 349.82 (1).

3',4'-Difluoroacetophenone, CA 369-33-5: 348.15 (1).

$C_8H_6F_2O_2$

  3,4-Difluorophenylacetic acid: 383.15 (1).

$C_8H_6F_3NO_3$

  4-Methoxy-3-nitrobenzotrifluoride, CA 394-25-2: Above 385.93 (1).

$C_8H_6N_2$

  Cinnoline, CA 253-66-7: Above 385.93 (1).

  Quinazoline, CA 253-82-7: 379.82 (1).

  Quinoxaline, CA 91-19-0: 371.48 (1).

C_8H_6O

  2,3-Benzofuran, CA 271-89-6: 329.26 (1).

$C_8H_6O_2$

  2-Coumaranone, CA 553-86-6: Above 385.93 (1).

  Phenylglyoxal, CA 1074-12-0: Below 296.15 (7).

  Phthalic dicarboxaldehyde, CA 643-79-8: Above 385.93 (1).

$C_8H_6O_3$

  Piperonal, CA 120-57-0: Above 385.93 (1).

$C_8H_6O_4$

  Phthalic acid, CA 88-99-3: 440.93 (6); 441.15* (5).

  Terephthalic acid, CA 100-21-0: 533.15* (6).

C_8H_6S

  Thianaphthene, CA 95-15-8: Above 385.93 (1).

$C_8H_6S_2$

  2,2'-Bithiophene, CA 492-97-7: Above 385.93 (1).

C_8H_7Br

  beta-Bromostyrene, mixed isomers, CA 103-64-0: 352.59 (1).

  2-Bromostyrene, CA 2039-88-5: 359.26 (1).

  3-Bromostyrene, CA 2039-86-4: 340.93 (1).

  4-Bromostyrene, CA 2039-82-9: 348.71 (1).

C_8H_7BrO

  2-Bromoacetophenone, CA 70-11-1: Above 385.93 (1).

(continues)

C_8H_7BrO *(continued)*

2'-Bromoacetophenone, CA 2142-69-0: Above 385.93 (1).

3'-Bromoacetophenone, CA 2142-63-4: Above 385.93 (1).

4'-Bromoacetophenone, CA 99-90-1: Above 385.93 (1).

C_8H_7Cl

2-Chlorostyrene, CA 2039-87-4: 332.04 (1).

3-Chlorostyrene, CA 2039-85-2: 335.93 (1).

4-Chlorostyrene, CA 1073-67-2: 333.15 (1); 341.15 (3).

$C_8H_7ClF_3NS$

2-(2-Chloro-1,1,2-trifluoroethylthio)aniline, CA 81029-02-9: Above 385.93 (1).

C_8H_7ClO

2-Chloroacetophenone, CA 532-27-4: 391.15 (5,6).

2'-Chloroacetophenone, CA 2142-68-9: 362.04 (1); 365.15 (3).

3'-Chloroacetophenone, CA 99-02-5: 378.15 (1).

4'-Chloroacetophenone, CA 99-91-2: 363.15 (1); 390.93 (7).

Phenylacetyl chloride, CA 103-80-0: 375.93 (1).

2-Toluoyl chloride, CA 933-88-0: 349.82 (1).

3-Toluoyl chloride, CA 1711-06-4: 349.82 (1).

4-Toluoyl chloride, CA 874-60-2: 355.37 (1).

C_8H_7ClOS

Benzyl chlorothioformate, CA 37734-45-5: 391.48 (2).

$C_8H_7ClO_2$

Benzyl chloroformate, practical, CA 501-53-1: 353.15* (4); 364.82 (1); 381.05 (4).

2-Methoxybenzoyl chloride, CA 21615-34-9: 357.59 (1).

3-Methoxybenzoyl chloride, CA 1711-05-3: 365.37 (1).

4-Methoxybenzoyl chloride, CA 100-07-2: 360.93 (1).

Methyl 3-chlorobenzoate, CA 2905-65-9: 377.59 (1).

Methyl 4-chlorobenzoate, CA 1126-46-1: 379.82 (1).

Phenoxyacetyl chloride, CA 701-99-5: 381.48 (1).

C_8H_7F

3-Fluorostyrene, CA 350-51-6: 302.59 (1).

4-Fluorostyrene, CA 405-99-2: 299.82 (1); 308.15 (3).

C_8H_7FO

2'-Fluoroacetophenone, CA 445-27-2: 334.82 (1).

3'-Fluoroacetophenone, CA 455-36-7: 353.71 (1).

4'-Fluoroacetophenone, CA 403-42-9: 344.26 (1).

$C_8H_7FO_2$

3-Fluoro-*para*-anisaldehyde, CA 351-54-2: Above 385.93 (1).

3-Fluorophenylacetic acid, CA 331-25-9: 383.15 (1).

$C_8H_7F_3O$

3-(Trifluoromethyl)anisole, CA 454-90-0: 322.04 (1).

2-(Trifluoromethyl)benzyl alcohol, CA 346-06-5: 365.37 (1).

3-(Trifluoromethyl)benzyl alcohol, CA 349-75-7: 357.59 (1).

4-(Trifluoromethyl)benzyl alcohol, CA 349-95-1: 373.71 (1).

$C_8H_7F_5Si$

Dimethylpentafluorophenylsilane, CA 13888-77-2: 313.71 (1).

C_8H_7N

Benzyl cyanide, CA 140-29-4: 374.82 (1,3); 385.93* (6).

Benzyl isocyanide, CA 10340-91-7: 352.59 (1).

Indole, CA 120-72-9: Above 385.93 (1).

2-Methylbenzonitrile, CA 529-19-1: 355.15 (3); 357.59 (1).

3-Methylbenzonitrile, CA 620-22-4: 359.82 (1).

4-Methylbenzonitrile, CA 104-85-8: 358.15 (1); 365.15 (3).

C_8H_7NO

Benzyl isocyanate, CA 3173-56-6: 318.15 (1).

Mandelonitrile, technical, CA 532-28-5: 370.37 (1).

2-Methoxybenzonitrile, CA 6609-56-9: Above 385.93 (1).

3-Methoxybenzonitrile, CA 1527-89-5: 378.15 (1).

2-Methylbenzoxazole, CA 95-21-6: 348.15 (1).

ortho-Tolyl isocyanate, CA 614-68-6: 365.37 (1).

meta-Tolyl isocyanate, CA 621-29-4: 338.71 (1).

para-Tolyl isocyanate, CA 622-58-2: 339.82 (1).

C_8H_7NOS

2-Methoxyphenyl isothiocyanate, CA 3288-04-8: Above 385.93 (1).

4-Methoxyphenyl isothiocyanate, CA 2284-20-0: Above 385.93 (1).

$C_8H_7NO_2$

2-Methoxyphenyl isocyanate, CA 700-87-8: Above 385.93 (1).
3-Methoxyphenyl isocyanate, CA 18908-07-1: 369.26 (1).
4-Methoxyphenyl isocyanate, CA 5416-93-3: 371.48 (1).
3-Nitrostyrene, CA 586-39-0: 380.37 (1).

$C_8H_7NO_2S$

4-Toluenesulfonyl cyanide, technical, CA 19158-51-1: Above 385.93 (1).

$C_8H_7NO_3$

2'-Nitroacetophenone, CA 577-59-3: Above 385.93 (1).

$C_8H_7NO_3S$

Cyanomethyl benzenesulfonate, technical, CA 10531-13-2: Above 385.93 (1).
4-Toluenesulfonyl isocyanate, CA 4083-64-1: Above 385.93 (1).

$C_8H_7NO_4$

Methyl 2-nitrobenzoate, CA 606-27-9: Above 385.93 (1).
2-Nitrophenyl acetate, CA 610-69-5: Above 385.93 (1).

C_8H_7NS

Benzyl isothiocyanate, CA 622-78-6: Above 385.93 (1).
Benzyl thiocyanate, CA 3012-37-1: Above 385.93 (1).
2-Methylbenzothiazole, CA 120-75-2: 375.37 (1).
2-Tolyl isothiocyanate, CA 614-69-7: 380.93 (1).
4-Tolyl isothiocyanate, CA 622-59-3: 380.93 (1).

$C_8H_7NS_2$

2-(Methylthio)benzothiazole, CA 615-22-5: Above 385.93 (1).

C_8H_7NSe

2-Methylbenzselenazole, CA 2818-88-4: Above 385.93 (1).

C_8H_8

1,3,5,7-Cyclooctatetraene, CA 629-20-9: Below 295.15 (7); 295.37 (1,3).
Styrene, CA 100-42-5: 304.26 (1,6,7); 304.55* (4); 305.15 (3,5);
307.55* (4).

C_8H_8BrClO

4-Bromophenyl 2-chloroethyl ether, CA 55162-34-0: Above 385.93 (1).
4-Chlorophenyl 2-bromoethyl ether, CA 2033-76-3: 383.15 (1).

$C_8H_8Br_2$

Dibromoethylbenzene, mixed isomers: Nonflammable (2).

$C_8H_8Br_2O$

2-Bromoethyl 4-bromophenyl ether, CA 18800-30-1: Above 385.93 (1).

$C_8H_8ClNO_2$

2-Methyl-3-nitrobenzyl chloride, CA 60468-54-4: Above 385.93 (1).

4-Methyl-3-nitrobenzyl chloride: Above 385.93 (1).

5-Methyl-2-nitrobenzyl chloride, CA 66424-91-7: Above 385.93 (1).

$C_8H_8Cl_2$

Dichloroethylbenzene, mixed isomers, CA 1331-29-9: 369.26 (7).

2-Dichloromethyltoluene, CA 612-12-4: 380.93 (1).

3-Dichloromethyltoluene, CA 626-16-4: Above 385.93 (1).

$C_8H_8Cl_2O$

2,6-Dichlorobenzyl methyl ether, CA 33486-90-7: Above 385.93 (1).

2,4-Dichloro-*alpha*-methylbenzyl alcohol ±, CA 1475-13-4: 382.04 (1).

$C_8H_8F_3N$

2-(Trifluoromethyl)benzylamine, CA 3048-01-9: 342.59 (1).

3-(Trifluoromethyl)benzylamine, CA 2740-83-2: 344.26 (1).

4-(Trifluoromethyl)benzylamine, CA 3300-51-4: 348.15 (1).

$C_8H_8F_3NO$

5-Methoxy-3-(trifluoromethyl)aniline, CA 349-55-3: Above 385.93 (1).

$C_8H_8N_2$

4-Aminobenzyl cyanide, CA 3544-25-0: Above 385.93 (1).

C_8H_8O

Acetophenone, CA 98-86-2: 333.15 (3); 349.82 (6); 355.37 (1,2,4); 355.37* (7); 366.15* (4); 278.15 (5).

2,3-Dihydrobenzofuran, CA 496-16-2: 339.82 (1).

2-Methylbenzaldehyde, CA 529-20-4: 340.37 (1); 350.15 (3).

3-Methylbenzaldehyde, CA 620-23-5: 351.48 (1); 356.15 (3).

4-Methylbenzaldehyde, CA 104-87-0: 353.15 (1); 358.15 (3).

Phenylacetaldehyde, CA 122-78-1: 344.26 (6); 359.82 (1,3).

Phthalan, CA 496-14-0: 336.48 (1).

C_8H_8O *(continued)*

Styrene oxide, CA 96-09-3:   347.04* (6,7); 352.59 (1,3,5).

Styrene oxide, (R)-(-), CA 20780-53-4:   353.15 (3).

Styrene oxide, (S)-(+), CA 20780-54-5:   353.15 (3).

C_8H_8OS

Cyclopropyl 2-thienyl ketone, CA 6193-47-1:   365.37 (1).

4-Keto-4,5,6,7-tetrahydrothianaphthene, CA 13414-95-4:   Above 385.93 (1).

4-(Methylthio)benzaldehyde, CA 3446-89-7:   Above 385.93 (1).

S-Phenyl thioacetate, CA 934-87-2:   352.59 (1).

Phenyl vinyl sulfoxide, CA 20451-53-0:   Above 385.93 (1).

$C_8H_8O_2$

1,4-Benzodioxan, CA 493-09-4:   360.93 (1).

2'-Hydroxyacetophenone, CA 118-93-4:   Above 385.93 (1).

2-Methoxybenzaldehyde, CA 135-02-4:   390.93 (1); 390.93* (6).

3-Methoxybenzaldehyde, CA 591-31-1:   383.15 (1).

4-Methoxybenzaldehyde, CA 123-11-5:   382.04 (1).

Methyl benzoate, CA 93-58-3:   355.93 (1,3,5,6,7).

3,4-(Methylenedioxy)toluene, CA 7145-99-5:   349.26 (1).

Phenyl acetate, CA 122-79-2:   349.82 (1); 353.15 (6,7); 367.15 (3).

Phenylacetic acid, CA 103-82-2:   Above 373.15 (6).

$C_8H_8O_3$

2-Hydroxy-4-methoxybenzaldehyde, CA 673-22-3:   Above 385.93 (1).

2-Hydroxy-5-methoxybenzaldehyde, CA 672-13-9:   Above 385.93 (1).

Methyl salicylate, CA 119-36-8:   369.26 (6); 374.15 (5,7); Above 385.93 (1).

Piperonyl alcohol, CA 495-76-1:   Above 385.93 (1).

Resorcinol monoacetate, technical, CA 102-29-4:   Above 385.93 (1)

Tetrahydrophthalic anhydride, CA 85-43-8:   430.37* (7).

ortho-Vanillin, CA 148-53-8:   Above 385.93 (1).

$C_8H_8O_3S$

Phenyl vinylsulfonate, CA 26914-43-2:   383.15 (1).

$C_8H_8O_4$

5-Acetoxymethyl-2-furaldehyde, CA 10551-58-3:   380.93 (1).

Dehydroacetic acid, CA 520-45-6:   430.37* (6).

C_8H_8S

Phenyl vinyl sulfide, CA 1822-73-7:   318.71 (1); 340.15 (3).

C_8H_9Br

(1-Bromoethyl)benzene, CA 585-71-7:   354.82 (1).

(2-Bromoethyl)benzene, CA 103-63-9:   362.59 (1); 369.15 (3).

1-Bromo-2-ethylbenzene, CA 1973-22-4:   323.15 (3); 344.26 (1).

1-Bromo-4-ethylbenzene, CA 1585-07-5:   337.04 (1).

Bromoethylbenzene, mixed isomers: 366.15 (2).

2-(Bromomethyl)toluene, CA 89-92-9:   355.37 (1).

3-(Bromomethyl)toluene, CA 620-13-3:   355.37 (1).

4-(Bromomethyl)toluene, CA 104-81-4:   370.93 (1).

2-Bromo-*meta*-xylene, CA 576-22-7:   347.04 (1).

2-Bromo-*para*-xylene, CA 553-94-6:   352.59 (1).

3-Bromo-*ortho*-xylene, CA 576-23-8:   353.71 (1).

4-Bromo-*ortho*-xylene, technical, CA 583-71-1:   353.71 (1)

4-Bromo-*meta*-xylene, technical, CA 583-70-0:   352.04 (1).

5-Bromo-*meta*-xylene, CA 556-96-7:   360.37 (1).

C_8H_9BrO

4-Bromophenethyl alcohol, CA 4654-39-1:   Above 385.93 (1).

2-Bromophenetole, CA 589-10-6:   338.71 (1).

4'-Bromophenetole, CA 588-96-5:   376.48 (1).

4-Bromophenyl methyl carbinol, CA 5391-88-8:   336.48 (1).

$C_8H_9BrO_2$

1-Bromo-2,4-dimethoxybenzene, CA 17715-69-4:   Above 385.93 (1).

1-Bromo-2,5-dimethoxybenzene, CA 25245-34-5:   Above 385.93 (1).

2-(4-Bromophenoxy)ethanol, CA 34743-88-9:   Above 385.93 (1).

4-Bromoveratrole, CA 2859-78-1:   382.59 (1).

C_8H_9Cl

(2-Chloroethyl)benzene, CA 622-24-2:   377.15 (3); 339.82 (1).

1-Chloro-4-ethylbenzene, CA 622-98-0:   337.15 (5,6).

2-(Chloromethyl)toluene, CA 552-45-4:   347.04 (1).

3-(Chloromethyl)toluene, CA 620-19-9:   348.71 (1).

4-(Chloromethyl)toluene, CA 104-82-5:   348.71 (1).

2-Chloro-*meta*-xylene, CA 6781-98-2:   336.48 (1).

2-Chloro-*para*-xylene, CA 95-72-7:   330.37 (1); 340.15 (3).

4-Chloro-*ortho*-xylene, CA 615-60-1:   339.82 (1).

C_8H_9ClO

Benzyl chloromethyl ether, CA 3587-60-8: 364.26 (1).
4-Chloro-3-ethylphenol, CA 14143-32-9: Above 385.93 (1).
2-Chloroethyl phenyl ether, CA 622-86-6: 380.37 (6).
2-Chlorophenethyl alcohol, CA 19819-95-5: Above 385.93 (1).
3-Chlorophenethyl alcohol, CA 5182-44-5: Above 385.93 (1).
4-Chlorophenethyl alcohol, CA 1875-88-3: Above 385.93 (1).
3-Methoxybenzyl chloride, CA 824-98-6: 374.82 (1).
4-Methoxybenzyl chloride, CA 824-94-2: 382.59 (1).

C_8H_9ClOS

4-Chloro-2'-butyrothienone, CA 43076-59-1: 383.15 (1).

$C_8H_9ClO_2$

2-Chloro-1,4-dimethoxybenzene, CA 2100-42-7: 390.37 (6).
5-Chloro-1,3-dimethoxybenzene, CA 7051-16-3: 385.37 (1).

C_8H_9F

3-Fluoro-*ortho*-xylene, CA 443-82-3: 309.26 (1).

C_8H_9FO

2-Fluorophenethyl alcohol, CA 50919-06-7: 371.48 (1).
4-Fluorophenethyl alcohol, CA 7589-27-7: 377.59 (1).

C_8H_9N

N-Benzylidenemethylamine, CA 622-29-7: Above 385.93 (1).
2,3-Cyclopentenopyridine, CA 533-37-9: 340.37 (1).
Indoline, CA 496-15-1: 365.93 (1).
2-Methyl-5-vinylpyridine, CA 140-76-1: 347.04* (2).
5-Norbornene-2-carbonitrile, mixed isomers, CA 95-11-4: 338.71 (1).

C_8H_9NO

Acetanilide, CA 103-84-4: 442.59* (6); 447.04* (1,5,7).
5-Acetyl-2-methylpyridine, CA 36357-38-7: 377.04 (1).
2'-Aminoacetophenone, CA 551-93-9: Above 385.93 (1).
N-Methylformanilide, CA 93-61-8: 399.82 (1).

$C_8H_9NO_2$

1,4-Benzodioxan-6-amine, CA 22013-33-8: Above 385.93 (1).
Ethyl isonicotinate, CA 1570-45-2: 360.93 (1).

(continues)

$C_8H_9NO_2$ *(continued)*

Ethyl nicotinate, CA 614-18-6: 366.48 (1).

4-Ethylnitrobenzene, CA 100-12-9: Above 385.93 (1).

Ethyl picolinate, CA 2524-52-9: 380.37 (1).

Methyl anthranilate, CA 134-20-3: Above 373.15 (6); 377.59 (1).

Methyl 6-methylnicotinate, CA 5470-70-2: 376.48 (1).

Methyl 2-pyridylacetate, CA 1658-42-0: Above 385.93 (1).

2-Nitro-*meta*-xylene, CA 81-20-9: 360.93 (1).

3-Nitro-*ortho*-xylene, CA 83-41-0: 380.15 (3); 380.93 (1).

4-Nitro-*ortho*-xylene, CA 99-51-4: Above 385.93 (1).

4-Nitro-*meta*-xylene, CA 89-87-2: 380.37 (1).

Piperonylamine, CA 2620-50-0: Above 385.93 (1).

$C_8H_9NO_3$

2-Methyl-3-nitroanisole, CA 4837-88-1: Above 385.93 (1).

3-Methyl-2-nitroanisole, CA 5345-42-6: Above 385.93 (1).

3-Methyl-4-nitroanisole, CA 5367-32-8: Above 385.93 (1).

4-Methyl-3-nitroanisole, CA 17484-36-5: Above 385.93 (1).

3-Methyl-2-nitrobenzyl alcohol, CA 80866-76-8: Above 385.93 (1).

4-Methyl-3-nitrobenzyl alcohol, CA 40870-59-5: Above 385.93 (1).

2-Nitrophenethyl alcohol, CA 15121-84-3: Above 385.93 (1).

3-Nitrophenethyl alcohol, CA 52022-77-2: Above 385.93 (1).

$C_8H_9N_3O_4$

2,4-Dinitro-*N*-ethylaniline, CA 3846-50-2: Above 385.93 (1).

C_8H_{10}

1,2-Dimethylbenzene, CA 95-47-6: 290.15 (7); 303.15 (5); 304.15 (2,3); 305.37 (1,4,6).

1,3-Dimethylbenzene, CA 108-38-3: 298.15 (1,3,5,7); 300.37 (6); 302.15 (2,4).

1,4-Dimethylbenzene, CA 106-42-3: 298.15 (3,5,7); 300.37 (1,2,4,6).

Dimethylbenzene, mixed isomers, CA 1330-20-7: 302.59 (1).

6,6-Dimethylfulvene, CA 2175-91-9: 316.48 (1).

Ethylbenzene, CA 100-41-0: 288.15 (2,3,4,7); 294.26 (6); 295.37 (1); 296.15 (5).

1-Ethynylcyclohexene, CA 931-49-7: 301.48 (1).

5-Methylene-2-norbornene, technical, CA 694-91-7: 277.59 (1).

1,7-Octadiyne, CA 871-84-1: 296.48 (1).

$C_8H_{10}BrN$

   4-Bromo-N,N-dimethylaniline, CA 586-77-6: Above 385.93 (1).

   4-Bromo-2,6-dimethylaniline, CA 24596-19-8: Above 385.93 (1).

   4-Bromophenethylamine, CA 73918-56-6: Above 385.93 (1).

$C_8H_{10}ClN$

   2-(2-Chlorophenyl)ethylamine, CA 13078-80-3: 382.04 (1).

   2-(4-Chlorophenyl)ethylamine, CA 156-41-2: 379.26 (1).

$C_8H_{10}Cl_2$

   3,4-Dichlorobicyclo(3,2,1)oct-2-ene, technical, mixed isomers, CA 57615-42-6: 375.93 (1).

$C_8H_{10}O$

   2,4-Dimethylphenol, CA 105-67-9: Above 385.93 (1).

   2,6-Dimethylphenol, CA 576-26-1: 347.04 (1); 351.48 (1); 360.93* (2).

   2-Ethylphenol, CA 90-00-6: 351.48 (1).

   3-Ethylphenol, CA 620-17-7: 367.59 (1).

   4-Ethylphenol, CA 123-07-9: 373.71 (1); 377.04 (6).

   2-Methylanisole, CA 578-58-5: 324.82 (1).

   3-Methylanisole, CA 100-84-5: 325.15 (3); 327.59 (1).

   4-Methylanisole, CA 104-93-8: 326.48 (1); 332.15 (3); 333.15 (6).

   2-Methylbenzyl alcohol, CA 89-95-2: 377.59 (1).

   3-Methylbenzyl alcohol, CA 587-03-1: 378.71 (1).

   3-Methylene-2-norbornanone, CA 5597-27-3: 328.15 (1).

   5-Norbornene-2-carboxaldehyde, CA 5453-80-5: 321.15 (3); 323.71 (1).

   Phenetole, CA 103-73-1: 330.37 (1,3); 335.93 (6,7).

   1-Phenylethanol ±, CA 13323-81-4: 358.15 (1); 359.15 (3); 366.48 (6); 369.26* (7).

   1-Phenylethanol, (S)-(-), CA 1445-91-6: 358.15 (1); 359.15 (3).

   1-Phenylethanol, (R)-(+), CA 1517-69-7: 358.15 (1); 359.15 (3).

   2-Phenylethanol, CA 60-12-8: 369.26 (6); 375.37 (1,5,7).

$C_8H_{10}OS$

   3-Acetyl-2,5-dimethylthiophene, CA 2530-10-1: 372.04 (1).

   2-(Ethylthio)phenol, CA 29549-60-8: 366.48 (1).

   Methoxymethyl phenyl sulfide, CA 13865-50-4: 368.15 (1).

   1-Methoxy-4-(methylthio)benzene, CA 3517-99-5: 380.93 (1).

   4-Methoxy-$alpha$-toluenethiol, technical, CA 6258-60-2: Above 385.93 (1).

   4-(Methylthio)benzyl alcohol, CA 3446-90-0: Above 385.93 (1).

   2-(Phenylthio)ethanol, CA 699-12-7: Above 385.93 (1).

$C_8H_{10}O_2$

3-Acetyl-2,5-dimethylfuran, CA 10599-70-9: 352.04 (1).

Allyl glycidyl ether, CA 106-92-3: 318.15 (3).

1,3-Dimethoxybenzene, CA 151-10-0: 360.93 (1).

2-Ethoxyphenol, CA 94-71-3: 364.26 (1).

2-Hydroxyphenethyl alcohol, CA 7768-28-7: Above 385.93 (1).

3-Hydroxyphenethyl alcohol, CA 13398-94-2: Above 385.93 (1).

2-Methoxybenzyl alcohol, CA 612-16-8: Above 385.93 (1).

3-Methoxybenzyl alcohol, CA 6971-51-3: Above 385.93 (1).

4-Methoxybenzyl alcohol, CA 105-13-5: Above 385.93 (1).

2-Methoxy-4-methylphenol, CA 93-51-6: 372.59 (1).

2-Phenoxyethanol, CA 122-99-6: 383.15 (1); 394.26 (5,6,7); 394.26* (4,6); 399.82 (2).

Veratrole, CA 91-16-7: 345.15 (3); 360.37 (1).

$C_8H_{10}O_2S$

Ethyl 2-thiopheneacetate, CA 57382-97-5: 377.59 (1).

Ethyl 3-thiopheneacetate, CA 37784-63-7: 376.48 (1).

4-(2-Thienyl)butyric acid, CA 4653-11-6: Above 385.93 (1).

$C_8H_{10}O_3$

2-Acetyl-1,3-cyclohexanedione, CA 4056-73-9: Above 385.93 (1).

Crotonic anhydride, CA 623-68-7: 366.15 (3); 383.71 (1).

1,2-Cyclohexanedicarboxylic anhydride, cis, CA 13149-00-3: Above 385.93 (1).

4,4-Dimethoxy-2,5-cyclohexadien-1-one, CA 935-50-2: 374.26 (1).

2,3-Dimethoxyphenol, CA 5150-42-5: 382.04 (1).

2,6-Dimethoxyphenol, CA 91-10-1: Above 385.93 (1).

3,5-Dimethoxyphenol, CA 500-99-2: 351.48 (1).

Methacrylic anhydride, CA 760-93-0: 357.59 (1).

Methyl 2,5-dimethyl-3-furoate, CA 6148-34-1: 348.15 (3); 353.71 (1).

$C_8H_{10}O_3S$

4-Ethylbenzenesulfonic acid, technical, CA 98-69-1: Above 385.93 (1).

Methyl para-toluenesulfonate, CA 80-48-8: Above 385.93 (1); 425.37* (6).

$C_8H_{10}O_4$

Diethyl acetylenedicarboxylate, CA 762-21-0: 367.59 (1).

Ethylene diacrylate, CA 2274-11-5: As 70% solution, 373.71 (1).

$C_8H_{10}S$

Benzyl methyl sulfide, CA 766-92-7: 346.48 (1).

2,4-Dimethylthiophenol, CA 13616-82-5: 331.48 (1).

2,5-Dimethylthiophenol, CA 4001-61-0: 356.48 (1).

3,4-Dimethylthiophenol, CA 18800-53-8: 363.15 (1).

Ethyl phenyl sulfide, CA 622-38-8: 344.15 (3); 347.04 (1).

Methyl *para*-tolyl sulfide, CA 623-13-2: 358.15 (1).

Phenethyl mercaptan, CA 4410-99-5: 363.71 (1).

$C_8H_{11}Cl$

3-Chlorobicyclo(3,2,1)oct-2-ene, mixed isomers, CA 35242-17-2: 332.04 (1).

$C_8H_{11}ClSi$

Chlorodimethylphenylsilane, CA 768-33-2: 334.82 (1); 345.15 (3).

$C_8H_{11}N$

N-Benzylmethylamine, CA 103-67-3: 348.15 (3); 350.93 (1).

2,3,6-Collidine, CA 1462-84-6: 327.15 (3).

2,4,6-Collidine, CA 108-75-8: 327.15 (3); 330.37 (1).

Collidine, mixed isomers, CA 29611-84-5: 315.93 (1).

1-Cyclohexenylacetonitrile, CA 6975-71-9: As 92% solution, 357.04 (1).

N,*N*-Dimethylaniline, CA 121-69-7: 335.93 (1,3,5,6,7).

2,3-Dimethylaniline, CA 87-59-2: 369.26 (1); 370.15 (5,6).

2,4-Dimethylaniline, CA 95-68-1: 363.71 (1).

2,5-Dimethylaniline, CA 95-78-3: 367.04 (1).

2,6-Dimethylaniline, CA 87-62-7: 364.26 (1).

3,4-Dimethylaniline, CA 95-64-7: 371.48 (1).

3,5-Dimethylaniline, CA 108-69-0: 366.48 (1).

Dimethylaniline, mixed isomers, CA 1300-73-8: 364.26 (1); 369.82 (7).

N-Ethylaniline, CA 103-69-5: 358.15 (1,3,5); 358.15* (6,7).

2-Ethylaniline, CA 578-54-1: 358.15* (7); 370.93 (1).

3-Ethylaniline, CA 587-02-0: 358.15 (1).

4-Ethylaniline, CA 589-16-2: 358.15 (1).

5-Ethyl-2-methylpyridine, CA 104-90-5: 333.15 (3); 339.26 (1); 341.83* (6); 347.04* (7); 347.15 (5).

alpha-Methylbenzylamine ±, CA 618-36-0: 343.15 (3); 352.59 (1); 352.59* (6).

alpha-Methylbenzylamine, (*S*)-(-), CA 2627-86-3: 338.15 (3); 352.59 (1).

alpha-Methylbenzylamine, (*R*)-(+), CA 3886-69-9: 343.15 (3); 352.59 (1).

2-Methylbenzylamine, CA 89-93-0: 357.04 (1).

(continues)

$C_8H_{11}N$ *(continued)*

   3-Methylbenzylamine, CA 100-81-2: 353.71 (1).

   4-Methylbenzylamine, CA 104-84-7: 348.15 (1).

   2-Norbornanecarbonitrile, mixed isomers, CA 2234-26-6: 352.04 (1).

   Phenethylamine, CA 64-04-0: 343.15 (3); 363.71 (1).

   2-Propylpyridine, CA 622-39-9: 329.26 (1).

$C_8H_{11}NO$

   meta-Amino-*alpha*-methylbenzyl alcohol, CA 2454-37-7: 430.37* (6).

   2-Aminophenethyl alcohol, CA 5339-89-5: Above 385.93 (1).

   2-Anilinoethanol, CA 122-98-5: Above 385.93 (1); 424.82* (6,7);
   425.15 (5).

   2-Ethoxyaniline, CA 94-70-2: 388.15* (6).

   3-Ethoxyaniline, CA 621-33-0: Above 385.93 (1).

   4-Ethoxyaniline, CA 156-43-4: 388.71 (1); 389.15 (5,6).

   2-Methoxybenzylamine, CA 6850-57-3: 373.15 (1).

   3-Methoxybenzylamine, CA 5071-96-5: 383.15 (1).

   4-Methoxybenzylamine, CA 2393-23-9: Above 385.93 (1).

   2-Methoxy-5-methylaniline, CA 120-71-8: Above 385.93 (1).

   4-Methoxy-2-methylaniline, CA 102-50-1: Above 385.93 (1).

   N-Methyl-*para*-anisidine, CA 5961-59-1: Above 385.93 (1).

   3-Pyridinepropanol, CA 2859-67-8: 383.15 (1).

$C_8H_{11}NO_2$

   2,4-Dimethoxyaniline, CA 2735-04-8: Above 385.93 (1).

   2,5-Dimethoxyaniline, CA 102-56-7: 423.15* (6).

   3,5-Dimethoxyaniline, CA 10272-07-8: Above 385.93 (1).

   Methyl 1,5-dimethyl-2-pyrrolecarboxylate, CA 73476-31-0: 374.82 (1).

   Methyl 1-methyl-2-pyrroleacetate, CA 51856-79-2: 379.82 (1).

$C_8H_{11}NO_2S$

   Ethyl 4-methyl-5-thiazoleacetate: Above 385.93 (1).

$C_8H_{11}NO_3$

   Ethyl (ethoxymethylene)cyanoacetate, CA 94-05-3: Above 385.93 (1).

$C_8H_{11}N_3$

   3,3-Dimethyl-1-phenyltriazene, CA 7227-91-0: 380.37 (1).

$C_8H_{11}O_2P$

   Dimethyl phenylphosphonite, CA 2946-61-4: Above 385.93 (1).

$C_8H_{11}P$

Dimethylphenylphosphine, CA 672-66-2: 323.15 (1).

C_8H_{12}

1,3-Cyclooctadiene, mixed isomers, CA 1700-10-3: 298.15 (3).

1,3-Cyclooctadiene, *cis,cis*, CA 3806-59-5: 297.59 (1).

1,5-Cyclooctadiene, mixed isomers, CA 111-78-4: 304.82 (1); 308.15 (6); 310.93 (3).

1,5-Cyclooctadiene, *cis,cis*, CA 1552-12-1: 305.15 (3).

4-Vinyl-1-cyclohexene, CA 100-40-3: 288.71* (7); 289.15 (3,6); 293.15 (1); 294.15 (2,5).

$C_8H_{12}Cl_2O_2$

Suberoyl chloride, CA 10027-07-3: Above 385.93 (1).

$C_8H_{12}Cl_2O_6$

Triethylene glycol, bis(chloroformate), CA 17134-17-7: Above 385.93 (1).

$C_8H_{12}N_2$

2-(4-Aminophenyl)ethylamine, CA 13472-00-9: Above 385.93 (1).

1,6-Dicyanohexane, CA 629-40-3: Above 385.93 (1).

2,3-Diethylpyrazine, CA 15707-24-1: 343.71 (1).

1,6-Diisocyanohexane, CA 929-57-7: 373.15 (3).

N,N-Dimethyl-1,4-phenylenediamine, CA 99-98-9: 363.71 (1).

2-(2-Methylaminoethyl)pyridine, CA 5638-76-6: 369.82 (1).

2-Methyl-3-propylpyrazine, CA 15986-80-8: 342.04 (1).

N-Phenylethylenediamine, CA 1664-40-0: Above 385.93 (1).

meta-Xylylenediamine, CA 1477-55-0: Above 385.93 (1).

$C_8H_{12}N_2O$

2-Isopropyl-3-methoxypyrazine, CA 25773-40-4: 339.82 (1).

2-Methyl-6-propoxypyrazine, CA 67845-28-7: 348.15 (1).

$C_8H_{12}N_2O_2$

1,6-Diisocyanatohexane, CA 822-06-0: 373.15 (3); 413.15 (1); 413.15* (4).

$C_8H_{12}O$

1-Acetyl-1-cyclohexene, CA 932-66-1: 338.71 (1).

1-Acetyl-2-methyl-1-cyclopentene, CA 3168-90-9: 339.82 (1).

3,5-Dimethyl-2-cyclohexen-1-one, CA 1123-09-7: 352.59 (1).

(continues)

$C_8H_{12}O$ *(continued)*

4,4-Dimethyl-2-cyclohexen-1-one, CA 1073-13-8: 337.59 (1).

1-Ethynyl-1-cyclohexanol, CA 78-27-3: 335.93 (1).

5-Norbornene-2-methanol, mixed isomers, CA 95-12-5: 359.82 (1).

Vinylcyclohexane monoxide, CA 106-86-5: 330.93 (7).

$C_8H_{12}O_2$

2-Acetylcyclohexanone, CA 874-23-7: 352.59 (1).

1,2,5,6-Diepoxycyclooctane, CA 27035-39-8: 378.15 (1).

3,6-Dihydro-4,6,6-trimethyl-2H-pyran-2-one, CA 22954-83-2: 372.59 (1).

3-Ethoxy-2-cyclohexen-1-one, CA 5323-87-5: 380.37 (1).

Ethyl sorbate, CA 2396-84-1: 342.59 (1).

3-Hydroxy-2,2,4-trimethyl-3-pentenoic acid, *beta*-lactone, CA 3173-79-3: 335.93 (1).

Methyl 1-cyclohexene-1-carboxylate, CA 18448-47-0: 344.15 (3); 347.04 (1).

2-Norbornyl formate, *exo*, CA 41498-71-9: 327.04 (1).

2-Octynoic acid, CA 5663-96-7: 374.26 (1).

Tetrahydro-2-(2-propynyloxy)-2H-pyran, CA 6089-04-9: 332.59 (1).

Vinyl cyclohexene dioxide, CA 106-87-6: 383.15 (7).

$C_8H_{12}O_3$

1,4-Dioxaspiro[4,5]decan-2-one, CA 4423-79-4: 378.15 (1).

3-Ethyl-2-methylglutaric anhydride, CA 6970-57-6: 368.15 (1).

Ethyl 2-oxocyclopentanecarboxylate, CA 611-10-9: 350.15 (3); 350.93 (1).

$C_8H_{12}O_4$

Diethyl fumarate, CA 623-91-6: 364.82 (1); 367.15 (3); 377.59 (6,7).

Diethyl maleate, CA 141-05-9: 366.48 (1,2,3,5); 394.26* (6,7).

Dimethyl 3-methylglutaconate, mixed isomers, CA 52313-87-8: 370.93 (1).

Ethyl diacetoacetate, CA 603-69-0: 358.15 (1).

Methallylidene diacetate, CA 10476-95-6: 356.48 (1).

$C_8H_{12}O_5$

Dimethyl acetylsuccinate ±, CA 10420-33-4: Above 385.93 (1).

Dimethyl 3-oxoadipate, CA 5457-44-3: Above 385.93 (1).

Methyl 2,5-dihydro-2,5-dimethoxy-2-furancarboxylate, CA 62435-72-7: 377.59 (1).

$C_8H_{12}O_6$

1,1,2-Triacetoxyethane, CA 2983-35-9: Above 385.93 (1).

$C_8H_{12}Si$

Dimethylphenylsilane, CA 766-77-8: 308.15 (1); 311.15 (3).

$C_8H_{13}BrO_4$

Diethyl 2-bromo-2-methylmalonate, CA 29263-94-3: Above 385.93 (1).

$C_8H_{13}ClO$

3-Cyclopentylpropionyl chloride, CA 104-97-2: 357.59 (1).
2,2,3,3-Tetramethylcyclopropanecarboxylic acid, CA 24303-61-5: 339.26 (1).

$C_8H_{13}N$

Cycloheptyl cyanide, CA 32730-85-1: 358.15 (1).
2,4-Dimethyl-3-ethylpyrrole, CA 517-22-6: 344.82 (1).
1-Ethynylcyclohexylamine, CA 30389-18-5: 315.37 (1).

$C_8H_{13}NO$

2-Oxooctanenitrile, technical, CA 80997-84-8: 352.59 (1).
Tropinone, CA 532-24-1: 363.15 (1).

$C_8H_{13}NO_2$

N-Acetylcaprolactam, CA 1888-91-1: Above 385.93 (1).

$C_8H_{13}NO_3$

1-Carbethoxy-4-piperidone, CA 29976-53-2: 360.93 (1).

$C_8H_{13}NO_5$

Diethyl formamidomalonate, CA 6326-44-9: Above 385.93 (1).

C_8H_{14}

Allylcyclopentane, CA 3524-75-2: 285.37 (1).
Cyclooctene, mixed isomers, CA 931-88-4: 294.15 (3).
Cyclooctene, *cis*, CA 931-87-3: 298.15 (1).
1,3-Dimethyl-1-cyclohexene, CA 2808-76-6: 285.93 (1).
1,4-Dimethyl-1-cyclohexene, CA 70688-47-0: 284.82 (1).
2,5-Dimethyl-1,5-hexadiene, CA 627-58-7: 286.48 (1); 292.15 (3).
2,5-Dimethyl-2,4-hexadiene, CA 764-13-6: 298.15 (3); 302.59 (1).
Ethylidenecyclohexane, CA 1003-64-1: 297.04 (1).
1,7-Octadiene, CA 3710-30-3: 282.59 (1,3); 298.15 (7).
1-Octyne, CA 629-05-0: 289.15 (3); 290.93 (1).

(continues)

164

C_8H_{14} *(continued)*

4-Octyne, CA 1942-45-6: 293.15 (1).

Vinylcyclohexane, CA 695-12-5: 289.15 (3); 294.26 (1).

$C_8H_{14}F_6O_6S_2Si$

Diisopropylsilyl bis(trifluoromethanesulfonate), CA 85272-30-6: 383.15 (3).

$C_8H_{14}N_2$

1-Piperidinepropionitrile, CA 3088-41-3: 375.93 (1).

1-Pyrrolidinebutyronitrile, CA 35543-25-0: 372.59 (1).

$C_8H_{14}N_2O_4$

Diisopropyl azodicarboxylate, CA 2446-83-5: 379.26 (1).

$C_8H_{14}O$

Cyclohexyl methyl ketone, CA 823-76-7: 330.15 (3); 335.37 (1).

Cyclooctanone, CA 502-49-8: 345.93 (1,3).

Cyclooctene oxide, CA 286-62-4: 329.26 (1).

2,6-Dimethylcyclohexanone, mixed isomers, CA 2816-57-1: 324.26 (1,3).

3,5-Dimethyl-1-hexyn-3-ol ±, CA 107-54-0: 317.59 (1); 330.15 (5); 330.37* (6).

1,2-Epoxy-7-octene, CA 19600-63-6: 311.48 (1).

4-Ethylcyclohexanone, CA 5441-51-0: 336.48 (1).

2-Ethyl-2-hexenal, CA 645-62-5: 341.15 (5); 341.48* (6,7).

2-Ethyl-2-hexenal, *trans*, CA 64344-45-2: 326.48 (1).

6-Methyl-5-hepten-2-one, CA 110-93-0: 323.71 (1); 328.15 (3); 330.37 (6).

2-Norbornanemethanol ±, CA 5240-72-2: 357.59 (1).

2-Octenal, *trans*, CA 2548-87-0: 338.71 (1).

1-Octyn-3-ol, CA 818-72-4: 337.04 (1); 339.15 (3).

$C_8H_{14}O_2$

1,4-Butanediol, divinyl ether, CA 3891-33-6: 335.93 (2).

2-Butyn-1-al, diethyl acetal, CA 2806-97-5: 328.71 (1).

Butyl methacrylate, CA 97-88-1: 314.15 (5); 323.71 (1); 325.37* (6,7); 327.15 (3); 339.15* (2,4).

Cycloheptanecarboxylic acid, CA 1460-16-8: Above 385.93 (1).

Cyclohexyl acetate, CA 622-45-7: 326.15 (3); 330.93 (1,5,6); 337.15* (2).

Cyclohexylacetic acid, CA 5292-21-7: Above 385.93 (1).

3-Cyclopentylpropionic acid, CA 140-77-2: 319.82 (1).

1,2,7,8-Diepoxyoctane, CA 2426-07-5: 370.93 (1).

2-Ethyl-2-hexenoic acid, CA 5309-52-4: 438.71* (6).

(continues)

$C_8H_{14}O_2$ *(continued)*

Isobutyl methacrylate, CA 97-86-9: 314.82 (1); 319.15 (3); 322.15* (2,4).

Methyl cyclohexanecarboxylate, CA 4630-82-4: 333.15 (1).

1-Methyl-1-cyclohexanecarboxylic acid, CA 1123-25-7: 373.71 (1).

6-Methyl-2,4-heptanedione, CA 3002-23-1: 332.04 (1).

gamma-Octanoic lactone, CA 104-50-7: Above 385.93 (1).

2,2-Pentamethylene-1,3-dioxolane, CA 177-10-6: 342.04 (1).

$C_8H_{14}O_3$

Butyl acetoacetate, CA 591-60-6: 358.15 (7); 358.15* (6).

tert-Butyl acetoacetate, CA 1694-31-1: 333.71 (1).

Butyric anhydride, CA 106-31-0: 355.37 (6); 360.15 (3); 360.93 (1,2,5).

Diethylene glycol, divinyl ether, CA 764-99-8: 383.15* (2).

2-(2-Ethoxyethoxy)ethanol, CA 111-90-0: 367.04* (7).

2-Ethoxyethyl methacrylate, CA 2370-63-0: 344.82 (1).

Ethyl 4-acetylbutyrate, CA 13984-57-1: 342.59 (1).

Ethyl butyrylacetate, CA 3249-68-1: 351.48 (1).

Ethyl 2-ethylacetoacetate, CA 607-97-6: 342.15 (3).

Ethyl isobutyrylacetate, CA 7152-15-0: 317.15 (3); 326.48 (1).

Isobutyric anhydride, CA 97-72-3: 332.59 (1,6,7); 341.15 (3).

Methyl 4,4-dimethyl-3-oxopentanoate, CA 55107-14-7: 350.37 (1).

$C_8H_{14}O_4$

Adipic acid, monoethyl ester, CA 626-86-8: Above 385.93 (1).

1,4-Bis(2-hydroxyethoxy)-2-butyne, technical, CA 1606-85-5: Above 385.93 (1).

1,3-Butanediol diacetate, technical, CA 1117-31-3: 358.15 (1).

Diethyl methylmalonate, CA 609-08-5: 349.82 (1); 355.15 (3).

Diethyl succinate, CA 123-25-1: 363.71 (6); 369.15 (3); 383.15 (1,5).

Dimethyl adipate, CA 627-93-0: 380.37 (1); 413.71 (2).

Ethylene glycol diglycidyl ether, technical, CA 2224-15-9: Above 385.93 (1).

Isosorbide dimethyl ether, CA 5306-85-4: 381.48 (1).

$C_8H_{14}O_4S$

Dimethyl 3,3'-thiodipropionate, CA 4131-74-2: Above 385.93 (1).

$C_8H_{14}O_5$

Diethyl 1,3-acetonedicarboxylate, CA 105-50-0: 344.15 (3).

Diethylene glycol diacetate, CA 628-68-2: 397.04 (6); 408.15* (6).

$C_8H_{14}O_6$

  Diethyl tartrate, *levo*, CA 87-91-2: 366.48 (1,6).

  Diethyl tartrate, *dextro,* CA 13811-71-7: 366.48 (1).

$C_8H_{14}Si$

  2,4-Cyclopentadien-1-yltrimethylsilane, CA 3559-74-8: 302.59 (1).

$C_8H_{15}Br$

  8-Bromo-1-octene, CA 2695-48-9: 351.48 (1).

$C_8H_{15}BrO_2$

  2-Bromooctanoic acid ±, CA 70610-87-6: Above 385.93 (1).

  8-Bromooctanoic acid, CA 17696-11-6: 383.15 (1).

  Ethyl 2-bromohexanoate, CA 615-96-3: 368.71 (1).

  Ethyl 6-bromohexanoate, CA 25542-62-5: 352.15 (3).

$C_8H_{15}ClO$

  2-Ethylhexanoyl chloride, CA 760-67-8: 342.59 (1).

  Octanoyl chloride, CA 111-64-8: 353.71 (1); 355.15 (5,6); 355.15* (2).

$C_8H_{15}N$

  3-Azabicyclo(3,2,2)nonane, CA 283-24-9: 337.04 (1).

  2-(1-Cyclohexenyl)ethylamine, CA 3399-73-3: 330.93 (1).

  Heptyl cyanide, CA 124-12-9: 347.04 (1); 353.15 (3).

  Perhydroindole, mixed isomers, CA 4375-14-8: 332.59 (1).

$C_8H_{15}NO$

  5,6-Dihydro-2,4,4,6-tetramethyl-4*H*-1,3-oxazine, CA 26939-18-4: 318.71 (1).

  1-Isopropyl-4-piperidone, CA 5355-68-0: 350.37 (1).

  1-Propyl-4-piperidone, CA 23133-37-1: 348.71 (1).

$C_8H_{15}NO_2$

  3-Cyanopropionaldehyde diethyl acetal, CA 18381-45-8: 363.71 (1).

  N,N-Diethylacetoacetamide, CA 2235-46-3: 367.15 (3); 394.26* (6,7).

  2-(Dimethylamino)ethyl methacrylate, CA 2867-47-2: 337.15 (3); 343.71 (1); 347.04* (6,7).

  Ethyl isonipecotate, CA 1126-09-6: 353.15 (1).

  Ethyl nipecotate ±, CA 71962-74-8: 358.15 (3); 363.71 (1).

  Ethyl pipecolinate, CA 15862-72-3: 319.26 (1).

  Ethyl 1-pyrrolidineacetate, CA 22041-19-6: 353.15 (1).

  N-(Isobutoxymethyl)acrylamide, CA 16669-59-3: 352.15 (3).

C_8H_{16}

Cyclooctane, CA 292-64-8: 301.15 (3); 303.15 (1).

Diisobutylene, CA 25167-70-8: Below 266.48 (2); 267.15 (3); 268.15 (6); 274.82* (7).

1,1-Dimethylcyclohexane, CA 590-66-9: 280.37 (1); 284.15 (3).

1,2-Dimethylcyclohexane, mixed isomers, CA 583-57-3: 282.15 (3); 288.71 (1).

1,2-Dimethylcyclohexane, cis, CA 2207-01-4: 284.15 (3); 285.37 (1); 288.71 (2); 289.71 (7).

1,2-Dimethylcyclohexane, trans, CA 6876-23-9: 278.48 (7); 279.26 (1); 280.15 (3); 283.71 (2).

1,3-Dimethylcyclohexane, mixed isomers, CA 591-21-9: 279.15 (7); 282.59 (1); 284.26 (6).

1,3-Dimethylcyclohexane, cis, CA 638-04-0: 278.71 (1).

1,4-Dimethylcyclohexane, mixed isomers, CA 589-90-2: 278.15 (3); 284.26 (6); 288.71 (1).

1,4-Dimethylcyclohexane, cis, CA 624-29-3: 279.26 (1,3); 288.71 (2,6); below 294.15 (5).

1,4-Dimethylcyclohexane, CA 2207-04-7: 277.59 (2); 279.15 (3); 283.71 (6); 289.15 (7).

2,5-Dimethyl-3-hexene, trans, CA 692-70-6: 271.15 (3).

Ethylcyclohexane, CA 1678-91-7: 292.04 (1,2); below 294.15 (5); 308.15 (6).

Isooctene, mixed isomers, CA 11071-47-9: Below 266.48 (6).

2-Methyl-1-heptene, CA 15870-10-7: 283.15 (1).

2-Methyl-2-heptene, CA 627-97-4: 285.15 (3).

5-Methyl-1-heptene, CA 13151-04-7: 283.15 (3).

1-Octene, CA 111-66-0: 281.15 (3); 294.26 (1,5,7); 294.26* (6).

2-Octene, mixed isomers, CA 111-67-1: 287.15 (7); 294.26 (2); 294.26* (6).

2-Octene, cis, CA 7642-04-8: 294.15 (5).

2-Octene, trans, CA 13389-42-9: 287.15 (3); 294.26 (1,5).

3-Octene, trans, CA 14919-01-8: 287.15 (3).

4-Octene, trans, CA 14850-23-8: 287.15 (3); 294.26 (1).

Octene, mixed isomers, CA 25377-83-7: 281.15 (7).

Propylocyclopentane, CA 2040-96-2: 291.15 (3).

2,3,4-Trimethyl-1-pentene, CA 565-76-4: Below 294.15 (5,6).

2,4,4-Trimethyl-1-pentene, CA 107-39-1: 267.04 (1); 269.15 (3).

2,4,4-Trimethyl-2-pentene, CA 107-40-4: 271.48 (1); 274.82* (6); 275.15 (2,3,5).

3,4,4-Trimethyl-2-pentene, CA 598-96-9: Below 294.15 (5,6).

$C_8H_{16}Br_2$

1,8-Dibromooctane, CA 4549-32-0: Above 385.93 (1).

168

$C_8H_{16}Cl_2O_3$

Bis[2-(2-chloroethoxy)ethyl] ether, CA 638-56-2:   Above 394.26 (6).

$C_8H_{16}I_2$

1,8-Diiodooctane, CA 24772-63-2:   Above 385.93 (1).

$C_8H_{16}NO_3P$

Diisopropyl cyanomethylphosphonate, CA 58264-04-3:   Above 385.93 (1).

$C_8H_{16}N_2O_2$

Ethyl 4-amino-1-piperidinecarboxylate, CA 58859-46-4:   351.48 (1).

$C_8H_{16}N_2O_4$

Diethyl 1,2-dimethyl-1,2-hydrazinedicarboxylate, CA 15429-36-4:   348.15 (3).

$C_8H_{16}O$

Cycloheptanemethanol, CA 4448-75-3:   365.93 (1).

1-Cyclohexylethanol ±, CA 1193-81-3:   345.93 (1).

2-Cyclohexylethanol, CA 4442-79-9:   359.15 (3); 359.82 (1).

Cyclooctanol, CA 696-71-9:   359.26 (1); 361.15 (3).

3-Cyclopentyl-1-propanol, CA 767-05-5:   355.37 (1).

2,3-Dimethylcyclohexanol, mixed isomers, CA 1502-24-5:   338.71 (1).

2,5-Dimethylcyclohexanol, CA 3809-32-3:   343.15 (1).

2,6-Dimethylcyclohexanol, mixed isomers, CA 5337-72-4:   328.15 (1); 343.15 (3).

3,4-Dimethylcyclohexanol, mixed isomers, CA 5715-23-1:   351.48 (1).

3,5-Dimethylcyclohexanol, mixed isomers, CA 5441-52-1:   346.48 (1); 347.15 (3).

1,2-Epoxyoctane, CA 2984-50-1:   310.37 (1).

2-Ethylcyclohexanol, mixed isomers, CA 3760-20-1:   341.48 (1).

4-Ethylcyclohexanol, mixed isomers, CA 4534-74-1:   350.93 (1).

2-Ethylhexanal, CA 123-05-7:   315.37 (1); 317.15 (3,6); 324.82* (7); 325.15 (5).

5-Methyl-3-heptanone, CA 541-85-5:   315.15 (3); 317.04 (1).

6-Methyl-5-hepten-2-ol ±, CA 4630-06-2:   340.93 (1).

1-Octanal, CA 124-13-0:   324.82 (1,3,5,6,7).

2-Octanone, CA 111-13-7:   324.82 (6); 329.15 (3); 335.93 (1,4); 344.26 (7).

3-Octanone, CA 106-68-3:   319.26 (1,4); 324.15 (3); 332.04 (7).

1-Octen-3-ol, CA 3391-86-4:   334.26 (1); 360.15 (3).

Octyl aldehyde, mixed isomers: 325.15 (5).

2,2,5,5-Tetramethyltetrahydrofuran, CA 15045-43-9:   277.04 (1).

$C_8H_{16}OSi$

1-(Trimethylsilyloxy)cyclopentene, CA 19980-43-9: 309.15 (3); 311.48 (1).

$C_8H_{16}O_2$

Acetic acid, 2-ethylbutyl ester, CA 10031-87-5: 327.15 (5); 327.59* (6).

Acetic acid, 4-methylpentyl ester, CA 628-95-5: 316.15 (5); 318.15 (6).

Acetic acid, 4-methyl-2-pentyl ester, CA 108-84-9: 316.48* (6); 318.15 (6); 318.15* (4,7).

Butyl butyrate, CA 109-21-7: 322.59 (1,5); 324.15 (2); 326.15 (3,4); 326.48* (6,7).

Butyraldol, CA 496-03-7: 347.04* (6).

1,2-Cyclohexanedimethanol, cis, CA 15753-50-1: Above 385.93 (1).

1,4-Cyclohexanedimethanol, mixed isomers, CA 105-08-8: 434.26 (1); 439.82 (6); 440.15* (4).

1,2-Cyclooctanediol, trans, technical, CA 42565-22-0: Above 385.93 (1).

2,5-Dimethyl-4-hydroxy-3-hexanone, technical, CA 815-77-0: 334.26 (1).

Ethyl caproate, CA 123-66-0: 322.59 (1,6); 326.15 (3); 327.59* (7).

Ethylene glycol, butyl vinyl ether, CA 4223-11-4: 413.15* (4).

2-Ethylhexanoic acid ±, CA 149-57-5: Above 385.93 (1); 391.48* (6); 399.82* (2,7); 400.15 (5).

Ethyl isocaproate, CA 25415-67-2: 316.15 (3).

Hexyl acetate, CA 142-92-7: 310.37 (1); 314.15 (3).

Isobutyl butyrate, CA 539-90-2: 323.15 (5,6).

Isobutyl isobutyrate, CA 97-85-8: 310.37 (1); 311.15* (4); 311.48 (2,6); 313.15 (3); 317.04* (2); 322.15 (4).

Isooctanoic acid, mixed isomers, CA 25103-52-0: 405.37* (6).

Methyl enanthate, CA 106-73-0: 325.93 (1).

Octanoic acid, CA 124-07-2: 383.15 (1); 405.15* (4).

7-Octene-1,2-diol ±, CA 90970-68-6: Above 385.93 (1).

Pentyl propionate, CA 624-54-4: 314.15 (5); 314.26* (6).

2-Propylpentanoic acid, CA 99-66-1: 384.26 (1).

$C_8H_{16}O_2S_2$

Ethyl bis(ethylthio)acetate, CA 20461-95-4: 298.15 (1,3).

$C_8H_{16}O_2Si$

1-Methoxy-3-trimethylsilyloxy-1,3-butadiene, CA 59414-23-3: 318.15 (1).

1-Methoxy-3-trimethylsilyloxy-1,3-butadiene, trans, CA 54125-02-9: 332.15 (3).

4-(Trimethylsilyloxy)-3-penten-2-one, mixed isomers, CA 13257-81-3: 308.15 (3); 332.04 (1).

170

$C_8H_{16}O_3$

tert-Butyl 2-hydroxyisobutyrate, CA 36293-63-7:   295.15 (3).

2,5-Diethoxytetrahydrofuran, mixed isomers, technical, CA 3320-90-9:
329.82 (1).

Diethylene glycol, ethyl vinyl ether: 364.25* (4).

2-Ethoxyethyl isobutyrate, CA 54396-97-3:   330.93 (1).

Ethylene glycol, monobutyl ether acetate, CA 112-07-2:   344.26 (2,6);
347.05 (4); 353.71* (2,4); 360.95 (4,7); 388.15* (4).

Ethyl 2-hydroxycaproate ±, CA 6946-90-3:   354.82 (1).

Pentyl lactate, CA 6382-06-5:   352.59 (2,6,7).

$C_8H_{16}O_3Si$

Methyl 3-trimethylsilyloxy-2-butenoate, mixed isomers, CA 62269-44-7:
332.15 (3); 337.04 (1).

$C_8H_{16}O_4$

12-Crown-4, CA 294-93-9:   Above 385.93 (1).

Diethylene glycol, ethyl ether acetate, CA 112-15-2:   380.15 (4,5); 380.37*
(2,4,6); 383.15* (4,7).

Ethyl diethoxyacetate, CA 6065-82-3:   346.48 (1).

$C_8H_{16}O_4Si$

Ethyl trimethylsilylmalonate, CA 18457-03-9:   323.15 (3).

$C_8H_{16}O_5$

2-[2-(2-Methoxyethoxy)ethoxy]-1,3-dioxolane, CA 74733-99-6:   381.48 (1).

$C_8H_{16}S_2$

Ethylcyclohexyl dimercaptan, CA 28679-10-9:   394.26* (2).

$C_8H_{17}Br$

1-Bromooctane, CA 111-83-1:   351.48 (1); 353.15 (3).

2-Ethylhexyl bromide, CA 18908-66-2:   342.59 (1).

$C_8H_{17}BrO$

8-Bromo-1-octanol, CA 50816-19-8:   Above 385.93 (1).

$C_8H_{17}Cl$

1-Chloro-2-ethylhexane, CA 2350-24-5:   333.15* (6,7); 410.93* (2).

1-Chlorooctane, CA 111-85-3:   327.59 (1); 336.15 (3); 343.15 (6).

$C_8H_{17}ClO_2$

2-(2-Butoxy)ethoxyethyl chloride, CA 1120-23-6: 360.93 (6).

$C_8H_{17}Cl_3Si$

Octyltrichlorosilane, CA 5283-66-9: 358.15 (3); 369.82 (1).

$C_8H_{17}I$

1-Iodooctane, CA 629-27-6: 360.15 (3); 368.15 (1).

$C_8H_{17}N$

1-Butylpyrrolidine, CA 767-10-2: 309.26 (1).

Cyclooctylamine, CA 5452-37-9: 335.93 (1); 340.15 (3).

N,N-Dimethylcyclohexylamine, CA 98-94-2: 312.15 (3); 315.37 (1).

2,3-Dimethylcyclohexylamine, mixed isotopes, CA 42195-92-6: 324.26 (1).

N-Ethylcyclohexylamine, CA 5459-93-8: 303.15* (6,7); 317.04 (1); 318.15 (3).

5-Ethyl-2-methylpiperidine, CA 104-89-2: 325.37* (6).

$C_8H_{17}NO$

1-Diethylamino-3-butanone, CA 3299-38-5: 316.48 (1).

N,N-Diisopropylacetamide, CA 759-22-8: 348.15 (3).

N-(2-Hydroxyethyl)cyclohexylamine, CA 2842-38-8: 393.71* (6).

4-Isobutylmorpholine, CA 10315-98-7: 318.15 (3).

$C_8H_{17}NO_2$

N,N-Diethylglycine, ethyl ester, CA 2644-21-5: 335.15 (3).

$C_8H_{17}NO_3$

2-Ethylhexyl nitrate, CA 27247-96-7: 348.71 (1).

Isooctyl nitrate, mixed isomers: 369.26* (6).

$C_8H_{17}O_5P$

Triethyl phosphonoacetate, CA 867-13-0: Above 385.93 (1).

C_8H_{18}

2,2-Dimethylhexane, CA 590-73-8: 269.82 (1); 273.15 (3).

2,3-Dimethylhexane, CA 584-94-1: 278.48 (7); 280.37* (6).

2,4-Dimethylhexane, CA 589-43-5: 276.15 (3); 283.15* (6,7).

2,5-Dimethylhexane, CA 592-13-2: 275.15 (3); 299.82 (1).

3-Ethyl-2-methylpentane, CA 609-26-7: Below 294.26 (5,6).

(continues)

C_8H_{18} *(continued)*

Hexamethylethane, CA 594-82-1: 277.59 (1).

2-Methylheptane, CA 592-27-8: 250.15 (3); 277.59 (1).

3-Methylheptane, CA 598-81-1: 250.15 (3).

4-Methylheptane, CA 589-53-7: 250.15 (3); 279.82 (1).

Octane, CA 111-65-9: 285.15 (5); 286.15 (3,4,6,7); 288.71 (1); 295.37 (2).

2,2,3-Trimethylpentane, CA 564-02-3: Below 294.15 (5,6).

2,2,4-Trimethylpentane, CA 540-84-1: 261.15 (5,6,7); 265.37 (1,2); 277.15 (3); 277.59* (6).

2,3,3-Trimethylpentane, CA 560-21-4: Below 294.15 (5,6).

2,3,4-Trimethylpentane, CA 565-75-3: 277.15 (3); 278.15 (1,2).

$C_8H_{18}BF_3O$

Boron trifluoride dibutyl etherate, CA 593-04-4: 343.15 (1).

$C_8H_{18}Br_2Sn$

Dibutyltin dibromide, CA 996-08-7: 383.15 (1).

$C_8H_{18}ClP$

Di-*tert*-butylchlorophosphine, CA 13716-10-4: 334.26 (1).

$C_8H_{18}Cl_2Si$

Di-*tert*-butyldichlorosilane, CA 18395-90-9: 258.15 (3).

$C_8H_{18}Cl_2Sn$

Dibutyltin dichloride, CA 683-18-1: Above 385.93 (1); 441.48* (7).

Di-*tert*-butyltin dichloride, CA 19429-30-2: Above 385.93 (1).

$C_8H_{18}F_3NOSi_2$

Bis(trimethylsilyl)trifluoroacetamide, CA 25561-30-2: 297.04 (1); 307.15 (3).

$C_8H_{18}NO$

Di-*tert*-butyl nitroxide, CA 2406-25-9: 318.71 (1).

$C_8H_{18}N_2$

1,3-Cyclohexanebis(methylamine), CA 2579-20-6: 379.26 (1).

N,N'-Diethyl-2-butene-1,4-diamine, CA 112-21-0: 347.04 (1).

N,N,N',*N*'-Tetramethyl-2-butene-1,4-diamine, CA 4559-79-9: 323.71 (1).

$C_8H_{18}N_2O$

4-[2-(Dimethylamino)ethyl]morpholine, CA 4385-05-1: 348.71 (1).

O-Methyl-N,N'-diisopropylisourea, CA 54648-79-2: 308.71 (1).

4-Methyl-1-piperazinepropanol, CA 5317-33-9: Above 385.93 (1).

$C_8H_{18}O$

Butyl ether, CA 142-96-1: 298.15 (1,4,5,6,7); 303.15 (3); 303.75 (2).

2-Ethyl-1-hexanol, CA 104-76-7: 346.48 (6); 350.37 (1); 354.15 (3,4,7); 355.15* (4); 358.15 (5); 358.15* (2).

2-Ethyl-4-methyl-1-pentanol, CA 106-67-2: 343.15 (6); 350.15 (4).

Isobutyl ether, CA 628-55-7: 282.04 (1); 289.15 (3).

Isooctyl alcohol, mixed isomers, CA 26952-21-6: 355.37* (2,4,6).

4-Methyl-3-heptanol, CA 14979-39-6: 327.59 (1).

6-Methyl-2-heptanol, CA 4730-22-7: 340.37 (1).

1-Octanol, CA 111-87-5: 353.15 (3); 354.26 (1,5,6); 363.71* (2).

2-Octanol ±, CA 4128-31-8: 344.26 (1); 349.15 (3); 361.15 (4,6).

2-Octanol, (R)-(-), CA 5978-70-1: 344.26 (1); 349.15 (3).

2-Octanol, (S)-(+), CA 6169-06-8: 344.26 (1); 349.15 (3).

sec-Octanol, mixed isomers, CA 25339-16-6: 358.15 (2).

3-Octanol ±, CA 20296-29-1: 338.71 (1); 341.15 (3).

4-Octanol, CA 589-62-8: 338.15 (3).

2-Propyl-1-pentanol, CA 58175-57-8: 344.82 (1).

2,4,4-Trimethyl-1-pentanol, CA 16325-63-6: 333.15 (1).

$C_8H_{18}OS$

Butyl sulfoxide, technical, CA 2168-93-6: 394.26 (1).

$C_8H_{18}O_2$

1-tert-Butoxy-2-ethoxyethane, CA 51422-54-9: 307.04 (1).

tert-Butyl peroxide, CA 110-05-4: 274.26 (1); 291.15 (3); 291.48* (6,7).

2,5-Dimethyl-2,5-hexanediol, CA 110-03-2: 399.82 (1).

2-(2-Ethylbutoxy)ethanol, CA 4468-93-3: 355.15 (5); 355.37* (4,6,7).

2-Ethyl-1,3-hexanediol, mixed isomers, CA 94-96-2: 383.15 (6); 399.82* (6,7); 400.15 (5); 402.59* (2).

2-(Hexyloxy)ethanol, CA 112-25-4: 363.71* (4,7).

1,2-Octanediol, CA 1117-86-8: Above 385.93 (1).

4,5-Octanediol ±, CA 22520-40-7: 383.15 (5).

2,2,4-Trimethyl-1,3-pentanediol, CA 144-19-4: 385.93 (7); 385.93* (4,6).

$C_8H_{18}O_2S$

Butyl sulfone, CA 598-04-9: 416.48 (1).

$C_8H_{18}O_2Si$

Ethyl 3-(trimethylsilyl)propionate, CA 17728-88-0: 327.04 (1).

1-Methoxy-2-methyl-1-trimethylsiloxypropene, CA 31469-15-5: 287.59 (1); 303.15 (3).

$C_8H_{18}O_3$

2-(2-Butoxyethoxy)ethanol, CA 112-34-5: 351.15 (4,5,6,7); 372.15 (3); 373.15 (1); 374.25 (2,4); 380.37* (2); 383.15* (6); 388.75 (4).

Diethylene glycol, monoisobutyl ether, CA 18912-80-6: 378.71 (6).

Dipropylene glycol, dimethyl ether, mixed isomers, CA 89399-28-0: 328.15 (1).

Dipropylene glycol, ethyl ether, mixed isomers, CA 15764-24-6: 364.15* (2,4).

Ethoxyacetaldehyde, diethyl acetal, CA 4819-77-6: 318.71 (1).

2-Ethoxyethyl ether, CA 112-36-7: 327.59 (1); 340.15 (3); 355.15 (5); 355.37* (2,4,6,7); 364.15* (2).

Triethyl orthoacetate, CA 78-39-7: 305.15 (3); 309.26 (1).

Trimethyl orthovalerate, CA 13820-09-2: 314.82 (1); 328.15 (3).

$C_8H_{18}O_3Si$

Triethoxyvinylsilane, CA 78-08-0: 292.15 (3); 307.59 (1).

$C_8H_{18}O_4$

Ethylene glycol, monomethyl ether acetal, CA 10143-67-6: 366.48* (6).

Triethylene glycol, dimethyl ether, CA 112-49-2: 383.71 (1,5); 384.15* (2,4,6).

Triethylene glycol, monoethyl ether, CA 112-50-5: 397.04 (2,4); 408.15 (5); 408.15* (4,6,7).

$C_8H_{18}O_5$

Tetraethylene glycol, CA 112-60-7: 447.15* (5); 449.82 (1); 455.37* (6,7); 458.15* (2); 464.15 (4).

$C_8H_{18}S$

Butyl sulfide, CA 544-40-1: 335.15 (3); 349.82 (1).

sec-Butyl sulfide, CA 626-26-6: 312.59 (1).

tert-Butyl sulfide, CA 107-47-1: 304.15 (3); 322.04 (1).

Methyl heptanethiol, CA 63834-87-7: 319.26 (7).

1-Octanethiol, CA 111-88-6: 341.15 (3); 342.04 (1); 342.04* (6); 352.59* (2).

tert-Octyl mercaptan: 319.26* (2,6).

$C_8H_{18}S_2$

Butyl disulfide, CA 629-45-8: 366.48 (1).

sec-Butyl disulfide, CA 5943-30-6: Above 385.93 (1).

tert-Butyl disulfide, CA 110-06-5: 335.37 (1); 337.15 (3).

$C_8H_{18}Si$

Cyclohexyldimethylsilane, CA 29681-56-9: 304.82 (1).

3-Methyl-1-trimethylsilyl-2-butene, CA 18293-99-7: 296.15 (3).

$C_8H_{18}Si_2$

Bis(trimethylsilyl)acetylene, CA 14630-40-1: 275.93 (1); 302.15 (3).

$C_8H_{19}ClO_2Si$

Chloromethyl(diisopropoxy)methylsilane, CA 2212-08-0: 327.59 (1).

$C_8H_{19}ClSi$

Di-tert-butylchlorosilane, CA 56310-18-0: 312.59 (1).

$C_8H_{19}N$

Dibutylamine, CA 111-92-2: 313.15 (3); 314.26 (1); 315.15 (2,5); 320.37 (6); 325.15* (4,7); 330.37* (2).

Di-sec-butylamine, CA 626-23-3: 297.15 (5); 297.15* (6,7).

Diisobutylamine, CA 110-96-3: 294.15 (7); 302.15 (5,6); 302.59* (2); 303.15 (3).

N,N-Diisopropylethylamine, CA 7087-68-5: 283.15 (3); 283.71 (1).

N,N-Dimethylhexylamine, CA 4385-04-0: 307.15 (3).

1,5-Dimethylhexylamine, CA 543-82-8: 322.04 (1).

2-Ethylhexylamine ±, CA 104-75-6: 325.37 (1); 330.37* (2); 333.15 (3,5); 333.15* (6,7).

N-Heptylmethylamine, CA 36343-05-2: 325.15 (3).

1-Methylheptylamine, CA 693-16-3: 323.71 (1).

Octylamine, CA 111-86-4: 333.15 (3,5,6,7); 335.93 (1).

tert-Octylamine, CA 107-45-9: 305.37 (1); 306.15 (5); 306.15* (6); 308.15* (2); 333.15 (3).

$C_8H_{19}NO$

2-(Diisopropylamino)ethanol, CA 96-80-0: 330.37 (1); 337.15 (3); 340.37 (2); 352.59* (6,7).

$C_8H_{19}NO_2$

4-Aminobutyraldehyde diethyl acetal, technical, CA 6346-09-4: 335.93 (1).

(continues)

$C_8H_{19}NO_2$ *(continued)*

N-Butyldiethanolamine, CA 102-79-4: 391.48* (6,7); 399.82 (1); 419.26 (2).

N-*tert*-Butyldiethanolamine, CA 2160-93-2: 413.71* (6).

2,2-Diethoxy-N,N-dimethylethylamine, CA 3616-56-6: 318.15 (1); 323.15 (3).

Di(2-ethoxyethyl)amine, CA 124-21-0: 358.15 (2).

2-(2-Diethylaminoethoxy)ethanol, CA 140-82-9: 369.15 (3).

$C_8H_{19}NSi_2$

1,3-Divinyl-1,1,3,3-tetramethyldisilazane, CA 7691-02-3: 315.15 (3).

$C_8H_{19}O_3P$

Dibutyl phosphite, CA 1809-19-4: 322.04 (6,7); 394.26* (1,2).

$C_8H_{19}O_4P$

Dibutyl phosphate, CA 107-66-4: Above 385.93 (1).

$C_8H_{20}N_2$

1,8-Diaminooctane, CA 373-44-4: 438.15 (1).

N,N'-Diethyl-1,3-butanediamine, CA 32280-46-9: 319.26* (6).

2-(Diisopropylamino)ethylamine, CA 121-05-1: 327.15 (3).

N,N'-Diisopropylethylenediamine, CA 4013-94-9: 324.82 (1).

N,N'-Dimethyl-1,6-hexanediamine, CA 13093-04-4: 353.15 (1); 359.15 (3).

2,5-Dimethyl-2,5-hexanediamine, CA 23578-35-0: 335.93 (1).

N,N,N',N'-Tetramethyl-1,3-butanediamine, CA 97-84-7: 316.15 (3); 318.15 (2); 319.15* (4).

N,N,N',N'-Tetramethyl-1,4-butanediamine, CA 111-51-3: 316.15 (3); 319.26 (1).

N,N,N'-Triethylethylenediamine, CA 105-04-4: 305.37 (1).

$C_8H_{20}N_2O_2S$

N,N,N',N'-Tetraethylsulfamide, CA 2832-49-7: Above 385.93 (1).

$C_8H_{20}O_3Si$

Ethyltriethoxysilane, CA 78-07-8: 302.15 (3); 313.15 (4).

$C_8H_{20}O_3Si_2$

Trimethylsilyl (trimethylsilyloxy)acetate, CA 33581-77-0: 308.15 (3); 316.48 (1).

$C_8H_{20}O_4Si$

Tetraethoxysilane, CA 78-10-4: 311.15 (3); 319.82 (1); 324.82* (6); 325.15 (5); 328.15 (4).

$C_8H_{20}O_4Ti$

Tetraethoxytitanium, practical, CA 3087-36-3:   301.15 (3); 302.04 (1).

$C_8H_{20}Pb$

Tetraethyl lead, CA 78-00-2:   353.15 (5); 366.48 (6,7).

$C_8H_{20}Si$

Tetraethylsilane, CA 631-36-7:   298.15 (3); 299.26 (1).

$C_8H_{20}Sn$

Tetraethyltin, CA 597-64-8:   326.48 (1).

$C_8H_{21}NOSi_2$

N,O-Bis(trimethylsilyl)acetamide, CA 10416-59-8:   284.82 (1); 317.15 (3).

$C_8H_{22}BN$

Borane-N,N-diisopropylethylamine complex: 313.15 (1).

$C_8H_{22}N_2O_3Si$

N-[3-(Trimethoxysilyl)propyl]ethylenediamine, technical, CA 1760-24-3:
Above 385.93 (1).

$C_8H_{22}N_4$

N,N'-Bis(3-aminopropyl)ethylenediamine, technical, CA 10563-26-5:   Above
385.93 (1).

$C_8H_{22}O_2Si_2$

1,2-Bis(trimethylsilyloxy)ethane, CA 7381-30-8:   319.26 (1); 324.15 (3).

$C_8H_{22}S_2Si_2$

1,2-Ethanedithiobis(trimethylsilane), CA 51048-29-4:   340.15 (3).

$C_8H_{23}N_5$

Tetraethylenepentamine, CA 112-57-2:   427.15 (4); 435.93* (2,6,7); 436.16
(5); 458.15 (1).

$C_8H_{24}O_2Si_3$

Octamethyltrisiloxane, CA 107-51-7:   302.59 (1).

$C_8H_{24}O_4Si_4$

Octamethylcyclotetrasiloxane, CA 556-67-2:   324.15 (3); 333.15 (1).

$C_9H_3ClF_6O$

  3,5-Bis(trifluoromethyl)benzoyl chloride, CA 785-56-8: 345.37 (1).

$C_9H_3Cl_3O_3$

  1,3,5-Benzenetricarboxylic acid chloride, CA 4422-95-1: Above 385.93 (1).

$C_9H_3F_6N$

  3,5-Bis(trifluoromethyl)benzonitrile, CA 27126-93-8: 345.93 (1).

$C_9H_4F_6O$

  3,5-Bis(trifluoromethyl)benzaldehyde, CA 401-95-6: 343.15 (1).

$C_9H_4O_5$

  Trimellitic anhydride, CA 552-30-7: 500.15 (4).

$C_9H_5BrF_6$

  3,5-Bis(trifluoromethyl)benzyl bromide, CA 32247-96-4: 299.26 (1).

$C_9H_5ClF_6$

  3,5-Bis(trifluoromethyl)benzyl chloride, CA 75462-59-8: 344.82 (1).

$C_9H_5F_5$

  Allylpentafluorobenzene, CA 1736-60-3: 315.93 (1).

C_9H_6BrN

  4-Bromoisoquinoline, CA 1532-97-4: Above 385.93 (1).

  4-Bromoquinoline, CA 5332-24-1: Above 385.93 (1).

C_9H_6ClN

  4-Chlorocinnamonitrile, mixed isomers, CA 28446-72-2: Above 385.93 (1).

  2-Chloroquinoline, CA 612-62-4: Above 385.93 (1).

  4-Chloroquinoline, CA 611-35-8: Above 385.93 (1).

$C_9H_6F_3N$

  2-(Trifluoromethyl)phenylacetonitrile, CA 3038-47-9: 362.59 (1).

  3-(Trifluoromethyl)phenylacetonitrile, CA 2338-76-3: 322.04 (1).

  4-(Trifluoromethyl)phenylacetonitrile, CA 2338-75-2: Above 385.93 (1).

$C_9H_6F_6O$

  3,5-Bis(trifluoromethyl)benzyl alcohol, CA 32707-89-4: 370.93 (1).

$C_9H_6N_2O_2$

2,4-Toluenediisocyanate, technical, CA 584-84-9: 394.26 (1); 399.82 (6); 405.15* (6,7).

2,6-Toluenediisocyanate, CA 91-08-7: Above 385.93 (1).

C_9H_6O

Phenylpropargyl aldehyde, CA 2579-22-8: 367.59 (1).

C_9H_7ClO

alpha-Chlorocinnamaldehyde, CA 18365-42-9: Above 385.93 (1).

Cinnamoyl chloride, mostly trans, CA 102-92-1: Above 385.93 (1).

$C_9H_7ClO_3$

Acetylsalicyloyl chloride, CA 5538-51-2: 383.15 (1).

$C_9H_7F_3O$

2'-(Trifluoromethyl)acetophenone, CA 17408-14-9: 357.59 (1).

3'-(Trifluoromethyl)acetophenone, CA 349-76-8: 352.15 (3); 357.04 (1).

4'-(Trifluoromethyl)acetophenone, CA 709-63-7: 357.59 (1).

$C_9H_7F_6N$

3,5-Bis(trifluoromethyl)benzylamine, technical: 351.48 (1).

C_9H_7N

Cinnamonitrile, mostly trans, CA 1885-38-7: Above 385.93 (1).

Isoquinoline, CA 119-65-3: 380.37 (1).

Quinoline, CA 91-22-5: 365.15 (3); 374.26 (1).

C_9H_7NO

Quinoline N-oxide hydrate, CA 1613-37-2: Above 385.93 (1).

$C_9H_7NO_2$

3,4-Methylenedioxyphenylacetonitrile, CA 4439-02-5: Above 385.93 (1).

C_9H_7NS

2-(3-Thienyl)pyridine, CA 21298-55-5: Above 385.93 (1).

C_9H_8

Indene, CA 95-13-6: 331.15 (3); 332.04 (1).

1-Phenyl-1-propyne, CA 673-32-5: 335.37 (1).

180

C_9H_8ClFO

3-Chloro-4'-fluoropropiophenone, CA 347-93-3:   365.37 (1).

$C_9H_8ClF_3$

1-(2-Chloroethyl)-2-(trifluoromethyl)benzene: 355.37 (1).

$C_9H_8Cl_2$

(2,2-Dichlorocyclopropyl)benzene, CA 2415-80-7:   377.59 (1).

$C_9H_8Cl_2O$

3,4'-Dichloropropiophenone, CA 3946-29-0:   354.26 (1).

3',4'-Dichloropropiophenone, CA 6582-42-9:   Above 385.93 (1).

$C_9H_8Cl_2O_2$

trans-3,6-Endomethylene-1,2,3,6-tetrahydrophthaloyl chloride, CA 4582-21-2:
383.15 (1).

$C_9H_8Cl_3NO$

Benzyl 2,2,2-trichloroacetimidate, CA 81927-55-1:   383.15 (1).

$C_9H_8N_2$

2-Methylquinoxaline, CA 7251-61-8:   380.37 (1).

1-Phenylpyrazole, CA 1126-00-7:   Above 385.93 (1).

C_9H_8O

Cinnamaldehyde, trans, CA 104-55-2:   344.26 (1).

1-Indanone, CA 83-33-0:   384.82 (1).

2-Indanone, CA 615-13-4:   373.15 (1).

2-Methylbenzofuran, CA 4265-25-2:   340.93 (1).

C_9H_8OS

Thiochroman-4-one, CA 3528-17-4:   Above 385.93 (1).

$C_9H_8O_2$

4-Chromanone, CA 491-37-2:   383.15 (1).

Dihydrocoumarin, CA 119-84-6:   Above 385.93 (1).

1-Phenyl-1,2-propanedione, CA 579-07-7:   357.59 (1).

$C_9H_8O_3$

4-Acetoxybenzaldehyde, CA 878-00-2:   Above 385.93 (1).

(continues)

$C_9H_8O_3$ *(continued)*

1,4-Benzodioxan-6-carboxaldehyde, CA 29668-44-8: Above 385.93 (1).

Methyl benzoylformate, CA 15206-55-0: Above 385.93 (1).

C_9H_9Br

Cinnamyl bromide, technical, CA 4392-24-9: 360.93 (1).

C_9H_9BrO

2-Bromo-4'-methylacetophenone, CA 619-41-0: Above 385.93 (1).

2-Bromopropiophenone, CA 2114-00-3: Above 385.93 (1).

3'-Bromopropiophenone, CA 19829-31-3: Above 385.93 (1).

4'-Bromopropiophenone, CA 10342-83-3: Above 385.93 (1).

$C_9H_9BrO_2$

Benzyl 2-bromoacetate, CA 5437-45-6: Above 385.93 (1).

2-Bromo-2'-methoxyacetophenone, CA 31949-21-0: Above 385.93 (1).

2-(3-Bromophenyl)-1,3-dioxolane, CA 17789-14-9: Above 385.93 (1).

C_9H_9Cl

4-Chloro-*alpha*-methylstyrene, technical, CA 1712-70-5: 347.04 (1).

Cinnamyl chloride, CA 2687-12-9: 352.59 (1).

Vinylbenzyl chloride, mixed isomers, CA 26446-61-7: 377.59* (6).

C_9H_9ClO

3-Chloropropiophenone, technical, CA 936-59-4: Above 385.93 (1).

3'-Chloropropiophenone, CA 34841-35-5: Above 385.93 (1).

4'-Chloropropiophenone, CA 6285-05-8: Above 385.93 (1).

4-Ethylbenzoyl chloride, CA 16331-45-6: Above 385.93 (1).

Hydrocinnamoyl chloride, CA 645-45-4: 381.48 (1).

$C_9H_9ClO_2$

Benzyloxyacetyl chloride, CA 19810-31-2: Above 385.93 (1).

4-Chlorophenyl 2,3-epoxypropyl ether, CA 2212-05-7: 380.37 (1).

3-Ethoxybenzoyl chloride, CA 61956-65-8: Above 385.93 (1).

4-Ethoxybenzoyl chloride, CA 16331-46-7: Above 385.93 (1).

$C_9H_9ClO_3$

3,5-Dimethoxybenzoyl chloride, CA 17213-57-9: Above 385.93 (1).

C$_9$H$_9$FO

(2-Fluorophenyl)acetone, CA 2836-82-0: 358.15 (1).

(4-Fluorophenyl)acetone, CA 459-03-0: 364.82 (1).

2'-Fluoropropiophenone, CA 446-22-0: 352.59 (1).

4'-Fluoropropiophenone, CA 456-03-1: 349.82 (1).

C$_9$H$_9$FO$_2$

Ethyl 3-fluorobenzoate, CA 451-02-5: 355.37 (1).

Ethyl 4-fluorobenzoate, CA 451-46-7: 354.26 (1).

C$_9$H$_9$F$_3$O

alpha-Methyl-2-(trifluoromethyl)benzyl alcohol, CA 79756-81-3: 362.59 (1).

2-(Trifluoromethyl)phenethyl alcohol: 375.93 (1).

3-(Trifluoromethyl)phenethyl alcohol, CA 455-01-6: 358.15 (1).

C$_9$H$_9$F$_3$O$_3$S

2,2,2-Trifluoroethyl *para*-toluenesulfonate, CA 433-06-7: Above 385.93 (1).

C$_9$H$_9$N

2,5-Dimethylbenzonitrile, technical, CA 13730-09-1: 365.37 (1).

Hydrocinnamonitrile, CA 645-59-0: Above 385.93 (1).

alpha-Methylbenzyl cyanide, technical, CA 1823-91-2: 367.04 (1).

2-Methylbenzyl cyanide, CA 22364-68-7: 383.15 (1).

3-Methylbenzyl cyanide, CA 2947-60-6: Above 385.93 (1).

4-Methylbenzyl cyanide, CA 2947-61-7: 379.26 (1).

1-Methylindole, CA 603-76-9: Above 385.93 (1).

4-Methylindole, CA 16096-32-5: Above 385.93 (1).

6-Methylindole, CA 3420-02-8: Above 385.93 (1).

Tripropargylamine, CA 6921-29-5: 330.37 (1).

C$_9$H$_9$NO

2,5-Dimethylbenzoxazole, CA 5676-58-4: 364.26 (1).

2,6-Dimethylphenyl isocyanate, CA 28556-81-2: 359.82 (1).

2-Ethoxybenzonitrile, CA 6609-57-0: Above 385.93 (1).

3-Ethoxybenzonitrile, CA 25117-75-3: Above 385.93 (1).

2-Ethylphenyl isocyanate, CA 40411-25-4: 352.04 (1).

1-Furfurylpyrrole, CA 1438-94-4: 365.93 (1).

(3-Methoxyphenyl)acetonitrile, CA 19924-43-7: 372.04 (1).

(4-Methoxyphenyl)acetonitrile, CA 104-47-2: Above 385.93 (1).

(continues)

C_9H_9NO *(continued)*

alpha-Methylbenzyl isocyanate, (S)-(-), CA 14649-03-7:   338.71 (1).
alpha-Methylbenzyl isocyanate, (R)-(+), CA 33375-06-3:   338.71 (1).

C_9H_9NOS

6-Methoxy-2-methylbenzothiazole, CA 2941-72-2:   Above 385.93 (1).

C_9H_9NOSe

5-Methoxy-2-methylbenzselenazole, CA 2946-17-0:   Above 385.93 (1).

$C_9H_9NO_2$

2,3-Dimethoxybenzonitrile, CA 5653-62-3:   Above 385.93 (1).
2-Ethoxyphenyl isocyanate, CA 5395-71-1:   375.37 (1).
5-Nitroindan, CA 7436-07-9:   Above 385.93 (1).

$C_9H_9NO_3$

2,4-Dimethoxyphenyl isocyanate: Above 385.93 (1).
2,5-Dimethoxyphenyl isocyanate, CA 56309-62-7:   Above 385.93 (1).

C_9H_9NS

2,5-Dimethylbenzothiazole, CA 95-26-1:   Above 385.93 (1).
alpha-Methylbenzyl isothiocyanate, CA 4478-92-6:   338.71 (1).
Phenethyl isothiocyanate, CA 2257-09-2:   Above 385.93 (1).

$C_9H_9N_3O_3$

4-Methoxybenzyloxycarbonyl azide, CA 25474-85-5:   Above 385.93 (1).

C_9H_{10}

Allylbenzene, CA 300-57-2:   306.48 (1); 313.15 (3).
Cyclopropylbenzene, CA 873-49-4:   317.04 (1).
Indan, CA 496-11-7:   323.15 (1); 327.15 (3).
alpha-Methylstyrene, CA 98-83-9:   318.71 (1,3); 327.04 (6); 330.95* (4); 331.15 (5).
beta-Methylstyrene, mixed isomers, CA 637-50-3:   333.15 (1).
beta-Methylstyrene, *trans*, CA 873-66-5:   325.93 (1); 331.15 (3).
2-Methylstyrene, CA 611-15-4:   331.15 (3).
3-Methylstyrene, CA 100-80-1:   324.26 (1); 325.15 (3).
4-Methylstyrene, CA 622-97-9:   318.71 (1,3).
Vinyltoluene, mixed isomers, CA 25013-15-4:   322.04 (6); 333.15 (3).

$C_9H_{10}Cl_2O_2$

Propylene glycol, 2,4-dichlorophenyl ester, CA 56927-95-8: 441.48* (2).

$C_9H_{10}N_2$

3-Anilinopropionitrile, CA 1075-76-9: Above 385.93 (1).

$C_9H_{10}O$

2-Allylphenol, CA 1745-81-9: 362.04 (1).

Allyl phenyl ether, CA 1746-13-0: 335.93 (1).

Cinnamyl alcohol, CA 104-54-1: Above 385.93 (1).

2,4-Dimethylbenzaldehyde, CA 15764-16-6: 362.04 (1).

2,5-Dimethylbenzaldehyde, CA 5779-94-2: 360.93 (1).

4-Ethylbenzaldehyde, CA 4748-78-1: 365.37 (1).

Hydrocinnamaldehyde, CA 104-53-0: 363.15 (1); 368.15 (3); 369.26 (6).

1-Indanol, CA 6351-10-6: Above 385.93 (1).

5-Indanol, CA 1470-94-6: Above 385.93 (1).

Isochroman, CA 493-05-0: 338.71 (1).

beta-Methoxystyrene, mixed isomers, technical, CA 4747-15-3: 350.37 (1).

2'-Methylacetophenone, CA 577-16-2: 348.71 (1); 357.15 (3).

3'-Methylacetophenone, CA 585-74-0: 355.15 (3); 358.15 (1).

4'-Methylacetophenone, technical, CA 122-00-9: 355.15 (3); 365.37 (1); 369.26 (6).

2-Phenylpropionaldehyde ±, CA 34713-70-7: 349.26 (1,3); as 80% solution, 342.59 (1).

2-Propenylphenol, mixed isomers, CA 6380-21-8: 363.71 (1).

Propiophenone, CA 93-55-0: 358.15 (2,4); 360.93 (1); 369.15* (4); 372.04* (6,7); 372.15 (5).

4-Vinylanisole, CA 637-69-4: 349.82 (1).

$C_9H_{10}O_2$

Benzyl acetate, CA 140-11-4: 363.71 (6); 375.37 (1,4,5,7).

4-Chromanol, CA 1481-93-2: Above 385.93 (1).

1,2-Epoxy-3-phenoxypropane ±, CA 122-60-1: Above 385.93 (1).

2-Ethoxybenzaldehyde, CA 613-69-4: 364.82 (1).

4-Ethoxybenzaldehyde, CA 10031-82-0: 348.15 (3); above 385.93 (1).

Ethyl benzoate, CA 93-89-0: 357.59 (1); 361.15 (5,6); 366.15 (3); above 368.71 (7).

Hydrocinnamic acid, CA 501-52-0: Above 385.93 (1).

2'-Hydroxy-5'-methylacetophenone, CA 1450-72-2: Above 385.93 (1).

2'-Hydroxypropiophenone, CA 610-99-1: Above 385.93 (1).

2-Methoxyacetophenone, CA 4079-52-1: 374.82 (1).

(continues)

$C_9H_{10}O_2$ *(continued)*

2'-Methoxyacetophenone, CA 579-74-8: 382.04 (1).

3'-Methoxyacetophenone, CA 586-37-8: 383.15 (1).

4'-Methoxyacetophenone, CA 100-06-1: Above 385.93 (1).

3-Methyl-*para*-anisaldehyde, CA 32723-67-4: 385.37 (1).

Methyl 2-methylbenzoate, CA 89-71-4: 355.37 (1).

Methyl 3-methylbenzoate, CA 99-36-5: 364.15 (3); 368.71 (1).

Methyl 4-methylbenzoate, CA 99-75-2: 362.15 (3); 368.15 (1).

Methyl phenylacetate, CA 101-41-7: 363.71 (1,6); 369.15 (3).

Phenoxy-2-propanone, CA 621-87-4: 358.15 (1).

Phenyl glycidol, CA 21915-53-7: Above 385.93 (1).

2-Phenylpropionic acid, CA 492-37-5: Above 385.93 (1).

para-Tolyl acetate, CA 140-39-6: 363.15 (1); 363.71 (6).

$C_9H_{10}O_3$

2,3-Dimethoxybenzaldehyde, CA 86-51-1: Above 385.93 (1).

2,5-Dimethoxybenzaldehyde, CA 93-02-7: Above 385.93 (1).

3,4-Dimethoxybenzaldehyde, CA 120-14-9: Above 385.93 (1).

3,5-Dimethoxybenzaldehyde, CA 7311-34-4: Above 385.93 (1).

2-Ethoxybenzoic acid, CA 134-11-2: Above 385.93 (1).

Ethyl salicylate, CA 118-61-6: 380.37 (1).

2'-Hydroxy-4'-methoxyacetophenone, CA 552-41-0: Above 385.93 (1).

2'-Hydroxy-5'-methoxyacetophenone, CA 705-15-7: 383.15 (1).

Methyl 4-(hydroxymethyl)benzoate, CA 6908-41-4: Above 385.83 (1).

Methyl mandelate ±, CA 4358-87-6: Above 385.93 (1).

Methyl 2-methoxybenzoate, CA 606-45-1: Above 385.93 (1).

Methyl 4-methoxybenzoate, CA 121-98-2: Above 385.93 (1).

Methyl phenoxyacetate, CA 2065-23-8: 383.15 (1).

$C_9H_{10}O_3S$

Methyl (phenylsulfinyl)acetate, CA 14090-83-6: Above 385.93 (1).

$C_9H_{10}O_4$

Ethyl *beta*-oxo-3-furanpropionate, CA 36878-91-8: 383.15 (1).

2-Hydroxyethyl salicylate, CA 87-28-5: Above 385.93 (1).

Methyl 4-methoxysalicylate, CA 5446-02-6: Above 385.93 (1).

$C_9H_{10}O_4S$

Methyl phenylsulfonylacetate, CA 34097-60-4: 351.15 (3); above 385.93 (1).

186

$C_9H_{10}S$

Cyclopropyl phenyl sulfide, CA 14633-54-6: 364.82 (1).

$C_9H_{11}Br$

Bromocumene, mixed isomers: 369.15 (2).

2-Bromomesitylene, CA 576-83-0: 369.26 (1).

1-Bromo-3-phenylpropane, CA 637-59-2: 374.82 (1).

2-Bromo-1-phenylpropane ±, CA 2114-39-8: 363.71 (1).

$C_9H_{11}BrO$

3-Phenoxypropyl bromide, CA 588-63-6: 369.26 (1).

$C_9H_{11}Cl$

1-Chloro-3-phenylpropane, CA 104-52-9: 360.37 (1).

2,5-Dimethylbenzyl chloride, CA 824-45-3: 338.71 (1).

3,4-Dimethylbenzyl chloride, technical, CA 102-46-5: 369.26 (1).

$C_9H_{11}ClO_2$

Propylene glycol, 2-chlorophenyl ether, CA 5335-22-8: 405.37* (2).

Propylene glycol, 4-chlorophenyl ether, CA 5335-23-9: 416.48* (2).

$C_9H_{11}ClO_2S$

2-Mesitylenesulfonyl chloride, CA 773-64-8: Above 385.93 (1).

$C_9H_{11}ClO_3$

Dimethyl 4-chlorophenyl orthoformate, CA 40923-81-7: 364.15 (3).

$C_9H_{11}ClO_3S$

2-Chloroethyl para-toluenesulfonate, CA 80-41-1: Above 385.93 (1).

$C_9H_{11}F$

2-Fluorocumene, CA 2022-67-5: 311.15 (3).

$C_9H_{11}N$

N-Allylaniline, CA 589-09-3: 362.59 (1).

1-Aminoindan, CA 34698-41-4: 367.59 (1).

5-Aminoindan, CA 24425-40-9: 383.15 (1).

2,3-Cyclohexenopyridine, CA 10500-57-9: 359.26 (1).

5-Ethyl-2-vinylpyridine, CA 5408-74-2: 366.48* (6).

2-Isopropenylaniline, CA 52562-19-3: 351.48 (1).   (continues)

$C_9H_{11}N$ *(continued)*

2-Methylindoline, CA 6872-06-6: 366.48 (1).

1,2,3,4-Tetrahydroisoquinoline, CA 91-21-4: 372.04 (1).

5,6,7,8-Tetrahydroisoquinoline, CA 36556-06-6: 373.15 (1).

1,2,3,4-Tetrahydroquinoline, CA 635-46-1: 373.71 (1).

$C_9H_{11}NO$

para-Acetotoluidine, CA 103-89-9: 441.15 (6,7); 441.15* (5).

N,N-Dimethylbenzamide, CA 611-74-5: 383.15 (1).

$C_9H_{11}NO_2$

Dimethyl anthranilate, CA 85-91-6: 363.71 (6).

Ethyl 2-aminobenzoate, CA 87-25-2: Above 385.93 (1).

Ethyl 3-aminobenzoate, CA 582-33-2: Above 385.93 (1).

N-Ethyl-3,4-(methylenedioxy)aniline, CA 32953-14-3: Above 385.93 (1).

Ethyl 2-pyridylacetate, CA 2739-98-2: Above 385.93 (1).

Ethyl 3-pyridylacetate, CA 39931-77-6: Above 385.93 (1).

1-Isopropyl-4-nitrobenzene, CA 1817-47-6: 372.15 (3).

2-Nitromesitylene, CA 603-71-4: 359.26 (1).

1-Nitro-2-propylbenzene, CA 7137-54-4: 378.15 (1).

C_9H_{12}

Bicyclo(4,3,0)nona-3,6(1)-diene, technical, CA 7603-37-4: 283.15 (1).

Cumene, CA 98-82-8: 304.15 (3,5); 308.71 (6); 317.04 (6,7); 319.26 (1).

5-Ethylidene-2-norbornene, mixed isomers, CA 16219-75-3: 311.48 (1).

2-Ethyltoluene, CA 611-14-3: 312.59 (1); 326.15 (3).

3-Ethyltoluene, CA 620-14-4: 310.93 (1); 317.15 (3).

4-Ethyltoluene, CA 622-96-8: 309.82 (1); 316.15 (3).

Mesitylene, CA 108-67-8: 316.15 (4); 317.59 (1); 323.15 (6); 330.15 (3).

1,8-Nonadiyne, CA 2396-65-8: 314.82 (1).

Propylbenzene, CA 103-65-1: 303.15 (6,7); 312.15 (5); 315.15 (3); 318.71 (2); 320.93 (1).

Pseudocumene, CA 95-63-6: 317.59 (6); 319.15 (4); 321.15 (3); 322.04 (1); 323.15 (5); 327.59 (6,7).

1,2,3-Trimethylbenzene, technical, CA 526-73-8: 321.48 (1); 324.15 (4); 326.15 (3,6).

5-Vinyl-2-norbornene, technical, CA 3048-64-4: 300.93 (1).

$C_9H_{12}BrN$

2-Bromo-*N,N*-dimethylbenzylamine, CA 1976-04-1: 377.59 (1).

$C_9H_{12}Cl_2$

2-(Dichloromethylene)bicyclo(3,3,0)octane: 373.15 (1).

$C_9H_{12}N_2$

2-Pyrrolidinopyridine, CA 2456-81-7: Above 385.93 (1).

$C_9H_{12}O$

2-Acetyl-5-norbornene, mixed isomers, CA 5063-03-6: 334.82 (1).

2,3-Dimethylanisole, CA 2944-49-2: 340.37 (1).

2,4-Dimethylanisole, CA 6738-23-4: 336.48 (1).

2,5-Dimethylanisole, CA 1706-11-2: 339.26 (1).

2,6-Dimethylanisole, CA 1004-66-6: 340.37 (1).

3,4-Dimethylanisole, CA 4685-47-6: 348.71 (1).

3,5-Dimethylanisole, CA 874-63-5: 338.71 (1).

2,4-Dimethylbenzyl alcohol, CA 16308-92-2: Above 385.93 (1).

2,5-Dimethylbenzyl alcohol, CA 53957-33-8: 380.37 (1).

3,5-Dimethylbenzyl alcohol, CA 27129-87-9: 379.82 (1).

4-Ethylbenzyl alcohol, CA 768-59-2: 377.59 (1).

2-Isopropylphenol, CA 88-69-7: 380.93 (1).

3-Isopropylphenol, technical, CA 618-45-1: As 60% solution, 370.37 (1).

2-Methylphenethyl alcohol, CA 19819-98-8: Above 385.93 (1).

3-Methylphenethyl alcohol, CA 1875-89-4: 382.59 (1).

4-Methylphenethyl alcohol, CA 699-02-5: 380.37 (1).

1-Phenyl-1-propanol ±, CA 93-54-9: 363.71 (1).

1-Phenyl-2-propanol, CA 14898-87-4: 358.15 (1).

2-Phenyl-1-propanol, CA 1123-85-9: 367.04 (1).

2-Phenyl-2-propanol, CA 617-94-7: 360.93 (1).

3-Phenyl-1-propanol, CA 122-97-4: 373.15 (6); 382.59 (1).

2-Propylphenol, CA 644-35-9: 366.48 (1).

4-Propylphenol, CA 645-56-7: 379.26 (1).

$C_9H_{12}OS$

1-Methoxy-2-methyl-4-(methylthio)benzene, CA 50390-78-8: 383.15 (1).

4-Methoxy-2-methyl-1-(methylthio)benzene, CA 22583-04-6: Above 385.93 (1).

$C_9H_{12}O_2$

2-Allyl-2-methyl-1,3-cyclopentanedione, CA 26828-48-8: 368.71 (1).

Benzaldehyde dimethyl acetal, CA 1125-88-8: 328.15 (3); 342.59 (1).

2-Benzyloxyethanol, CA 622-08-2: Above 385.93 (1); 402.15 (2,5); 402.15*
(6,7).

(continues)

$C_9H_{12}O_2$ *(continued)*

Cumene hydroperoxide, technical, CA 80-15-9: As 80% solution, 329.26 (1); 352.59 (6,7).

2,3-Dimethoxytoluene, CA 4463-33-6: 358.71 (1).

2,4-Dimethoxytoluene, CA 38064-90-3: 365.93 (1).

2,6-Dimethoxytoluene, CA 5673-07-4: 369.82 (1).

2-Ethoxybenzyl alcohol, CA 71672-75-8: Above 385.93 (1).

4-Ethoxybenzyl alcohol, CA 6214-44-4: Above 385.93 (1).

3-(4-Hydroxyphenyl)-1-propanol, CA 10210-17-0: Above 385.93 (1).

2-Isopropoxyphenol, CA 20920-83-6: 360.93 (1).

3-Isopropylcatechol, CA 2138-48-9: Above 385.93 (1).

2-Methoxy-*alpha*-methylbenzyl alcohol ±, CA 13513-82-1: Above 385.93 (1).

1-Methoxymethyl-5-norbornen-2-one, technical, CA 61855-77-4: 337.59 (1).

7-*syn*-Methoxymethyl-5-norbornen-2-one, technical, CA 52962-99-9: 364.82 (1).

2-Methoxyphenethyl alcohol, CA 7417-18-7: Above 385.93 (1).

3-Methoxyphenethyl alcohol, CA 5020-41-7: Above 385.93 (1).

4-Methoxyphenethyl alcohol, CA 702-23-8: Above 385.93 (1).

2-Methoxy-2-phenylethanol, CA 2979-22-8: 370.93 (1).

5-Norbornen-2-yl acetate, mixed isomers, CA 6143-29-9: 335.37 (1).

2-Phenyl-1,2-propanediol ±, CA 4217-66-7: Above 385.93 (1).

4-Propoxyphenol, CA 18979-50-5: Above 385.93 (1).

Propylene glycol, phenyl ether, CA 770-35-4: 311.15 (4); 399.82 (2).

$C_9H_{12}O_2S$

1,2-Dimethoxy-3-(methylthio)benzene, CA 51506-47-9: Above 385.93 (1).

1,4-Dimethoxy-2-(methylthio)benzene, technical, CA 2570-42-5: Above 385.93 (1).

$C_9H_{12}O_3$

2,3-Dimethoxybenzyl alcohol, CA 5653-67-8: Above 385.93 (1).

2,4-Dimethoxybenzyl alcohol, CA 7314-44-5: Above 385.93 (1).

2,5-Dimethoxybenzyl alcohol, CA 33524-31-1: Above 385.93 (1).

3,4-Dimethoxybenzyl alcohol, CA 93-03-8: Above 385.93 (1).

3,5-Dimethoxybenzyl alcohol, CA 705-76-0: Above 385.93 (1).

Hexahydro-4-methylphthalic anhydride, CA 19438-60-9: Above 385.93 (1).

Homovanillyl alcohol, CA 2380-78-1: Above 385.93 (1).

3-Phenoxy-1,2-propanediol, CA 538-43-2: 383.15 (1).

1,2,3-Trimethoxybenzene, CA 634-36-6: Above 385.93 (1).

1,2,4-Trimethoxybenzene, CA 135-77-3: Above 385.93 (1).

1,3,5-Trimethoxybenzene, CA 621-23-8: 358.71 (1).

190

$C_9H_{12}O_3S$

Ethyl *para*-toluenesulfonate, CA 80-40-0: 430.93 (1,5,6,7).

$C_9H_{12}S$

2-Isopropylthiophenol, technical, CA 6262-87-9: 359.26 (1).

3-Phenylpropyl mercaptan, CA 24734-68-7: 363.71 (1).

$C_9H_{13}Cl_3$

1-Trichloromethyl-*cis*-perhydropentalene, CA 18127-07-6: 383.15 (1).

$C_9H_{13}N$

Benzedrine, CA 300-62-9: Below 373.15* (7); above 373.15 (6); above 383.15 (5).

4-*tert*-Butylpyridine, CA 3978-81-2: 336.48 (1); 340.15 (3).

N,N-Dimethylbenzylamine, CA 103-83-3: 326.15 (3); 327.59 (1).

N,N-Dimethyl-*meta*-toluidine, CA 121-72-2: 358.15 (1).

N,N-Dimethyl-*para*-toluidine, CA 99-97-8: 356.48 (1).

N-Ethylbenzylamine, CA 14321-27-8: 339.82 (1); 342.15 (3).

N-Ethyl-N-methylaniline, CA 613-97-8: 347.59 (1).

N-Ethyl-*meta*-toluidine, CA 102-27-2: 362.59 (1); 382.15 (3).

6-Ethyl-*ortho*-toluidine, CA 24549-06-2: 362.59 (1).

2-Isopropylaniline, CA 40085-42-5: 368.71 (1).

4-Isopropylaniline, CA 99-88-7: 365.37 (1).

beta-Methylphenethylamine, CA 582-22-9: 352.59 (1).

N-Methylphenethylamine, CA 589-08-2: 347.04 (1).

3-Phenyl-1-propylamine, CA 2038-57-5: 363.71 (1); 364.15 (3).

2-Propylaniline, CA 1821-39-2: 370.93 (1).

4-Propylaniline, CA 2696-84-6: 377.04 (1).

2-(*para*-Tolyl)ethylamine, CA 3261-62-9: 363.71 (1).

2,4,6-Trimethylaniline, CA 88-05-1: 369.26 (1).

$C_9H_{13}NO$

N-Benzylethanolamine, CA 104-63-2: Above 385.93 (1).

2-Butoxypyridine, CA 27361-16-6: 348.15 (1).

2-Ethoxybenzylamine, CA 37806-29-4: Above 385.93 (1).

2-Methoxyphenethylamine, CA 2045-79-6: Above 385.93 (1).

3-Methoxyphenethylamine, CA 2039-67-0: Above 385.93 (1).

4-Methoxyphenethylamine, CA 55-81-2: 345.15 (3); above 385.93 (1).

1-Methyl-2-phenoxyethylamine, CA 35205-54-0: Above 385.93 (1).

2-(Methylphenylamino)ethanol, CA 93-90-3: 410.93* (6); 411.15 (5).

(continues)

$C_9H_{13}NO$ *(continued)*

  6-Methyl-2-pyridinepropanol, CA 61744-43-2: 383.15 (1).

  Norephedrine, $(1R,2S)$-$(-)$, CA 492-41-1: Above 385.93 (1).

  2-Oxo-1-cyclohexanepropionitrile, CA 4594-78-9: Above 385.93 (1).

  2-(Toluidino)ethanol, CA 136-80-1: 416.48* (7).

$C_9H_{13}NO_2$

  Veratrylamine, CA 5763-61-1: Above 385.93 (1).

$C_9H_{13}NO_2S$

  N-Ethyl-4-toluenesulfonamide, CA 80-39-7: 400.15 (5,6).

$C_9H_{13}N_3$

  1-(2-Pyridyl)piperazine, CA 34803-66-2: Above 385.93 (1).

C_9H_{14}

  4,5,6,7-Tetrahydroindan, technical, CA 695-90-9: 318.71 (1).

$C_9H_{14}Cl_2O_2$

  Azelaoyl chloride, CA 123-98-8: 383.15 (1).

$C_9H_{14}N_2$

  Azelanitrile, CA 1675-69-0: Above 385.93 (1).

  2,3-Diethyl-5-methylpyrazine, CA 18138-04-0: 353.15 (1).

$C_9H_{14}N_2O$

  2-*sec*-Butyl-3-methoxypyrazine, CA 24168-70-5: 350.37 (1).

  2-Isobutyl-3-methoxypyrazine, CA 24683-00-9: 353.15 (1).

$C_9H_{14}O$

  4-Acetyl-1-methyl-1-cyclohexene ±, CA 70286-20-3: 347.15 (3); 349.82 (1).

  2,4-Dimethyl-2,6-heptadienal, mixed isomers: 337.59 (1).

  Isophorone, CA 78-59-1: 357.59 (1,6); 357.59* (7); 369.15 (3,4,5); 369.26* (2).

  4-Isopropyl-2-cyclohexen-1-one, mixed isomers, CA 500-02-7: 360.37 (1).

  2,4-Nonadienal, *trans,trans*, CA 5910-87-2: 358.71 (1).

  Nonadienal, *trans*-2,*cis*-6, CA 557-48-2: 355.93 (1).

  Phorone, CA 504-20-1: 352.59 (1); 358.15 (3,5); 358.15* (6,7).

  2,3,4,5-Tetramethyl-2-cyclopentenone, mixed isomers, CA 54458-61-6: 346.48 (1).

$C_9H_{14}O_2$

  2-Allyl-3-hydroxy-2-methylcyclopentanone, $(2S,3S)$-(+), CA 72345-34-7:
  Above 385.93 (1).

  cis-Bicyclo(3,3,0)octane-2-carboxylic acid: 383.15 (1).

  2-(3-Butynyloxy)tetrahydro-2H-pyran, CA 40365-61-5: 345.93 (1).

  Methyl 2-octynoate, CA 111-12-6: 360.93 (6); 362.04 (1).

  2-Norbornaneacetic acid, CA 1007-01-8: Above 385.93 (1).

$C_9H_{14}O_3$

  Bis(2-methylallyl) carbonate, CA 64057-79-0: 345.93 (1).

  Ethyl 2-cyclohexanonecarboxylate, CA 1655-07-8: 358.15 (1); 369.15 (3).

  2-Furaldehyde diethyl acetal, CA 13529-27-6: 337.59 (1).

  Methyl 2-oxo-1-cycloheptanecarboxylate, CA 52784-32-4: Above 385.93 (1).

$C_9H_{14}O_3SSi$

  Trimethylsilyl benzenesulfonate, CA 17882-06-3: 288.15 (3).

$C_9H_{14}O_4$

  Diethyl 1,1-cyclopropanedicarboxylate, CA 1559-02-0: 359.82 (1).

  Diethyl 1,2-cyclopropanedicarboxylate, CA 20561-09-5: 371.48 (1).

  Diethyl ethylidenemalonate, CA 1462-12-0: Above 385.93 (1).

  Diethyl glutaconate, mixed isomers, CA 2049-67-4: Above 385.93 (1).

  Methyl 4-acetyl-5-oxohexanoate, CA 13984-53-7: Above 385.93 (1).

$C_9H_{14}O_5$

  Diethyl 1,3-acetonedicarboxylate, CA 105-50-0: 359.26 (1).

  Diethyl oxalpropionate, CA 5965-53-7: 370.15 (3); above 385.93 (1).

  Ethyl 2-acetoxy-2-methylacetoacetate ±, CA 25409-39-6: 377.04 (1).

$C_9H_{14}O_6$

  Triacetin, CA 102-76-1: 410.93* (7); 411.15 (4,5,6); 416.48* (2); 422.04
  (1); 424.82 (2); 426.15* (4).

$C_9H_{14}SSi$

  Phenylthiotrimethylsilane, CA 4551-15-9: 303.71 (1); 306.15 (3).

$C_9H_{14}Si$

  Phenyltrimethylsilane, CA 768-32-1: 313.15 (3); 317.59 (1).

$C_9H_{15}Br_6O_4P$

  Tris(2,3-dibromopropyl) phosphate, CA 126-72-7: Above 317.59 (7).

$C_9H_{15}N$

l-Pyrrolidino-l-cyclopentene, CA 7148-07-4: 319.82 (1).

Triallylamine, CA 102-70-5: 312.59* (7).

$C_9H_{15}NO$

l-Morpholincyclopentene, CA 936-52-7: 333.15 (1); 337.15 (3).

$C_9H_{15}NO_2$

Diacetone acrylamide, CA 2873-97-4: Above 385.93 (1).

Ethyl l-methyl-1,2,3,6-tetrahydro-4-pyridinecarboxylate, technical,
CA 40175-06-2: 374.26 (1).

$C_9H_{15}NO_3$

Ethyl l-piperidineglyoxylate, CA 53074-96-7: Above 385.93 (1).

$C_9H_{15}NO_6$

Dimethyl 4-nitroheptanedioate, crude, CA 7766-83-8: 332.04 (1).

Triethyl nitrilotricarboxylate, CA 3206-31-3: Above 385.93 (1).

C_9H_{16}

Hydrindan, mixed isomers, CA 496-10-6: 315.15 (3).

l-Isopropyl-l-cyclohexene, CA 4292-04-0: 298.71 (1).

1,8-Nonadiene, CA 4900-30-5: 299.26 (1).

l-Nonyne, CA 3452-09-3: 307.04 (1).

1,3,5-Trimethyl-l-cyclohexene, mixed isomers, CA 3643-64-9: 296.48 (1).

$C_9H_{16}N_2$

N-(2-Cyanoethyl)cyclohexylamine: 397.04* (6).

1,8-Diazabicyclo(5,4,0)undec-7-ene, CA 6674-22-2: Above 385.93 (1).

$C_9H_{16}O$

Cyclononanone, CA 3350-30-9: 299.15 (3); 338.71 (1).

Cyclooctanecarboxyaldehyde, technical, CA 6688-11-5: 340.37 (1).

2,4-Dimethyl-2,6-heptadien-1-ol, mixed isomers, CA 80192-56-9: 351.48 (1).

3-Methyl-2-norbornanemethanol ±, CA 6968-75-8: 363.71 (1).

2-Nonenal, *trans*, CA 18829-56-6: 357.59 (1).

3-Nonen-2-one, CA 14309-57-0: 354.82 (1).

3-Nonyn-1-ol, CA 31333-13-8: 367.04 (1).

2,2,6-Trimethylcyclohexanone, CA 2408-37-9: 324.82 (1).

3,3,5-Trimethylcyclohexanone, CA 873-94-9: 337.15 (3).

3,5,5-Trimethyl-2-cyclohexen-1-ol, CA 470-99-5: 353.15 (1).

$C_9H_{16}O_2$

Allyl hexanoate, CA 123-68-2: 338.71 (6).

Cyclohexanepropionic acid, CA 701-97-3: Above 385.93 (1).

1,7-Dioxaspiro(5,5)undecane, CA 180-84-7: 337.04 (1).

2-Ethylbutyl acrylate, CA 3953-10-4: 324.82* (6,7); 325.15 (5).

Methyl cyclohexylacetate, CA 14352-61-5: 347.59 (1).

(Methylcyclohexyl) acetate, mixed isomers, CA 30232-11-2: 337.15 (5,6); 339.15 to 342.15* (2).

gamma-Nonanoic lactone, CA 104-61-0: Above 385.93 (1).

$C_9H_{16}O_3$

Butyl levulinate, CA 2052-15-5: 364.82 (1).

Ethyl 2-isopropylacetoacetate, CA 1522-46-9: 349.15 (3).

Ethyl pivaloylacetate, CA 17094-34-7: 351.15 (3).

7-Oxononanoic acid, CA 20356-92-7: Above 385.93 (1).

$C_9H_{16}O_4$

Diethyl dimethylmalonate, CA 1619-62-1: 344.26 (1).

Diethyl ethylmalonate, CA 133-13-1: 361.48 (1); 365.15 (3).

Diethyl glutarate, CA 818-38-2: 369.26 (1).

Dimethyl diethylmalonate, CA 27132-23-6: 353.15 (1).

Dimethyl pimelate, CA 1732-08-7: Above 385.93 (1).

Ethyl 3,3-diethoxyacrylate, CA 32002-24-7: 375.93 (1).

Suberic acid monomethyl ester, CA 3946-32-5: Above 385.93 (1).

2,2'-Trimethylenebis-1,3-dioxolane, CA 6543-04-0: Above 385.93 (1).

$C_9H_{16}O_5$

Diethyl 3-hydroxyglutarate, CA 32328-03-3: Above 385.93 (1).

$C_9H_{16}O_6$

Diethyl bis(hydroxymethyl)malonate, CA 20605-01-0: Above 385.93 (1).

$C_9H_{17}BrO_2$

Ethyl 2-bromoheptanoate ±, CA 5333-88-0: 377.59 (1).

$C_9H_{17}Cl$

1-Chloro-3-cyclohexylpropane, CA 1124-62-5: 363.15 (3).

$C_9H_{17}ClO$

Nonanoyl chloride, CA 764-85-2: 368.15 (1).

$C_9H_{17}ClOS$

Octyl chlorothioformate, CA 13889-96-8: 399.82 (2).

$C_9H_{17}ClO_2$

2-Ethylhexyl chloroformate ±, CA 24468-13-1: 354.82 (1); 359.15 (4).
Octyl chloroformate, CA 7452-59-7: 348.71 (1).

$C_9H_{17}N$

Allylcyclohexylamine, CA 6628-00-8: 326.48 (1).
Decahydroquinoline, mixed isomers, CA 2051-28-7: 341.48 (1).
Octyl cyanide, CA 2243-27-8: 354.26 (1); 355.15 (3).
Perhydroisoquinoline, mixed isomers, CA 6329-61-9: 345.93 (1).
1,1,3,3-Tetramethylbutyl isocyanate, CA 14542-93-9: 324.82 (1).

$C_9H_{17}NO$

1-(3-Methylbutyryl)pyrrolidine, CA 60026-17-7: 377.59 (1).

$C_9H_{17}NO_2$

2-(Diethylamino)ethyl acrylate, CA 2426-54-2: 341.15 (3); 363.71* (6).
Ethyl 1-methylnipecotate, CA 5166-67-6: 341.48 (1); 355.15 (3).
Ethyl 1-methylpipecolinate, CA 30727-18-5: 347.04 (1).

$C_9H_{17}NS$

tert-Octyl isothiocyanate, CA 17701-76-7: 343.15 (1).

C_9H_{18}

2,6-Dimethyl-3-heptene, CA 2738-18-3: 288.59 (7).
Isopropylcyclohexane, CA 696-29-7: 308.71 (1,2); 351.15 (3).
1-Nonene, CA 124-11-8: 297.15 (3); 298.71* (6); 319.26 (1,2).
2-Nonene, trans, CA 6634-78-2: 305.37 (1)
3-Nonene, trans, CA 20063-92-7: 305.37 (1).
4-Nonene, mixed isomers, CA 2198-23-4: 300.37 (1); 303.15 (3).
Propylene trimer, mixed isomers, CA 13987-01-4: 297.04* (6).
1,1,3-Trimethylcyclohexane, CA 3073-66-3: 291.15 (3).
1,2,4-Trimethylcyclohexane, mixed isomers, CA 2234-75-5: 292.15 (3).
1,3,5-Trimethylcyclohexane, mixed isomers, CA 1839-63-0: 292.15 (3).

$C_9H_{18}B_2O_6$

Trimethylene borate, CA 20905-35-5: 383.15 (1).

$C_9H_{18}Br_2$

1,9-Dibromononane, CA 4549-33-1: Above 385.93 (1).

$C_9H_{18}ClN_2OP$

2-Cyanoethyl N,N-diisopropylchlorophosphoramidochloride, CA 89992-70-1: 383.15 (1).

$C_9H_{18}Cl_2$

1,9-Dichlorononane, CA 821-99-8: Above 385.93 (1).

$C_9H_{18}F_3NOSi$

N-tert-Butyldimethylsilyl-N-methyltrifluoroacetamide, CA 77377-52-7: 318.15 (3); 325.93 (1).

$C_9H_{18}N_2$

1,3-Di-tert-butylcarbodiimide, CA 691-24-7: 308.15 (1,3).

1,2,5,6-Tetrahydro-2,2,4,6,6-pentamethylpyrimidine, CA 556-72-9: 326.48 (1).

$C_9H_{18}N_2Si$

1-(tert-Butyldimethylsilyl)imidazole, CA 54925-64-3: 354.15 (3); 372.59 (1).

$C_9H_{18}O$

Cyclooctanemethanol, CA 3637-63-6: 366.48 (1).

2,6-Dimethyl-4-heptanone, technical, CA 108-83-8: 322.04 (1,2,3,4,5,6); 322.15* (4); 328.15* (2); 333.15 (4,7).

5-Methyl-2-octanone, CA 58654-67-4: 333.15 (2,6); 344.26* (2).

2-Nonanone, CA 821-55-6: 337.59 (1,4); 341.15 (3); 344.26 (7).

3-Nonanone, CA 925-78-0: 339.15 (3); 340.93 (1).

5-Nonanone, CA 502-56-7: 333.71 (1); 338.15 (3).

3-Nonen-1-ol, cis, CA 10340-23-5: 335.37 (1).

Nonyl aldehyde, CA 124-19-6: 337.04 (1); 350.15 (3).

2,2,4,4-Tetramethyl-3-pentanone, CA 815-24-7: 305.93 (1); 307.15 (3).

3,3,5-Trimethylcyclohexanol, CA 116-02-9: 347.15 (5); 360.93* (6,7).

Trimethylcyclohexanol, mixed isomers, CA 1321-60-4: 347.04* (6).

$C_9H_{18}OSi$

1-Trimethylsiloxycyclohexene, CA 6651-36-1: 314.15 (3); 315.37 (1).

$C_9H_{18}O_2$

Butyl 2-methylbutyrate, CA 51115-64-1: 331.15 (3).   *(continues)*

$C_9H_{18}O_2$ *(continued)*

Butyl isovalerate, CA 109-19-3:   326.15 (5,6).

Ethyl heptanoate, CA 106-30-9:   339.26 (1); 347.15 (3).

Heptyl acetate, CA 112-06-1:   340.93 (1).

Isopentyl butyrate, CA 106-27-4:   332.15 (5,6).

Methyl octanoate, CA 111-11-5:   342.15 (3); 345.93 (1).

Nonanoic acid, CA 112-05-0:   373.15 (1).

Pentyl butyrate, CA 540-18-1:   330.37 (6).

$C_9H_{18}O_3$

3-(2-Ethylbutoxy)propionic acid, CA 10213-74-8:   410.93* (6,7).

$C_9H_{18}O_4$

Ethyl 3,3-diethoxypropionate, CA 10601-80-6:   357.59 (1).

$C_9H_{18}O_5$

Triethylene glycol, monomethyl ether acetate, CA 3610-27-3:   399.82* (6,7).

$C_9H_{19}Br$

1-Bromononane, CA 693-58-3:   363.15 (1,3).

$C_9H_{19}BrO$

Bromoethyl octyl ether, CA 96384-68-8:   381.48 (1).

$C_9H_{19}Cl$

1-Chlorononane, CA 2473-01-0:   347.59 (1).

$C_9H_{19}ClO$

Chloromethyl octyl ether, CA 24566-90-3:   362.04 (1).

$C_9H_{19}F$

1-Fluorononane, CA 463-18-3:   322.04 (1).

$C_9H_{19}F_3O_3SSi$

Dimethylhexylsilyl trifluoromethanesulfonate: 360.93 (1).

$C_9H_{19}I$

1-Iodononane, technical, CA 4282-42-2:   358.15 (1).

$C_9H_{19}N$

N-Isopropylcyclohexylamine, CA 1195-42-2:   307.04 (1); 307.04* (6); 319.15 (3).

2,2,6,6-Tetramethylpiperidine, CA 768-66-1:   297.59 (1).

3,3,5-Trimethylhexahydroazepine ±, CA 35466-89-8:   340.37 (1).

$C_9H_{19}NO$

N,N-Dibutylformamide, CA 761-65-9:   373.71 (1).

5-Diethylamino-2-pentanone, CA 105-14-6:   338.71 (1).

N,N-Diisobutylformamide, CA 2591-76-6:   363.15 (3).

$C_9H_{19}N_2OP$

Bis(pyrrolidino)methoxyphosphine, CA 89983-14-2:   338.15 (1).

$C_9H_{19}O_4P$

Dimethyl (2-oxoheptyl)phosphonate, technical, CA 36969-89-8:   Above 385.93 (1).

$C_9H_{19}O_5P$

Triethyl 2-phosphonopropionate, CA 3699-66-9:   362.04 (1).

C_9H_{20}

3,3-Diethylpentane, CA 1067-20-5:   Below 294.15 (7).

2,5-Dimethylheptane, CA 2216-30-0:   297.15 (7).

3,5-Dimethylheptane, CA 926-82-9:   296.21 (7).

4,4-Dimethylheptane, CA 1068-19-5:   294.15 (7).

3-Ethyl-2,3-dimethylpentane, CA 16747-33-4:   281.15 (7).

3-Ethyl-4-methylhexane, CA 3074-77-9:   297.15 (5,6).

4-Ethyl-2-methylhexane, CA 3074-75-7:   Below 294.15 (5,6).

2-Methyloctane, CA 3221-61-2:   299.15 (3).

3-Methyloctane, CA 2216-33-3:   295.15 (3).

4-Methyloctane, CA 2216-34-4:   293.15 (3).

Nonane, CA 111-84-2:   304.26 (1,3,5,6,7).

2,2,3,3-Tetramethylpentane, CA 7154-79-2:   Below 294.15 (5,6).

2,2,4,4-Tetramethylpentane, CA 1186-53-4:   Below 294.15 (5,6).

2,2,4-Trimethylhexane, CA 16747-26-5:   286.15 (3).

2,2,5-Trimethylhexane, CA 3522-94-9:   286.15 (2,5); 286.15* (6).

2,3,3-Trimethylhexane, CA 16747-28-7:   299.15 (5).

2,3,4-Trimethylhexane, CA 921-47-1:   300.15 (5).

2,3,5-Trimethylhexane, CA 1069-53-0:   286.15 (3).

3,3,4-Trimethylhexane, CA 16747-31-2:   299.15 (5).

$C_9H_{20}N_2$

1-(3-Aminopropyl)-2-pipecoline, CA 25560-00-3: 361.48 (1).

4-Amino-2,2,6,6-tetramethylpiperidine, CA 36768-62-4: 345.37 (1).

5-Amino-2,2,4-trimethyl-1-cyclopentanemethylamine, CA 67907-32-8: 370.37 (1).

N-Cyclohexyl-1,3-propanediamine, CA 3312-60-5: 352.59* (6); 374.26 (1).

$C_9H_{20}N_2O$

Tetraethylurea, CA 1187-03-7: 354.15 (3).

$C_9H_{20}O$

2,6-Dimethyl-4-heptanol, technical, CA 108-82-7: 339.26 (1); 347.04 (6,7).

1-Nonanol, CA 143-08-8: 348.71 (1); 353.15* (2).

2-Nonanol, CA 628-99-9: 355.37 (1); 369.15 (3).

3-Nonanol, CA 624-51-1: 369.15 (3).

5-Nonanol, CA 623-93-8: 350.15 (3).

3,5,5-Trimethyl-1-hexanol, technical, CA 3452-97-9: 354.26 (1); 366.15 (5); 366.48* (2,6).

$C_9H_{20}O_2$

2-Butyl-2-ethyl-1,3-propanediol, CA 115-84-4: Above 385.93 (1); 410.93* (6).

Dibutoxymethane, CA 2568-90-3: 333.15 (5,6).

1,9-Nonanediol, CA 3937-56-2: Above 385.93 (1).

$C_9H_{20}O_2Si$

tert-Butyl trimethylsilylacetate, CA 41108-81-0: 315.15 (3).

$C_9H_{20}O_3$

1-(Butoxyethoxy)-2-propanol, CA 124-16-3: 394.26* (6,7).

Diethylene glycol, *tert*-butyl methyl ether, CA 52788-79-1: 345.93 (1).

Dipropylene glycol, isopropyl ether, CA 54518-03-5: 363.71* (2).

3-Ethoxypropionaldehyde diethyl acetal, CA 7789-92-6: 326.48 (1).

Triethyl orthoproprionate, CA 155-80-0: 263.15 (1); 312.15 (3).

$C_9H_{20}O_3Si$

Allyltriethoxysilane, CA 2550-04-1: 294.26 (1).

$C_9H_{20}O_4$

Tetraethyl orthocarbonate, CA 78-09-1: 305.15 (3); 325.93 (1).

Tripropylene glycol, CA 1638-16-0: 413.71 (2,6,7); 413.71* (2).

$C_9H_{20}O_4Si_2$

Bis(trimethylsilyl) malonate, CA 18457-04-0: Below 293.15 (3); 297.04 (1).

$C_9H_{20}O_5Si$

3-Glycidoxypropyltrimethoxysilane, CA 2530-83-8: 383.15 (1).

$C_9H_{20}S$

Nonyl mercaptan, CA 1455-21-6: 352.04 (1).
tert-Nonyl mercaptan, CA 25360-10-5: 338.71* (2); 340.93* (6).

$C_9H_{20}S_2$

1,9-Nonanedithiol, CA 3489-28-9: 374.82 (1).

$C_9H_{21}BO_3$

Triisopropyl borate, CA 5419-55-6: 290.37 (1); 301.15 (5,6,7).
Tripropyl borate, CA 688-71-1: 305.37 (1); 341.48* (7).

$C_9H_{21}BO_6$

Tris(2-methoxyethyl) borate, CA 14987-42-7: 360.37 (1).

$C_9H_{21}ClO_3Ti$

Chlorotitanium triisopropoxide, CA 20717-86-6: 295.37 (1).

$C_9H_{21}ClSi$

Chlorotripropylsilane, CA 995-25-5: 336.15 (3); 343.15 (1).
Triisopropylsilyl chloride, CA 13154-24-0: 335.93 (1); 337.15 (3).

$C_9H_{21}N$

N,N-Dibutylmethylamine, CA 3405-45-6: 324.82* (7).
3-Dimethylamino-2,4-dimethylpentane, CA 54561-96-5: 307.15 (3).
Nonylamine, CA 112-20-9: 335.93 (1); 347.15 (3).
Tripropylamine, CA 102-69-2: 302.04 (2); 309.82 (1); 313.71* (6,7);
314.15 (3).

$C_9H_{21}NO_2$

3-Diisopropylamino-1,2-propanediol, CA 60302-96-7: 383.15 (1).
N,N-Dimethylformamide diisopropyl acetal, CA 18503-89-4: 296.48 (1);
298.15 (3).
N,N-Dimethylformamide dipropyl acetal, CA 6006-65-1: 310.93 (1);
319.15 (3).
3-Dipropylamino-1,2-propanediol, CA 60302-96-7: 383.15 (1).

$C_9H_{21}NO_3$

Triisopropanolamine, CA 122-20-3:   383.15 (1); 424.82* (2); 425.15 (5); 433.15* (2,6,7).

$C_9H_{21}N_3$

1,3,5-Triethylhexahydro-1,3,5-triazine, CA 7779-27-3:   343.15 (1).

$C_9H_{21}O_3P$

Triisopropyl phosphite, technical, CA 116-17-6:   347.04 (1); 347.04* (2).

$C_9H_{21}O_4P$

Tripropyl phosphate, CA 513-08-6:   Above 385.93 (1).

$C_9H_{21}O_5PSi$

Diethyl (trimethylsilyloxycarbonylmethyl)-phosphonate, CA 66130-90-3: 291.15 (3); 317.59 (1).

$C_9H_{22}N_2$

2-Amino-5-diethylaminopentane, CA 140-80-7:   341.48 (1).

1,9-Diaminononane, CA 646-24-2:   Above 385.93 (1).

$C_9H_{22}N_2O$

tert-Butoxybis(dimethylamino)methane, CA 5815-08-7:   314.15 (3).

$C_9H_{22}O_3Si$

Propyltriethoxysilane, CA 2550-02-9:   323.15 (4).

$C_9H_{22}Si$

Triisopropylsilane, CA 6485-79-6:   310.37 (1).

Tripropylsilane, CA 998-29-8:   316.48 (1).

$C_9H_{23}NO_3Si$

3-Aminopropyltriethoxysilane, CA 919-30-2:   371.15 (3); 377.59 (1).

$C_9H_{23}NSi$

N,N-Diisopropyltrimethylsilylamine, CA 17425-88-6:   301.48 (1).

$C_9H_{24}N_4$

N,N'-Bis(3-aminopropyl)-1,3-propanediamine, CA 38983-98-1:   383.15 (1).

$C_9H_{27}O_4PSi$

Tris(trimethylsilyl) phosphate, CA 10497-05-9: 297.59 (1); 312.15 (3).

$C_{10}F_{18}$

Perfluorodecalin, mixed isomers, CA 306-94-5: 313.71 (1).

$C_{10}H_5F_6N$

3,5-Bis(trifluoromethyl)phenylacetonitrile: Above 385.93 (1).

$C_{10}H_6F_6O$

3',5'-Bis(trifluoromethyl)acetophenone, CA 30071-93-3: 355.37 (1).

$C_{10}H_7Br$

1-Bromonaphthalene, CA 90-11-9: Above 385.93 (1).
2-Bromonaphthalene, CA 580-13-2: Above 385.93 (1).

$C_{10}H_7Cl$

1-Chloronaphthalene, CA 90-13-1: 394.26 (1,6).

$C_{10}H_7F$

1-Fluoronaphthalene, CA 321-38-0: 338.71 (1).

$C_{10}H_7F_3O_2$

4,4,4-Trifluoro-1-phenyl-1,3-butanedione, CA 326-06-7: 372.04 (1).

$C_{10}H_7I$

1-Iodonaphthalene, CA 90-14-2: Above 385.93 (1).

$C_{10}H_7NO$

4-Quinolinecarboxaldehyde, CA 4363-93-3: Above 385.93 (1).

$C_{10}H_7NO_2$

1-Nitronaphthalene, CA 86-57-7: 437.15 (5,6,7).

$C_{10}H_8$

Naphthalene, CA 91-20-3: 352.04 (1,4,6); 352.04* (7); 353.15 (5).

$C_{10}H_8ClN$

4-Chloroquinaldine, CA 4295-06-1: Above 385.93 (1).

$C_{10}H_8F_3NO$

alpha-Methoxy-alpha-(trifluoromethyl)phenylacetonitrile ±, CA 80866-87-1: 340.37 (1).

$C_{10}H_8N_2$

3,3'-Dipyridyl, CA 581-46-4:   Above 385.93 (1).

3-Indolylacetonitrile, CA 771-51-7:   Above 385.93 (1).

1,3-Phenylenediacetonitrile, CA 626-22-2:   Above 385.93 (1).

4-Phenylpyrimidine, CA 3438-48-0:   Above 385.93 (1).

$C_{10}H_8N_2O_2$

meta-Xylylene diisocyanate, CA 3634-83-1:   315.15 (3).

$C_{10}H_8O$

1,4-Epoxy-1,4-dihydronaphthalene, CA 573-57-9:   367.04 (1).

2-Naphthol, CA 135-19-3:   426.15 (5,6,7); 434.15 (4).

4-Phenyl-3-butyn-2-one, CA 1817-57-8:   368.15 (1).

$C_{10}H_8O_2S$

3-Acetoxythianaphthene, CA 24434-82-0:   Above 385.93 (1).

$C_{10}H_9ClO$

4-Chlorophenyl cyclopropyl ketone, CA 6640-25-1:   Above 385.93 (1).

2-Phenyl-1-cyclopropanecarboxylic acid chloride, trans, CA 939-87-7: Above 385.93 (1).

$C_{10}H_9ClO_3$

O-Acetylmandelic acid chloride, CA 1638-63-7:   Above 385.93 (1).

$C_{10}H_9Cl_2N$

2-Chloro-3-(3-chloro-ortho-tolyl)propionitrile, CA 21342-85-8:   Above 385.93 (1).

$C_{10}H_9FO_2$

gamma-(4-Fluorophenyl)-gamma-butyrolactone, CA 51787-96-3:   Above 385.93 (1).

$C_{10}H_9FO_3$

Methyl 4-fluorobenzoylacetate, CA 63131-29-3:   Above 385.93 (1).

$C_{10}H_9F_3O$

3-(Trifluoromethyl)phenylacetone, CA 21906-39-8:   362.04 (1).

$C_{10}H_9F_3O_3$

alpha-Methoxy-alpha-(trifluoromethyl)phenylacetic acid, CA 56135-03-6: Above 385.93 (1).

$C_{10}H_9N$

1-Aminonaphthalene, CA 134-32-7: Above 385.93 (1); 430.15 (4,5,6,7).

1-Methylisoquinoline, CA 1721-93-3: Above 385.93 (1).

4-Methylquinoline, CA 491-35-0: Above 385.93 (1).

6-Methylquinoline, CA 91-62-3: Above 385.93 (1).

7-Methylquinoline, CA 612-60-2: 383.15 (1).

8-Methylquinoline, CA 611-32-5: 378.15 (1).

1-Phenyl-1-cyclopropanecarbonitrile, CA 935-44-4: Above 385.93 (1).

Quinaldine, CA 91-63-4: 352.59 (1).

$C_{10}H_9NO$

4-Methoxycinnamonitrile, mixed isomers, CA 28446-68-6: Above 385.93 (1).

6-Methoxyquinoline, CA 5263-87-6: Above 385.93 (1).

2-Phenylcyclopropyl isocyanate, trans, technical, CA 63009-74-5: 380.37 (1).

$C_{10}H_9NO_2$

Ethyl 4-cyanobenzoate, CA 7153-22-2: Above 385.93 (1).

$C_{10}H_9NO_3$

Ethyl 4-isocyanatobenzoate, CA 30806-83-8: Above 385.93 (1).

$C_{10}H_9NS$

4-Benzylisothiazole, CA 36412-26-7: 383.15 (1).

$C_{10}H_{10}$

1,2-Dihydronaphthalene, technical, CA 447-53-0: 340.37 (1).

1,3-Divinylbenzene, CA 108-57-6: 347.15 (5).

Divinyl benzene, mixed isomers, technical, CA 1321-74-0: 334.82 (1); 349.26* (6).

$C_{10}H_{10}BrClO$

4'-Bromo-4-chlorobutyrophenone, CA 4559-96-0: Above 385.93 (1).

$C_{10}H_{10}ClFO$

4-Chloro-4'-fluorobutyrophenone, CA 3874-54-2: 383.15 (1).

$C_{10}H_{10}ClNO_2$

4'-Chloroacetoacetanilide, CA 101-92-8: 433.15 (7); 449.82* (7).

$C_{10}H_{10}Cl_2$

(2,2-Dichloro-1-methylcyclopropyl)benzene, CA 3591-42-2: 368.15 (1).

$C_{10}H_{10}N_2$

3-Methyl-1-phenylpyrazole, CA 1128-54-7: Above 385.93 (1).

$C_{10}H_{10}O$

Cyclopropyl phenyl ketone, CA 3481-02-5: 363.71 (1).

1,4-Epoxy-1,2,3,4-tetrahydronaphthalene, CA 35185-96-7: 363.15 (1).

alpha-Methylcinnamaldehyde, CA 101-39-3: 352.59 (1).

4-Phenyl-3-buten-2-one, trans, CA 122-57-6: 338.71 (1).

2-Phenyl-3-butyn-2-ol, CA 127-66-2: 369.26 (1).

alpha-Tetralone, CA 529-34-0: Above 385.93 (1); 403.15* (4).

beta-Tetralone, CA 530-93-8: Above 385.93 (1).

$C_{10}H_{10}O_2$

1,2-Diacetylbenzene, CA 704-00-7: Above 385.93 (1).

1,3-Diacetylbenzene, CA 6781-42-6: 383.15 (1).

Isosafrole, CA 120-58-1: Above 385.93 (1).

2-Methoxycinnamaldehyde, CA 1504-74-1: Above 385.93 (1).

Methyl cinnamate, trans, CA 103-26-4: Above 385.93 (1).

gamma-Phenyl-gamma-butyrolactone, CA 1008-76-0: Above 385.93 (1).

Safrole, CA 94-59-7: 370.93 (1); 373.15 (3,6).

$C_{10}H_{10}O_2S_2$

Furfuryl disulfide, CA 4437-20-1: Above 385.93 (1).

$C_{10}H_{10}O_3$

Ethyl benzoylformate, CA 1603-79-4: 383.15 (1).

Methyl-5-norbornene-2,3-dicarboxylic anhydride, CA 25134-21-8: Above 385.93 (1).

$C_{10}H_{10}O_4$

Dimethyl isophthalate, CA 1459-93-4: 410.93 (6).

Dimethyl phthalate, CA 131-11-3: 419.26 (1,5,6,7); 430.15* (4); 433.15 (2).

Dimethyl terephthalate, CA 120-61-6: 419.15 to 420.15* (4); 426.15* (4,6).

Resorcinol diacetate, CA 108-58-7: 383.15 (1).

$C_{10}H_{11}BrO$

  2-Bromoisobutyrophenone, CA 10409-54-8: Above 385.93 (1).

$C_{10}H_{11}ClO$

  4-Chlorobutyrophenone, CA 939-52-6: Above 385.93 (1).

  4-Propylbenzoyl chloride, CA 52710-27-7: Above 385.93 (1).

$C_{10}H_{11}ClO_3$

  2,5-Dimethoxyphenylacetyl chloride, CA 52711-92-9: 380.93 (1).

  3,4-Dimethoxyphenylacetyl chloride, CA 10313-60-7: Above 385.93 (1).

$C_{10}H_{11}F_7O_2$

  2,2-Dimethyl-6,6,7,7,8,8,8-heptafluoro-3,5-octanedione, CA 17587-22-3: 311.48 (1); 326.15 (3).

$C_{10}H_{11}N$

  1,2-Dimethylindole, CA 875-79-6: Above 385.93 (1).

  2-Phenylbutyronitrile ±, CA 769-68-6: 378.15 (1).

  4-Phenylbutyronitrile, CA 2046-18-6: Above 385.93 (1).

$C_{10}H_{11}NO_2$

  Acetoacetanilide, CA 102-01-2: 435.93 (1); 458.15* (6,7).

$C_{10}H_{12}$

  Dicyclopentadiene, CA 77-73-6: 299.82 (1); 305.15 (3); 305.37* (6,7).

  2,5-Dimethylstyrene, CA 2039-89-6: 337.04 (1).

  2-Methyl-1-phenyl-1-propene, CA 768-49-0: 332.04 (1); 355.15 (3).

  1-Phenyl-2-butene, CA 1560-06-1: 344.15 (2,5); 344.26* (6).

  1,2,3,4-Tetrahydronaphthalene, CA 119-64-2: 344.15 (4,6); 348.15 (3); 350.15 (1,5); 350.37* (6).

$C_{10}H_{12}ClN_2O_5P$

  4-Nitrophenyl 4-morpholinylphosphonochloridate, CA 79838-05-4: Above 385.93 (1).

$C_{10}H_{12}N_2$

  3-(Benzylamino)propionitrile, CA 706-03-6: Above 385.93 (1).

  1-Ethyl-2-methylbenzimidazole, CA 5805-76-5: Above 385.93 (1).

$C_{10}H_{12}N_2O$

  Cotinine (-), CA 79838-05-4: Above 385.93 (1).

$C_{10}H_{12}O$

4-Allylanisole, CA 140-67-0: 354.26 (1).

2-Allyl-4-methylphenol, technical, CA 6628-06-4: 374.82 (1).

2-Allyl-6-methylphenol, CA 3354-58-3: 367.59 (1).

Anethole, *trans*, CA 4180-23-8: 363.71 (1).

Benzylacetone, CA 2550-26-7: 371.48 (1).

Butyrophenone, CA 495-40-9: 362.04 (1); 370.15 (3).

Cyclopropyl phenyl carbinol, CA 31729-66-5: 385.37 (1).

3',4'-Dimethylacetophenone, CA 3637-01-2: 375.93 (1).

4-Ethoxystyrene, technical, CA 5459-40-5: 360.93 (1).

3'-Ethylacetophenone, CA 22699-70-3: 356.48 (1).

4'-Ethylacetophenone, CA 937-30-4: 363.71 (1).

Isobutyrophenone, CA 611-70-1: 357.59 (1); 362.15 (3).

4-Isopropylbenzaldehyde, CA 122-03-2: 366.48 (1); 370.15 (3).

Mesitaldehyde, CA 487-68-3: 378.71 (1).

2-Methyl-3-phenyl-2-propen-1-ol, *trans*, CA 1504-55-8: Above 385.93 (1).

4'-Methylpropiophenone, CA 5337-93-9: 369.26 (1).

1-Phenyl-2-butanone, CA 1007-32-5: 363.71 (1).

3-Phenylbutyraldehyde ±, CA 16251-77-7: 369.82 (1).

1-Phenyl-1-cyclopropanemethanol, CA 1007-03-0: Above 385.93 (1).

1,2,3,4-Tetrahydro-1-naphthol, CA 529-33-9: Above 385.93 (1).

5,6,7,8-Tetrahydro-2-naphthol, CA 1125-78-6: Above 385.93 (1).

$C_{10}H_{12}O_2$

3,4-Dimethoxystyrene, technical, CA 6380-23-0: Above 385.93 (1).

2,3-Dimethyl-*para*-anisaldehyde, CA 38998-17-3: Above 385.93 (1).

2,5-Dimethyl-*para*-anisaldehyde, CA 6745-75-1: Above 385.93 (1).

Ethyl 2-methylbenzoate, CA 87-24-1: 364.82 (1).

Ethyl 3-methylbenzoate, CA 120-33-2: 374.26 (1).

Ethyl 4-methylbenzoate, CA 94-08-6: 372.59 (1).

Ethyl phenylacetate, CA 101-97-3: 350.93 (1); 372.04 (6).

Eugenol, CA 97-53-0: Above 385.93 (1).

Isoeugenol, mixed isomers, CA 97-54-1: Above 373.15 (6); above 385.93 (1).

Isopropyl benzoate, CA 939-48-0: 372.04 (6); 372.04* (7).

1-Methoxy-2-indanol, CA 56175-44-1: Above 385.93 (1).

2-Methoxyphenylacetone, CA 5211-62-1: Above 385.93 (1).

4-Methoxyphenylacetone, CA 122-84-9: 374.82 (1).

4'-Methoxypropiophenone, CA 121-97-1: 334.26 (1).

alpha-Methylbenzyl acetate ±, CA 50373-55-2: 363.71 (6); 364.26 (1).

2-Methyl-3-phenylglycidol, CA 4426-63-5: Above 385.93 (1).

(continues)

$C_{10}H_{12}O_2$ *(continued)*

Phenethyl acetate, CA 103-45-7: 374.82 (1); 383.15 (5,7); 383.15* (6).

Phenylacetaldehyde ethylene acetal, CA 101-49-5: 380.93 (1).

2-Phenylbutyric acid, CA 90-27-7: Above 385.93 (1).

4-Phenylbutyric acid, CA 1821-12-1: Above 385.93 (1).

4-Phenyl-1,3-dioxane, CA 772-00-9: Above 385.93 (1).

Thymoquinone, CA 490-91-5: 377.04 (1).

$C_{10}H_{12}O_3$

2',4'-Dimethoxyacetophenone, CA 829-20-9: Above 385.93 (1).

2',5'-Dimethoxyacetophenone, CA 1201-38-3: Above 385.93 (1).

3',4'-Dimethoxyacetophenone, CA 1131-62-0: Above 385.93 (1).

3',5'-Dimethoxyacetophenone, CA 39151-19-4: Above 385.93 (1).

2,4-Dimethoxy-3-methylbenzaldehyde, CA 7149-92-0: Above 385.93 (1).

2,3-Epoxypropyl 4-methoxyphenyl ether, CA 2211-94-1: Above 385.93 (1).

3-Ethoxy-4-methoxybenzaldehyde, CA 1131-52-8: Above 385.93 (1).

Ethyl mandelate ±, CA 4358-88-7: Above 385.93 (1).

Methyl 4-methoxyphenylacetate, CA 23786-14-3: 309.82 (1).

2-Phenoxyethyl acetate, CA 6192-44-5: 416.45* (4).

$C_{10}H_{12}O_4$

Diallyl fumarate, CA 2807-54-7: 347.15* (4).

Diallyl maleate, technical, CA 999-21-3: Above 385.93 (1); 396.15* (4).

Ethyl isodehydracetate, CA 3385-34-0: Above 385.93 (1).

Methyl 3,5-dimethoxybenzoate, CA 2150-37-0: Above 385.93 (1).

2,3,4-Trimethoxybenzaldehyde, CA 2103-57-3: Above 385.93 (1).

$C_{10}H_{12}O_4S$

Glycidyl tosylate, CA 6746-81-2: Above 385.93 (1).

$C_{10}H_{12}O_5$

3,4-Bis(acetoxymethyl)furan, CA 30614-73-4: Above 385.93 (1).

Diethyl 3,4-furandicarboxylate, CA 30614-77-8: 355.37 (1).

$C_{10}H_{13}Br$

1-Bromo-4-*tert*-butylbenzene, CA 3972-65-4: 370.37 (1).

$C_{10}H_{13}BrO$

4-Phenoxybutyl bromide, CA 1200-03-9: Above 385.93 (1).

$C_{10}H_{13}BrO_2$

1-(4-Bromophenoxy)-1-ethoxyethane, technical: 379.26 (1).

$C_{10}H_{13}Cl$

1-Chloro-2-methyl-2-phenylpropane, CA 515-40-0: 365.37 (1).
2,4,5-Trimethylbenzyl chloride, CA 1585-16-6: 379.26 (1).

$C_{10}H_{13}ClO$

4-*tert*-Butyl-2-chlorophenol, CA 98-28-2: 380.37 (6).
4-Phenoxybutyl chloride, CA 2651-46-9: Above 385.93 (1).

$C_{10}H_{13}FN_2$

1-(4-Fluorophenyl)piperazine, CA 2252-63-3: Above 385.93 (1).

$C_{10}H_{13}N$

1-Amino-5,6,7,8-tetrahydronaphthalene, CA 2217-41-6: Above 385.93 (1).
2,3-Cycloheptenopyridine, CA 7197-96-8: 366.48 (1).
3-Methyl-5,6,7,8-tetrahydroquinoline, CA 28712-62-1: 370.93 (1).
1,2,3,4-Tetrahydro-1-naphthylamine, CA 2217-40-5: Above 385.93 (1).

$C_{10}H_{13}NO$

N-Ethylacetanilide, CA 529-65-7: 325.37 (6).
4-Phenylmorpholine, CA 92-53-5: 377.59* (6); above 385.93 (1).

$C_{10}H_{13}NO_2$

2-Nitro-*para*-cymene, technical, CA 943-15-7: 383.15 (1).

$C_{10}H_{13}NO_3$

4-(4-Nitrophenyl)-1-butanol, CA 79524-20-2: Above 385.93 (1).

$C_{10}H_{13}NO_4$

2,5-Diethoxynitrobenzene, CA 119-23-3: Above 385.93 (1).

$C_{10}H_{14}$

Butylbenzene, CA 104-51-8: 327.15 (3); 332.59 (1); 344.26 (2); 344.26* (6,7).
sec-Butylbenzene, CA 135-98-8: 318.71 (1); 325.15 (2,3,5,6); 325.15* (7).
tert-Butylbenzene, CA 98-06-6: 307.59 (1); below 328.15 (5); 333.15 (2,3); 333.15* (6,7).
ortho-Cymene, CA 527-84-4: 323.71 (1).

(continues)

$C_{10}H_{14}$ *(continued)*

meta-Cymene, CA 535-77-3:   320.93 (1).

para-Cymene, CA 99-87-6:   320.37 (1,3,5,6,7).

1,2-Diethylbenzene, CA 135-01-3:   322.59 (1); 328.15 (3,5,); 330.37 (6).

1,3-Diethylbenzene, CA 141-93-5:   323.71 (1); 328.15 (3); 329.26 (6).

1,4-Diethylbenzene, CA 105-05-5:   328.15 (3); 329.15 (5,6); 329.82 (1).

Diethylbenzene, mixed isomers, CA 25340-17-4:   329.82 (1).

Isobutylbenzene, CA 538-93-2:   321.15 (3); below 328.15 (5); 328.15 (6,7); 333.15 (2).

1,2,3,4-Tetramethylbenzene, CA 488-23-3:   341.48 (1); 346.15 (4); 347.59 (6).

1,2,3,5-Tetramethylbenzene, technical, CA 527-53-7:   336.48 (1); 341.15 (4); 344.26 (6).

1,2,4,5-Tetramethylbenzene, CA 95-93-2:   327.59 (6); 340.15 (4); 347.04 (1).

$C_{10}H_{14}ClN$

N-(2-Chloroethyl)-*N*-ethylaniline, CA 92-49-9:   Above 385.93 (1).

$C_{10}H_{14}Cl_2O_2$

Ethyl 3-(2,2-dichlorovinyl)-2,2-dimethyl-l-cyclopropanecarboxylate, CA 59609-49-3:   383.15 (1).

$C_{10}H_{14}N_2$

Anabasine ±, technical, CA 13078-04-1:   366.48 (1).

Nicotine, (*S*)-(-), CA 54-11-5:   374.82 (1).

l-Phenylpiperazine, CA 92-54-6:   Above 385.93 (1); 413.71 (7).

$C_{10}H_{14}N_2O$

N,*N*-Diethylnicotinamide, CA 59-26-7:   Above 385.93 (1).

$C_{10}H_{14}O$

2-*sec*-Butylphenol, CA 89-72-5:   380.37 (7); 385.37 (1).

2-*tert*-Butylphenol, CA 88-18-6:   383.15 (1); above 385.93 (1).

3-*tert*-Butylphenol, CA 585-34-2:   382.04 (1).

4-*sec*-Butylphenol, CA 99-71-8:   388.71 (1,7).

Butyl phenyl ether, CA 1126-79-0:   355.37 (1); 355.37* (2,6,7).

Carvacrol, CA 499-75-2:   379.82 (1).

Carvone, (*S*)-(+), CA 2244-16-8:   362.04 (1); 371.15 (3).

Carvone, (*R*)-(-), CA 6485-40-1:   362.04 (1); 371.15 (3).

2-Cyclopenten-l-yl ether, CA 15131-55-2:   348.15 (1).

alpha-Ethylphenethyl alcohol ±, CA 701-70-2:   373.15 (1).

(continues)

$C_{10}H_{14}O$ *(continued)*

4-Isopropylbenzyl alcohol, CA 536-60-7: Above 385.93 (1).

8-Ketotricyclo[5,2,1,0^{2,6}]decane, CA 13380-94-4: 374.26 (1).

2-Methyl-1-phenyl-1-propanol, CA 611-69-8: 359.82 (1).

2-Methyl-1-phenyl-2-propanol, CA 100-86-7: 354.26 (1).

Myrtenal, (1R)-(-), CA 564-94-3: 352.04 (1).

Perillaldehyde, (S)-(-), CA 18031-40-8: 368.71 (1).

2-Phenyl-2-butanol ±, CA 1565-75-9: 363.71 (1).

3-Phenyl-1-butanol, CA 2722-36-3: Above 385.93 (1).

4-Phenol-1-butanol, CA 3360-41-6: Above 385.93 (1).

Thymol, CA 89-83-8: 375.37 (1).

Verbenone, (1S)-(-), CA 80-57-9: 358.15 (1).

$C_{10}H_{14}O_2$

4-*tert*-Butylcatechol, CA 98-29-3: 403.15 (6,7); 424.82 (1).

(2,2-Dimethoxyethyl)benzene, CA 101-48-4: 357.04 (1); 362.15 (3).

1-(2-Methoxyphenyl)-2-propanol, CA 15541-26-1: Above 385.93 (1).

3-(4-Methoxyphenyl propanol, CA 5406-18-8: Above 385.93 (1).

$C_{10}H_{14}O_3$

4-Carbethoxy-3-methyl-2-cyclohexen-1-one, technical, CA 487-51-4: Above 385.93 (1).

Diethylene glycol, phenyl ether, CA 104-68-7: 433.15* (4).

2,4-Dimethoxy-3-methylbenzyl alcohol, CA 78647-61-7: 383.15 (1).

3,4-Dimethoxyphenethyl alcohol, CA 7417-21-2: Above 385.93 (1).

4-Ethoxy-3-methoxybenzyl alcohol, CA 61813-58-9: Above 385.93 (1).

2,4,6-Trimethoxytoluene, technical, CA 14107-97-2: Above 385.93 (1).

3,4,5-Trimethoxytoluene, CA 6443-69-2: 347.59 (1).

Trimethyl orthobenzoate, CA 707-07-3: 355.37 (1).

$C_{10}H_{14}O_4$

Ethylene glycol dimethacrylate, CA 97-90-5: 346.15 (3); 373.71 (1); 386.15* (4).

Hydroquinone, bis(2-hydroxyethyl), CA 104-38-1: 497.04 (6).

2,3,4-Trimethoxybenzyl alcohol, CA 71989-96-3: 373.15 (1).

3,4,5-Trimethoxybenzyl alcohol, CA 3840-31-1: Above 385.93 (1).

$C_{10}H_{15}N$

N-Butylaniline, CA 1126-78-9: 380.37 (3); 380.37* (6).

4-Butylaniline, CA 104-13-2: 374.82 (1).

(continues)

$C_{10}H_{15}N$ *(continued)*

4-*sec*-Butylaniline, CA 30273-11-1: 380.93 (1).

4-*tert*-Butylaniline, CA 769-92-6: 374.82 (1).

N,N-Diethylaniline, CA 91-66-7: 358.15 (6); 361.15 (3); 370.93 (1).

2,6-Diethylaniline, CA 579-66-8: 388.15 (3); 396.48 (1).

3,7-Dimethyl-2,6-octadienenitrile, mixed isomers, CA 5146-66-7: Above 385.93 (1).

N,N-Dimethyl-1-phenethylamine, CA 2449-49-2: 337.59 (1); 352.59* (6).

N-Ethyl-2,3-xylidine, CA 41115-23-5: 344.26 (1).

N-Isopropylbenzylamine, CA 102-97-6: 345.15 (3); 360.93 (1).

2-Isopropyl-6-methylaniline, CA 5266-85-3: 314.82 (1).

1-Methyl-3-phenylpropylamine, CA 22374-89-6: 370.93 (1).

4-Phenylbutylamine, CA 13214-66-9: 374.82 (1).

N,N,3,5-Tetramethylaniline, CA 4913-13-7: 363.71 (1).

$C_{10}H_{15}NO$

2-Aminoethanol, *N*-ethyl-*N*-phenyl, CA 92-50-2: 405.15 (5); 405.37* (6).

N-Benzyl-*N*-methylethanolamine, CA 101-98-4: Above 385.93 (1).

4-Butoxyaniline, CA 4344-55-2: Above 385.93 (1).

Ephedrine, (1*R*,2*S*)-(-), CA 299-42-3: 358.71 (1).

Ephedrine, (1*S*,2*R*)-(+), CA 321-98-2: Above 385.93 (1).

3-Methoxy-*N,N*-dimethylbenzylamine, CA 15184-99-3: 369.26 (1).

$C_{10}H_{15}NO_2$

2,5-Dimethoxyphenethylamine, CA 3600-86-0: Above 385.93 (1).

2-(3,4-Dimethoxyphenyl)ethylamine, CA 120-20-7: Above 385.93 (1).

N-Phenyldiethanolamine, CA 120-07-0: 464.15* (5); 469.26* (6).

$C_{10}H_{15}NO_4$

Diethyl 2-(2-cyanoethyl)malonate, CA 17216-62-5: Above 385.93 (1).

$C_{10}H_{15}NO_5$

4-Ethyl 1-methyl-3-oxopiperidine-1,4-dicarboxylate ±, CA 76508-78-6: Above 385.93 (1).

$C_{10}H_{15}N_2P$

1,3-Dimethyl-2-phenyl-1,3,2-diazaphospholidine, CA 22429-12-5: 378.15 (3).

$C_{10}H_{15}N_3$

1-Azidoadamantane, CA 24886-73-5: 358.15 (1).

$C_{10}H_{15}O_2P$

Diethyl phenylphosphonite, CA 1638-86-4: Above 385.93 (1).

$C_{10}H_{16}$

Camphene, CA 5794-03-6: 307.15 (3); 309.82 (1).

2-Carene (+), CA 4497-92-1: 311.15 (3).

3-Carene (+), CA 498-15-7: 328.15 (3).

1,5-Cyclododecadiene, mixed isomers, CA 1124-78-3: 348.15 (3).

1,5-Dimethyl-1,5-cyclooctadiene, CA 3760-14-3: 328.71 (1).

2,6-Dimethyl-2,4,6-octatriene, mixed isomers, CA 673-84-7: 342.04 (1).

Dipentene, technical, CA 138-86-3: 315.15 (3); 315.93 (1); 318.15 (6); 323.15 (5).

1,2-Divinylcyclohexane, cis, CA 1004-84-8: 318.15 (3).

Isolimonene, trans, CA 5113-87-1: 312.15 (3).

Limonene, (S)-(-), CA 5989-54-8: 316.15 (3); 321.48 (1).

Limonene, (R)-(+), CA 5989-27-5: 316.15 (3); 321.48 (1); 323.15 (5).

Myrcene, technical, CA 123-35-3: 312.59 (1).

1,2,3,4,5-Pentamethylcyclopentadiene, CA 4045-44-7: 317.59 (1,3).

alpha-Phellandrene (-), CA 4221-98-1: 315.15 (3).

beta-Phellandrene, CA 555-10-2: 322.15 (5,6).

alpha-Pinene, (S)-(-), CA 7785-26-4: 305.37 (1); 306.15 (3,4,5,6).

alpha-Pinene, (R)-(+), CA 7785-70-8: 295.15 (3); 305.37 (1).

beta-Pinene, (S)-(-), CA 18172-67-3: 305.93 (1); 309.15 (3).

Sabinene (+), CA 3387-41-5: 309.82 (1).

alpha-Terpinene, CA 99-86-5: 319.26 (1); 323.15 (3).

gamma-Terpinene, CA 99-85-4: 323.15 (3); 324.82 (1).

Terpinolene, CA 586-62-9: 310.93 (7).

Tricyclo[5,2,1,0^{2,6}]heptane, CA 6004-38-2: 313.71 (1).

$C_{10}H_{16}Cl_2O_2$

Sebacoyl chloride, CA 111-19-3: Above 385.93 (1).

$C_{10}H_{16}N_2$

N,N-Diethyl-1,4-phenylenediamine, CA 93-05-0: Above 385.93 (1).

N,N,N',N'-Tetramethyl-1,4-phenylenediamine, CA 100-22-1: Above 385.93 (1).

$C_{10}H_{16}O$

Camphor ±, CA 21368-68-3: 337.59 (1,3); 339.15 (5,6,7).

Carveol (-), mixed isomers, CA 99-48-9: 371.48 (1).

Citral, mixed isomers, CA 5392-40-5: 363.71 (6); 368.15 (3); 374.82 (1).

(continues)

214

$C_{10}H_{16}O$ *(continued)*

2,4-Decadienal, *trans,trans*, CA 25152-84-5: 371.15 (3); 374.26 (1).

1-Decalone, *trans*, CA 21370-71-8: 364.26 (1).

1-Decalone, mixed isomers, CA 4832-16-0: 360.93 (1).

2-Decalone, mixed isomers, CA 4832-17-1: 374.82 (1).

Dihydrocarvone (+), mixed isomers, CA 7764-50-3: 354.26 (1).

Fenchone (-), CA 1195-79-5: 325.93 (1); 341.15 (3).

Fenchone (+), CA 7787-20-4: 339.15 (3).

Limonene oxide, mixed isomers, CA 74347-40-3: 338.71 (1).

Myrtenol (-), CA 515-00-4: 341.15 (3); 362.59 (1).

Perillyl alcohol, (S)-(-), CA 536-59-4: Above 385.93 (1).

alpha-Pinene oxide, CA 1686-14-2: 338.71 (1).

beta-Pinene oxide, CA 6931-54-0: 339.26 (1).

Pulegone (+), technical, CA 89-82-7: 328.15 (3); 355.37 (1).

Thujone, mixed isomers, technical, CA 1125-12-8: 337.59 (1).

$C_{10}H_{16}O_2$

Cyclohexyl methacrylate, CA 101-43-9: 314.15 (3).

5-(1-Hydroxy-1-methylethyl)-2-methyl-2-cyclohexen-1-one, (S)-(+),
CA 60593-11-5: Above 385.93 (1).

$C_{10}H_{16}O_3$

Ethyl 2-cyclohexanoneacetate, CA 24731-17-7: Above 385.93 (1).

Ethyl 4-methyl-2-cyclohexanone-1-carboxylate, CA 13537-82-1: Above
385.93 (1).

$C_{10}H_{16}O_4$

Diethyl allylmalonate, CA 2049-80-1: 365.93 (1).

Diethyl isopropylidenemalonate, CA 6802-75-1: Above 353.15 (3); 382.59 (1).

Diisopropyl maleate, CA 10099-70-4: 377.59* (6).

Dimethyl 1,4-cyclohexanedicarboxylate, mixed isomers, CA 94-60-0:
383.15 (1).

Ethyl 4-acetyl-5-oxohexanoate, CA 2832-10-2: Above 385.93 (1).

3,3,6,6-Tetramethoxy-1,4-cyclohexadiene, CA 15791-03-4: 379.26 (1).

$C_{10}H_{16}O_5$

Diethyl acetylsuccinate, CA 1115-30-6: Above 385.93 (1).

Diethyl ethoxymethylenemalonate, CA 87-13-8: 428.15 (1).

$C_{10}H_{16}O_6$

Triethyl methanetricarboxylate, CA 6279-86-3: Above 385.93 (1).

$C_{10}H_{16}SSi$

(Phenylthiomethyl)trimethylsilane, CA 17873-08-4: 312.59 (1); 365.15 (3).

$C_{10}H_{16}SeSi$

(Phenylselenomethyl)trimethylsilane, CA 56253-60-2: 368.15 (3).

$C_{10}H_{16}Si$

Benzyltrimethylsilane, CA 770-09-2: 335.15 (3).

$C_{10}H_{17}Cl$

Geranyl chloride, CA 5389-87-7: 363.15 (1).

$C_{10}H_{17}ClO_4$

Diethyl (3-chloropropyl)malonate, CA 18719-43-2: Above 385.93 (1).

$C_{10}H_{17}N$

1-Pyrrolidino-1-cyclohexene, technical, CA 1125-99-1: 312.59 (1).

$C_{10}H_{17}NO$

N-Butylacetanilide, CA 91-49-6: 414.15 (5).

1-Cyclohexyl-2-pyrrolidinone, CA 6837-24-7: 363.15 (3); above 385.93 (1).

1-Morpholino-1-cyclohexene, CA 670-80-4: 341.48 (1).

$C_{10}H_{17}NO_6$

Dimethyl 4-methyl-4-nitroheptanedicarboxylate, CA 10499-89-5: 356.48 (1).

$C_{10}H_{18}$

Cyclodecene, cis, CA 935-31-9: 335.37 (1,3).

1,9-Decadiene, CA 1647-16-1: 314.82 (1).

Decahydronaphthalene, cis, CA 493-01-6: 331.48 (1,3); 334.15 (5).

Decahydronaphthalene, trans, CA 493-02-7: 325.93 (1); 327.15 (5,6); 331.15 (3).

Decahydronaphthalene, mixed isomers, CA 91-17-8: 330.37 (1); 330.93 (4,6,7).

1-Decyne, CA 764-93-2: 321.15 (3); 323.15 (1).

5,7-Dimethyl-1,6-octadiene (+), CA 6874-43-7: 293.15 (3).

4-Isopropyl-1-methylcyclohexene (+), CA 1195-31-9: 308.15 (3).

Pinane, CA 473-55-2: 321.15 (3).

Vinylcyclooctane, CA 19780-51-9: 327.04 (1).

$C_{10}H_{18}BN$

Borane-*N*,*N*-diethylaniline complex, CA 13289-97-9: 294.26 (1).

$C_{10}H_{18}F_6O_6S_2Si$

Di-*tert*-butylsilyl bis(trifluoromethanesulfonate), CA 85272-31-7: 363.71 (1).

$C_{10}H_{18}O$

Borneol, CA 507-70-0: 338.71 (1,5,6,7).

2-*tert*-Butylcyclohexanone, CA 1728-46-7: 345.37 (1).

4-*tert*-Butylcyclohexanone, CA 98-53-3: 369.26 (1).

Chrysanthemyl alcohol ±, CA 18383-59-0: 358.15 (1).

Cineole, CA 470-82-6: 322.15 (3); 323.15 (1).

Citronellal (+), CA 106-23-0: 346.15 (3); 347.04 (6).

Cyclododecanone, CA 1502-06-3: 355.37 (1).

Decahydro-2-naphthol, mixed isomers, CA 825-51-4: Above 385.93 (1).

Dihydrocarveol, mixed isomers, CA 619-01-2: 364.82 (1).

4-Ethyl-1-octyn-3-ol, CA 5877-42-9: 356.15* (4).

Fenchyl alcohol, CA 1632-73-1: 347.04 (1).

Geraniol, CA 106-24-1: 349.82 (1); above 373.15 (6).

Isopinocampheol ±, CA 51152-11-5: 366.48 (1).

Isopulegol, technical, CA 89-79-2: 351.48 (1).

Linalool ±, CA 78-70-6: 328.15 (3); 344.26 (6); 349.26 (1).

para-Menth-1-en-9-ol (+), CA 13835-75-1: 376.48 (1).

Menthone (-), mixed isomers, CA 10458-14-7: 342.59 (1); 351.15 (3).

Myrtanol (-), *cis*, CA 51152-12-6: 341.15 (3).

Myrtanol (-), *trans*, CA 53369-17-8: 359.15 (3); 377.04 (1).

Nerol, CA 106-25-2: 349.82 (1); 373.15 (3).

Terpinen-4-ol, CA 562-74-3: 352.59 (1).

alpha-Terpineol, CA 98-55-5: 362.59 (1,3); 363.71 (6).

3,3,5,5-Tetramethylcyclohexanone, CA 14376-79-5: 346.48 (1); 363.15 (3).

$C_{10}H_{18}OS$

8-Mercaptomenthone, technical, CA 38462-22-5: 381.48 (1).

$C_{10}H_{18}O_2$

Citronellic acid ±, CA 57030-77-0: 383.15 (1).

Cyclohexanebutyric acid, CA 4441-63-8: Above 385.93 (1).

2-Cyclohexylethyl acetate, CA 21722-83-8: 354.26 (1).

gamma-Decanolactone, CA 706-14-9: Above 385.93 (1).

(continues)

$C_{10}H_{18}O_2$ *(continued)*

delta-Decanolactone ±, CA 705-86-2:   Above 385.93 (1).

3,6-Dimethyl-4-octyne-3,6-diol ±, CA 78-66-0:   Above 385.93 (1).

Ethyl cyclohexylacetate, CA 5452-75-5:   353.15 (1); 357.15 (3).

Ethyl 2,2,3,3-tetramethylcyclopropanecarboxylate, CA 771-10-8:   322.04 (1).

Hexyl methacrylate, CA 142-09-6:   355.37* (6).

Methyl cyclohexanepropionate, CA 20681-51-0:   362.04 (1).

Pinanediol: Above 385.93 (1).

Vinyl 2-ethylhexanoate, CA 94-04-2:   338.15 (1); 347.04* (6,7).

$C_{10}H_{18}O_3$

2,2-Diethylacetoacetic acid, ethyl ester, CA 1619-57-4:   349.82 (6).

2(3)-(Tetrahydrofurfuryloxy)tetrahydropyran, CA 710-14-5:   370.93 (1,3).

Trimethylacetic anhydride, CA 1538-75-6:   330.37 (1).

Valeric anhydride, CA 2082-59-9:   372.15 (3); 374.26 (1).

$C_{10}H_{18}O_4$

Azelaic acid, monomethyl ester, CA 2104-19-0:   Above 385.93 (1).

1,4-Butanediol diglycidyl ether, CA 2425-79-8:   Above 385.93 (1).

Dibutyl oxalate, CA 2050-60-4:   377.15 (5,6); 382.04 (1); 402.49* (6).

Diethyl adipate, CA 141-28-6:   Above 385.93 (1).

Diethyl isopropylmalonate, CA 759-36-4:   361.15 (3).

Diethyl propylmalonate, CA 2163-48-6:   364.82 (1).

Dimethyl suberate, CA 1732-09-8:   Above 385.93 (1).

Methyl 3-(dimethoxymethyl)-2,2-dimethylcyclopropanecarboxylate, CA 59829-77-5:   359.15 (3); 367.59 (1).

$C_{10}H_{18}O_4Si_2$

Bis(trimethylsilyl) acetylenedicarboxylate, CA 76734-92-4:   331.15 (3); 340.37 (1).

$C_{10}H_{18}O_5$

Di-*tert*-butyl dicarbonate, CA 24424-99-5:   310.37 (1).

Diethylene glycol, dipropionate, CA 6942-59-2:   399.82 (6).

$C_{10}H_{18}O_6$

Diisopropyl tartrate, *dextro*, CA 62961-64-2:   Above 385.93 (1).

Diisopropyl tartrate, *levo*, CA 2217-15-4:   382.59 (1).

Triethylene glycol diacetate, CA 111-21-7:   419.82 (2); 447.04* (4,6).

$C_{10}H_{19}Cl$

Menthyl chloride, *levo*, CA 16052-42-9: 352.04 (1).

$C_{10}H_{19}ClO$

Decanoyl chloride, CA 112-13-0: 379.26 (1).

$C_{10}H_{19}N$

3-Azaspiro(5,5)undecane, technical, CA 180-44-9: 359.82 (1).

Myrtanylamine (-), *cis*, CA 38235-68-6: 351.48 (1).

1,3,3-Trimethyl-6-azabicyclo(3,2,1)octane, CA 53460-46-1: 348.15 (1).

$C_{10}H_{19}NO_2$

Ethyl 1-piperidinepropionate, CA 19653-33-9: 360.93 (1).

Methacrylic acid, 2-(*tert*-butylamino)ethyl ester, CA 3775-90-4: 284.15* (4); 369.26* (6,7).

$C_{10}H_{19}O_5P$

Triethyl 4-phosphonocrotonate, technical, CA 20345-62-4: Above 385.93 (1).

$C_{10}H_{20}$

Butylcyclohexane, CA 1678-93-9: 314.26 (1); 328.15 (3).

tert-Butylcyclohexane, CA 3178-22-1: 315.37 (1,3).

Cyclodecane, CA 293-96-9: 338.15 (1).

1-Decene, CA 872-05-9: 317.15 (3); 320.93 (1); 322.04 (2).

4-Decene, *trans*, CA 19398-89-1: 318.15 (3).

5-Decene, *trans*, CA 7433-56-9: 319.26 (1).

Diamylene, mixed isomers, CA 25339-53-1: 320.93* (6).

Diethylcyclohexane, mixed isomers, CA 1331-43-7: 322.04 (6); 322.04* (7).

3,7-Dimethyl-1-octene, CA 4984-01-4: 325.15 (3).

$C_{10}H_{20}Br_2$

1,10-Dibromodecane, CA 4101-68-2: Above 385.93 (1).

$C_{10}H_{20}Cl_2$

1,10-Dichlorodecane, CA 2162-98-3: Above 385.93 (1).

$C_{10}H_{20}I_2$

1,10-Diiododecane, CA 16355-92-3: Above 385.93 (1).

$C_{10}H_{20}O$

2-*tert*-Butylcyclohexanol, mixed isomers, CA 13491-79-7:   352.59 (1).

4-*tert*-Butylcyclohexanol, mixed isomers, CA 98-52-2:   378.15 (1).

beta-Citronellol ±, CA 106-22-9:   352.59 (1); 369.26 (6); 372.15 (3).

4-Cyclohexyl-1-butanol, CA 4441-57-0:   382.04 (1).

2-Decanone, CA 693-54-9:   344.26 (1); 359.15 (3).

3-Decanone, CA 928-80-3:   299.82 (1); 354.15 (3).

4-Decanone, CA 624-16-8:   344.26 (1).

5-Decen-1-ol, *trans*, CA 56578-18-8:   335.37 (1).

9-Decen-1-ol, CA 13019-22-2:   372.04 (1).

Decyl aldehyde, CA 112-31-2:   356.15 (3); 358.71 (1).

Dihydromyrcenol, technical, CA 18479-58-8:   349.82 (1).

1,2-Epoxydecane, CA 2404-44-6:   351.48 (1).

2-Ethylhexyl vinyl ether, CA 103-44-6:   320.15 (5); 330.37* (2,6,7).

Isodecaldehyde, CA 3085-26-5:   358.15* (6).

Isooctyl vinyl ether, CA 37769-62-3:   333.15* (2,6).

Menthol ±, CA 89-78-1:   366.48 (1).

Neomenthol ±, CA 3623-51-6:   355.37 (1); 356.15 (3).

Rhodinol, CA 6812-78-8:   Above 373.15 (6).

$C_{10}H_{20}O_2$

Decanoic acid, CA 334-48-5:   Above 385.93 (1).

3,7-Dimethyl-7-hydroxyoctanal, CA 107-75-5:   Above 373.15 (6).

2-Ethylhexyl acetate, CA 103-09-3:   344.26 (2,6); 352.15* (2,4); 355.15 (5); 355.15* (4); 360.93* (7).

Ethyl octanoate, CA 106-32-1:   348.15 (1); 352.59 (6); 354.15 (3).

2-Hexenal diethyl acetal, mostly *trans*, CA 54306-00-2:   335.93 (1).

Isodecanoic acid, CA 28933-59-7:   422.04* (6).

Methyl nonanoate, CA 1731-84-6:   357.59 (1).

Octyl acetate, CA 112-14-1:   359.26 (1).

$C_{10}H_{20}O_2S$

2-Ethylhexyl thioglycolate, CA 7659-86-1:   329.15 (3).

$C_{10}H_{20}O_3Si$

Ethyl 2-(trimethylsilylmethyl)acetoacetate, CA 17906-77-3:   365.93 (1).

$C_{10}H_{20}O_4$

Butyl 2-butoxy-2-hydroxyacetate, technical, CA 68575-73-5:   347.59 (1).

(continues)

$C_{10}H_{20}O_4$ *(continued)*

Diethylene glycol, butyl ether acetate, CA 124-17-4:   389.15 (5,6); 389.15* (2,4,6,7).

Tetraethoxyethylene, CA 40923-93-1:   336.15 (3).

$C_{10}H_{20}O_5$

15-Crown-5, CA 33100-27-5:   Above 385.93 (1).

$C_{10}H_{20}O_5Si$

3-(Trimethoxysilyl)propyl methacrylate, CA 2530-85-0:   365.37 (1).

$C_{10}H_{20}S_2$

Limonene dimercaptan (+):   394.26* (2).

$C_{10}H_{21}Br$

1-Bromodecane, CA 112-29-8:   367.59 (1).

2-Bromodecane, technical, CA 39563-53-6:   369.26 (1).

$C_{10}H_{21}Cl$

1-Chlorodecane, CA 1002-69-3:   356.48 (1).

$C_{10}H_{21}F_3O_3SSi$

Triisopropylsilyl trifluoromethanesulfonate, CA 80522-42-5:   373.71 (1).

$C_{10}H_{21}I$

1-Iododecane, CA 2050-77-3:   Above 385.93 (1).

$C_{10}H_{21}N$

N-Butylcyclohexylamine, CA 10108-56-2:   366.48* (6,7).

N-tert-Butylcyclohexylamine, CA 51609-06-4:   323.15 (3).

N,N-Diethylcyclohexylamine, CA 91-65-6:   330.93 (1).

1,2,2,6,6-Pentamethylpiperidine, CA 79-55-0:   323.15 (1); 328.15 (3).

$C_{10}H_{21}NO$

3-Aminomethyl-3,5,5-trimethylcyclohexanol, mixed isomers, CA 15647-11-7: Above 385.93 (1).

N,N-Dibutylacetamide, CA 1563-90-2:   380.37 (6).

$C_{10}H_{21}NO_3$

Decyl nitrate, CA 2050-78-4:   385.93* (6).

$C_{10}H_{22}$

Decane, CA 124-18-5: 317.04 (2); 319.26 (1,2,5,6,7).

2,3-Dimethyloctane, CA 7146-60-3: Below 328.15 (5,6).

2,6-Dimethyloctane ±, CA 2051-30-1: 307.15 (3).

3,4-Dimethyloctane, CA 15869-92-8: Below 328.15 (6).

4,5-Dimethyloctane, CA 15869-96-2: Below 328.15 (5).

2-Methylnonane, CA 871-83-0: 319.15 (3).

3-Methylnonane, CA 5911-04-6: 319.15 (3).

4-Methylnonane, CA 17301-94-9: 317.15 (3).

2,5,5-Trimethylheptane, CA 1189-99-7: Below 328.15 (5,6).

$C_{10}H_{22}N_2$

1,8-Diamino-*para*-menthane, technical, CA 80-52-4: 366.48 (1); 375.15* (4).

$C_{10}H_{22}O$

l-Decanol, CA 112-30-1: 355.37 (1,3,5); 355.37* (6,7); 377.59 to 380.37* (2); 385.92* (2).

2-Decanol, CA 1120-06-5: 353.15 (3); 358.15 (1).

4-Decanol ±, CA 2051-31-2: 355.37 (1).

3,7-Dimethyl-1-octanol, CA 106-21-8: 370.15 (3).

Isodecyl alcohol, mixed isomers, CA 25339-17-7: 375.15* (4); 377.59* (2,6,7).

Isopentyl ether, CA 544-01-4: 318.71 (1).

Pentyl ether, CA 693-65-2: 330.37 (1,3,5); 330.37* (6,7); 336.48* (2).

$C_{10}H_{22}O_2$

1,2-Decanediol, CA 1119-86-4: Above 385.93 (1).

Ethylene glycol dibutyl ether, CA 112-48-1: 358.15 (6); 358.15* (2,4).

Ethylene glycol, 2-ethylhexyl ether, CA 1559-35-9: 383.15* (4,6).

$C_{10}H_{22}O_2Si_2$

2,3-Bis(trimethylsilyloxy)-1,3-butadiene, CA 31411-71-9: 297.59 (1).

1,2-Bis(trimethylsilyloxy)cyclobutene, CA 17082-61-0: 334.26 (1).

$C_{10}H_{22}O_3$

Diethylene glycol hexyl ether, CA 112-59-4: 413.15 (1); 413.75* (4,6).

Dipropylene glycol butyl ether, mixed isomers, CA 29911-28-2: 386.15* (2,4).

$C_{10}H_{22}O_4$

Triethylene glycol monobutyl ether, CA 143-22-6: 416.45* (4); 416.48 (6,7).

(continues)

$C_{10}H_{22}O_4$ *(continued)*

Tripropylene glycol, monomethyl ether, CA 20324-33-8: 383.15 (1); 387.04 (2); 394.15 (5,7); 399.82* (2).

$C_{10}H_{22}O_5$

Tetraethylene glycol dimethyl ether, CA 143-24-8: 413.71 (1,2,5); 414.15* (2,6).

$C_{10}H_{22}O_6$

Pentaethylene glycol, CA 4792-15-8: Above 385.93 (1).

$C_{10}H_{22}S$

1-Decanethiol, CA 143-10-2: 371.48 (1); 374.82* (2).

tert-Decylmercaptan, CA 11090-97-4: 360.93 (6).

Dipentyl sulfide, CA 872-10-6: 358.15 (5); 358.15* (6).

Dipentyl sulfide, mixed isomers: 358.15* (6).

$C_{10}H_{23}BO_2$

Butyldiisopropoxyborane, CA 86545-32-6: 302.59 (1).

$C_{10}H_{23}ClSi$

Chlorodimethyloctylsilane, CA 18162-84-0: 370.93 (1).

$C_{10}H_{23}N$

Decylamine, CA 2016-57-1: 358.71 (1); 362.15 (3); 372.15 (5,6,7).

N,N-Dimethyloctylamine, CA 7378-99-6: 338.15 (1).

Dipentylamine, CA 2050-92-2: 324.15 (5,6,7); 325.37 (1); 339.15 (3); 339.82 (2).

Dipentylamine, mixed isomers: 343.15* (2).

$C_{10}H_{23}NO$

2-(Dibutylamino)ethanol, CA 102-81-8: 363.71 (2); 366.15 (5); 366.48* (6); 368.15 (3); 377.59* (7).

$C_{10}H_{23}NO_2$

Diethylaminoacetaldehyde diethyl acetal, CA 3616-57-7: 338.15 (1).

Tetraethylammonium acetate tetrahydrate, CA 1185-59-7: 333.71 (1).

$C_{10}H_{23}NO_2Si_2$

Ethyl 2,2,5,5-tetramethyl-1,2,5-azadisiloliden-1-acetate, CA 78605-23-9: 356.15 (3).

$C_{10}H_{24}N_2$

2-(Dibutylamino)ethylamine, CA 3529-09-7: 355.15 (3).

N,N,N',N'-Tetraethylethylenediamine, CA 150-77-6: 332.04 (1).

N,N,N',N'-Tetramethyl-1,6-hexanediamine, CA 111-18-2: 347.04 (1).

$C_{10}H_{24}N_2O_2$

2,3-Dimethoxy-1,4-bis(dimethylamino)butane: 344.26 (1).

4,9-Dioxa-1,12-dodecanediamine, CA 7300-34-7: Above 385.93 (1).

$C_{10}H_{24}N_2O_4$

N,N,N',N'-Tetrakis(2-hydroxyethyl)ethylenediamine, CA 140-07-8: 383.15 (1).

$C_{10}H_{24}N_4$

1,4-Bis(3-aminopropyl)piperazine, CA 7209-38-3: 435.93 (1).

Tetrakis(dimethylamino)ethylene, CA 996-70-3: 326.48 (1).

$C_{10}H_{26}N_2Si$

Bis(diethylamino)dimethylsilane, CA 4669-59-4: 321.15 (3).

$C_{10}H_{26}N_4$

Spermine, CA 71-44-3: Above 385.93 (1).

$C_{10}H_{28}N_6$

Pentaethylenehexamine, technical, CA 4067-16-7: 383.15 (1); 439.15 (4); 448.15 (3).

$C_{10}H_{28}Si_3$

Tris(trimethylsilyl)methane, technical, CA 1068-69-5: 349.82 (1).

$C_{10}H_{30}O_3Si_4$

Decamethyltetrasiloxane, CA 141-62-8: 335.37 (1).

$C_{11}H_7ClO$

1-Naphthoyl chloride, CA 879-18-5: Above 385.93 (1).

2-Naphthoyl chloride, CA 2243-83-6: 383.15 (1).

$C_{11}H_7NO$

1-Naphthyl isocyanate, CA 1984-04-9: 383.15 (1).

$C_{11}H_8F_6O$

3'5'-Bis(trifluoromethyl)propiophenone: 349.26 (1).

$C_{11}H_8N_2O$

Di-2-pyridyl ketone, CA 19437-26-4: 383.15 (1).

$C_{11}H_8O$

1-Naphthaldehyde, CA 66-77-3: Above 385.93 (1).

$C_{11}H_9Br$

1-(Bromomethyl)naphthalene, CA 3163-27-7: Above 385.93 (1).

2-(Bromomethyl)naphthalene, CA 939-26-4: Above 385.93 (1).

1-Bromo-2-methylnaphthalene, technical, CA 2586-62-1: Above 385.93 (1).

1-Bromo-4-methylnaphthalene, CA 6627-78-7: Above 385.93 (1).

$C_{11}H_9Cl$

1-(Chloromethyl)naphthalene, CA 86-52-2: Above 385.93 (1); 405.37* (7).

$C_{11}H_9N$

2-Phenylpyridine, CA 1008-89-5: Above 385.93 (1).

3-Phenylpyridine, CA 1008-88-4: Above 385.93 (1).

$C_{11}H_{10}$

1-Methylnaphthalene, CA 90-12-0: 351.15 (3); 355.37 (1); 367.15 (5).

2-Methylnaphthalene, CA 91-57-6: 351.15 (3); 370.93 (1).

$C_{11}H_{10}O$

1-Methoxynaphthalene, CA 2216-69-5: Above 385.93 (1).

$C_{11}H_{10}O_2$

Ethyl phenylpropiolate, CA 2216-94-6: Above 385.93 (1).

$C_{11}H_{11}ClOS$

O-(5,6,7,8-Tetrahydro-2-naphthyl) chlorothioformate, CA 84995-63-1: Above 385.93 (1).

$C_{11}H_{11}FO_3$

Ethyl 2-fluorobenzoylacetate, CA 1479-24-9: Above 385.93 (1).

$C_{11}H_{11}N$

2,4-Dimethylquinoline, CA 1198-37-4: 383.15 (1).

2,8-Dimethylquinoline, CA 1463-17-8: Above 385.93 (1).

1-Naphthalenemethylamine, CA 118-31-0: Above 385.93 (1).

$C_{11}H_{11}NO_2$

Ethyl 3-(cyanomethyl)benzoate, CA 13288-86-3: Above 385.93 (1).

Ethyl phenylcyanoacetate, CA 4553-07-5: Above 385.93 (1).

$C_{11}H_{12}N_2$

1-Benzyl-2-methylimidazole, CA 13750-62-4: Above 385.93 (1).

$C_{11}H_{12}O$

1-Benzosuberone, CA 826-73-3: Above 385.93 (1).

Cyclobutyl phenyl ketone, CA 5407-98-7: Above 385.93 (1).

1-Methyl-2-tetralone, technical, CA 4024-14-0: Above 385.93 (1).

2-Methyl-1-tetralone, CA 1590-08-5: Above 385.93 (1).

4-Methyl-1-tetralone, CA 19832-98-5: Above 385.93 (1).

$C_{11}H_{12}O_2$

Cinnamyl acetate, CA 103-54-8: Above 385.93 (1).

Cyclopropyl 4-methoxyphenyl ketone, CA 7152-03-6: Above 385.93 (1).

4,7-Dihydro-2-phenyl-1,3-dioxepin, CA 2568-24-3: 383.15 (1).

Ethyl cinnamate, mostly *trans*, CA 103-36-6: Above 385.93 (1).

6-Methoxy-2-tetralone, CA 2472-22-2: Above 385.93 (1).

7-Methoxy-2-tetralone, CA 4133-34-0: Above 385.93 (1).

$C_{11}H_{12}O_3$

Benzyl acetoacetate, CA 5396-89-4: 345.15 (3); above 385.93 (1).

Ethyl benzoylacetate, technical, CA 94-02-0: 413.71* (1,6); 414.15 (5).

Ethyl-3-phenylglycidate ±, mixed isomers, CA 121-39-1: Above 385.93 (1).

$C_{11}H_{12}O_5$

Phthalic acid, mono(2-methoxyethyl) ester, CA 16501-01-2: 408.15 (7).

$C_{11}H_{13}BrO$

4'-Bromovalerophenone, CA 7295-44-5: Above 385.93 (1).

$C_{11}H_{13}ClO$

4-Butylbenzoyl chloride, CA 28788-62-7: Above 385.93 (1).

4-*tert*-Butylbenzoyl chloride, CA 1710-98-1: 360.93 (1).

$C_{11}H_{13}ClOS$

O-(3-*tert*-Butylphenyl) chlorothioformate, CA 97986-06-6: Above 385.93 (1).

226

$C_{11}H_{13}ClO_2$

4-Butoxybenzoyl chloride, CA 33863-86-4: Above 385.93 (1).

$C_{11}H_{13}F_3N_2$

Piperazine, 1-(3-trifluoromethyl)benzene, CA 15532-75-9: Above 385.93 (1).

$C_{11}H_{13}N$

N-Methyl-*N*-propargylbenzylamine, CA 555-57-7: 357.04 (1).
2,3,3-Trimethylindolenine, CA 1640-39-7: 366.48 (1).

$C_{11}H_{13}NO$

1-Benzyl-2-pyrrolidinone, CA 5291-77-0: Above 385.93 (1).
1-Benzyl-3-pyrrolidinone, CA 775-16-6: Above 385.93 (1).
4,4-Dimethyl-2-phenyl-2-oxazoline, CA 19312-06-2: 375.93 (1).
Homotryptophol, CA 3569-21-9: 383.15 (1).

$C_{11}H_{13}NO_2$

ortho-Acetoacetotoluidine, CA 93-68-5: 416.48 (1); 433.15 (5); 433.15* (7).

$C_{11}H_{13}NO_3$

ortho-Acetoacetylanisidine, CA 92-15-9: 435.93* (6,7).

$C_{11}H_{13}NO_4$

Butyl 4-nitrobenzoate, CA 120-48-9: 383.15 (1).

$C_{11}H_{14}$

2,4,6-Trimethylstyrene, CA 769-25-5: 348.15 (1).

$C_{11}H_{14}N_2$

2-Isopropylamino-4-methylbenzonitrile, CA 28195-00-8: Above 385.93 (1).
2,3-Dihydro-2-methylene-1,3,3-trimethyl-3*H*-pyrrolo(2,3-*b*)pyridine,
CA 41450-85-5: 376.48 (1).

$C_{11}H_{14}N_2O_2$

2-Ethyl-2-phenylmalonamide monohydrate, CA 80866-90-6: Above 385.93 (1).

$C_{11}H_{14}O$

2-Cyclopentylphenol, technical, CA 1518-84-9: Above 385.93 (1).
2,2-Dimethylpropiophenone, CA 938-16-9: 360.37 (1).
6-Methoxy-1,2,3,4-tetrahydronaphthalene, technical, CA 1730-48-9: Above
385.93 (1).   *(continues)*

$C_{11}H_{14}O$ *(continued)*

2',4',6'-Trimethylacetophenone, CA 1667-01-2: 383.15 (1).

Valerophenone, CA 1009-14-9: 375.93 (1).

$C_{11}H_{14}O_2$

4-Allyl-1,2-dimethoxybenzene, CA 93-15-2: 372.04 (6); above 385.93 (1).

4-Butoxybenzaldehyde, CA 5736-88-9: Above 385.93 (1).

Butyl benzoate, CA 136-60-7: 379.26 (1); 380.37* (6,7).

1,2-Dimethoxy-4-propenylbenzene, CA 93-16-3: Above 385.93 (1).

Ethyl hydrocinnamate, CA 2021-28-5: 371.15 (3); 380.93 (1).

Ethyl *ortho*-tolylacetate, CA 40291-39-2: 379.82 (1).

Ethyl *para*-tolylacetate, CA 14062-19-2: 382.59 (1).

4-(4-Methoxyphenyl)-2-butanone, CA 104-20-1: 344.15 (3); above 385.93 (1).

Methyl isoeugenol, CA 6379-72-2: Above 373.15 (6).

1-Phenylethyl propionate ±: 367.59 (1).

$C_{11}H_{14}O_3$

4-Allyl-2,6-dimethoxyphenol, technical, CA 6627-88-9: Above 385.93 (1).

tert-Butyl peroxybenzoate, CA 614-45-9: 265.93 (7); above 360.93* (6); 366.48 (1).

tert-Butyl phenyl carbonate, CA 6627-89-0: 374.26 (1).

2',4'-Dimethoxy-3'-methylacetophenone, CA 60512-80-3: Above 385.93 (1).

3,4-Dimethoxyphenylacetone, CA 776-99-8: Above 385.93 (1).

Ethyl 4-ethoxybenzoate, CA 23676-09-7: Above 385.93 (1).

Ethyl 4-methoxyphenylacetate, CA 14062-18-1: 319.26 (1).

$C_{11}H_{14}O_4$

Ethyl 3,4-dihydroxyhydrocinnamate, CA 3967-57-5: Above 385.93 (1).

Ethyl homovanillate, CA 60563-13-5: Above 385.93 (1).

2',3',4'-Trimethoxyacetophenone, CA 13909-73-4: Above 385.93 (1).

$C_{11}H_{15}Br$

4-(*tert*-Butyl)benzyl bromide, technical, CA 18880-00-7: 383.15 (1).

$C_{11}H_{15}Cl$

4-(*tert*-Butyl)benzyl chloride, CA 19692-45-6: 367.59 (1).

$C_{11}H_{15}ClO$

1-Adamantanecarboxylic acid chloride, CA 2094-72-6: Above 385.93 (1).

Chloro-4-*tert*-pentylphenol, mixed isomers: 380.37 (6).

228

$C_{11}H_{15}ClO_3$

Diethyl 4-chlorophenyl orthoformate, CA 25604-54-0:   355.15 (3).

$C_{11}H_{15}FO_2$

1-Adamantyl fluoroformate, technical, CA 62087-82-5:   362.04 (1).

$C_{11}H_{15}NO$

1 Benzyl-3-pyrrolidinol, CA 775-15-5:   383.15 (1).
4-(Diethylamino)benzaldehyde, CA 120-21-8:   Above 385.93 (1).

$C_{11}H_{15}NO_2$

N-Benzylglycine, ethyl ester, CA 6436-90-4:   Above 385.93 (1).
2-(Dimethylamino)ethyl benzoate, CA 2208-05-1:   Above 385.93 (1).
Ethyl 2-dimethylaminobenzoate, technical, CA 55426-74-9:   372.04 (1).

$C_{11}H_{15}NO_5$

Methyl 3,4,5-trimethoxyanthranilate, CA 5035-82-5:   Above 385.93 (1).

$C_{11}H_{15}O_5P$

Dimethyl (3-phenoxyacetonyl)phosphonate, CA 40665-68-7:   359.26 (1).

$C_{11}H_{16}$

4-tert-Butyltoluene, CA 98-51-1:   327.59 (1); 333.15 (3).
(2-Methylbutyl)benzene (+), CA 40560-30-3:   335.37 (1).
Pentamethylbenzene, CA 700-12-9:   364.26 (1); 366.48 (6).
Pentylbenzene, CA 538-68-1:   315.15 (3); 338.71 (1,5); 338.71* (6).
2-Phenylpentane, CA 3968-85-2:   341.48 (2).

$C_{11}H_{16}N_2$

1-Benzylpiperazine, CA 2759-28-6:   Above 385.93 (1); 396.15 (3).

$C_{11}H_{16}N_2O$

1-(2-Methoxyphenyl)piperazine, CA 35386-24-4:   Above 385.93 (1).

$C_{11}H_{16}N_2O_2S$

tert-Butyl S-(4,5-dimethylpyrimidin-2-yl) thiolcarbonate, CA 41840-28-2:
Above 385.93 (1).

$C_{11}H_{16}O$

3-Acetylnoradamantane: 370.37 (1).

(continues)

$C_{11}H_{16}O$ *(continued)*

Benzyl-*tert*-butanol, CA 103-05-9: Above 385.93 (1).

4-*tert*-Butylbenzyl alcohol, CA 877-65-6: Above 385.93 (1).

2-*tert*-Butyl-4-methylphenol, CA 2409-55-4: 319.82 (7); 373.15 (1).

2-*tert*-Butyl-5-methylphenol, CA 88-60-8: 376.71 (1).

2-*tert*-Butyl-6-methylphenol, CA 2219-82-1: 380.37 (1).

4-*tert*-Butyl-2-methylphenol, CA 98-27-1: 390.93 (6).

tert-Butyl-3-methylphenol, mixed isomers, CA 1333-13-7: 319.82 (6).

2,2-Dimethyl-3-phenyl-1-propanol, CA 13351-61-6: 382.59 (1).

4,4*a*,5,6,7,8-Hexahydro-4*a*-methyl-2(3*H*)-naphthalenone, (*R*)-(-),
CA 63975-59-7: Above 385.93 (1).

4,4*a*,5,6,7,8-Hexahydro-4*a*-methyl-2(3*H*)-naphthalenone, (*S*)-(+),
CA 4087-39-2: 383.15 (1).

Jasmone, *cis*, CA 488-10-8: 380.37 (1).

2-Pentylphenol, CA 136-81-2: 377.15 (5); 377.15* (6).

2-*sec*-Pentylphenol, CA 26401-74-1: 366.49* (7).

4-Pentylphenol, CA 14938-35-3: 377.04* (7); 385.15 (3).

4-*sec*-Pentylphenol, CA 25735-67-5: 405.37 (6); 405.37* (7).

4-*tert*-Pentylphenol, CA 80-46-6: 384.26* (6,7).

Pentyl phenyl ether, CA 2050-04-6: 358.15 (6); 358.15* (2).

1-Phenyl-2-pentanol ±, CA 705-73-7: 378.15 (1).

5-Phenyl-1-pentanol, CA 10521-91-2: Above 385.93 (1).

$C_{11}H_{16}OSi$

1-Phenyl-1-(trimethylsilyloxy)ethylene, CA 13735-81-4: 349.15 (3);
352.04 (1).

$C_{11}H_{16}O_2$

4-Butoxybenzyl alcohol, CA 6214-45-5: Above 385.93 (1).

1-Methoxybicyclo(2,2,2)oct-5-en-2-yl methyl ketone, CA 25489-00-3:
372.04 (1).

4-(4-Methoxyphenyl)-1-butanol, CA 22135-50-8: Above 385.93 (1).

Olivetol, CA 500-66-3: Above 385.93 (1).

$C_{11}H_{16}O_3$

Diethyl phenyl orthoformate, CA 14444-77-0: 330.37 (1); above 338.15 (3).

3-(3,4-Dimethoxyphenyl)-1-propanol, CA 3929-47-3: 383.15 (1).

4-Ethoxy-3-methoxyphenethyl alcohol, CA 77891-29-3: 383.15 (1).

Methyl 1-methoxybicyclo(2,2,2)oct-5-ene-2-carboxylate, CA 5259-50-7:
376.48 (1).

$C_{11}H_{16}O_4$

3,9-Divinyl-2,4,8,10-tetraoxaspiro(5,5)undecane, CA 78-19-3: 383.15 (1).

$C_{11}H_{17}N$

Benzyldiethylamine, CA 772-54-3: 350.15 (2,5); 350.15* (6).

N-Butylbenzylamine, CA 2403-22-7: 367.59 (1).

N-*tert*-Butylbenzylamine, CA 3378-72-1: 353.15 (1); 363.15 (3).

N,N,2,4,6-Pentamethylaniline, CA 13021-15-3: 352.04 (1).

4-*tert*-Pentylaniline, CA 2049-92-5: 374.82 (2,6).

$C_{11}H_{17}NO$

3-(N-Benzyl-N-methylamino)-1-propanol, CA 5814-42-6: Above 385.93 (1).

2-(N-Ethyl-*meta*-toluidino)ethanol, CA 91-88-3: Above 385.93 (1).

N-Methylpseudoephedrine, CA 51018-28-1: Above 385.93 (1).

4-Pentyloxyaniline, CA 39905-50-5: Above 385.93 (1).

$C_{11}H_{17}NO_2$

3-(N-Benzyl-N-methylamino)-1,2-propanediol, CA 60278-98-0: Above 385.93 (1).

(*meta*-Tolyl)diethanolamine, CA 91-99-6: 477.59* (6).

$C_{11}H_{17}NO_3$

4-[2-(2-Methoxyethoxy)ethoxy]aniline, CA 65673-48-5: Above 385.93 (1).

$C_{11}H_{17}O_3P$

Benzyl diethyl phosphite, CA 2768-31-2: Above 385.93 (1).

Diethyl benzylphosphonate, CA 1080-32-6: Above 385.93 (1).

$C_{11}H_{18}$

Ethyltetramethylcyclopentadiene, mixed isomers, CA 57693-77-3: 328.71 (1).

$C_{11}H_{18}N_2$

N'-Benzyl-N,N-dimethylethylenediamine, technical, CA 103-55-9: Above 385.93 (1).

$C_{11}H_{18}O$

2,4-Diethyl-2,6-heptadienal, mixed isomers: 359.26 (1).

Nopol, CA 128-50-7: 371.15 (3); 372.04 (1).

Patchenol ±, technical, CA 2226-05-3: 380.37 (1).

2,6,6-Trimethyl-1-cyclohexene-1-acetaldehyde, CA 472-66-2: 361.48 (1).

$C_{11}H_{18}O_2$

Geraniol formate, CA 105-86-2: 358.15 (6).

$C_{11}H_{18}O_3$

1,4-Cyclohexanedione mono-2,2-dimethyltrimethylene ketal, CA 69225-59-8:
Above 385.93 (1).

Methyl 1-methyl-2-oxocyclohexanepropionate, CA 53068-89-6: 383.15 (1).

$C_{11}H_{18}O_5$

Diethyl 2-acetylglutarate, CA 1501-06-0: Above 385.93 (1).

Diethyl 3-oxopimelate, CA 40420-22-2: Above 385.93 (1).

Diethyl 4-oxopimelate, CA 6317-49-3: Above 385.93 (1).

1,2:3,5-Di-O-isopropylidene-dextro-xylofuranose, CA 20881-04-3: Above
385.93 (1).

$C_{11}H_{18}O_6$

Triethyl 1,1,2-ethanetricarboxylate, CA 7459-46-3: Above 385.93 (1).

$C_{11}H_{19}ClO$

10-Undecenoyl chloride, CA 38460-95-6: 366.48 (1).

$C_{11}H_{19}ClO_2$

Menthyl chloroformate (-), CA 14602-86-9: 343.15 (1).

$C_{11}H_{19}N$

1-Adamantanemethylamine, CA 17768-41-1: 365.37 (1).

$C_{11}H_{19}NSi$

N-(Trimethylsilylmethyl)benzylamine, CA 53215-95-5: 369.26 (1).

$C_{11}H_{20}N_2O$

Bis(pentamethylene)urea, CA 5395-04-0: Above 385.93 (1).

$C_{11}H_{20}O$

4-tert-Amylcyclohexanone, CA 16587-71-6: 377.59 (1).

Cycloundecanone, CA 878-13-7: 369.26 (1).

Undecylenic aldehyde, CA 112-45-8: 365.93 (1).

$C_{11}H_{20}O_2$

Ethyl cyclohexanepropionate, CA 10094-36-7: 323.15 (1).

(continues)

$C_{11}H_{20}O_2$ (continued)

2-Ethylhexyl acrylate ±, CA 103-11-7:   352.59 (1); 355.15 (4,5); 355.15* (6,7); 364.15* (2,4).

2,2,6,6-Tetramethyl-3,5-heptanedione, CA 1118-71-4:   340.37 (1).

Undecanoic acid gamma-lactone, CA 104-67-6:   Above 385.93 (1).

Undecanoic acid delta-lactone, CA 710-04-3:   Above 385.93 (1).

10-Undecenoic acid, CA 112-38-9:   419.26* (7); 422.04 (1).

$C_{11}H_{20}O_4$

Di-tert-butylmalonate, CA 541-16-2:   362.04 (1).

Diethyl butylmalonate, CA 133-08-4:   367.04 (1).

Diethyl tert-butylmalonate, CA 759-24-0:   363.71 (1).

Diethyl diethylmalonate, CA 77-25-8:   367.59 (1).

Diethyl pimelate, CA 2050-20-6:   Above 385.93 (1).

Sebacic acid monomethyl ester, CA 818-88-2:   Above 385.93 (1).

$C_{11}H_{21}BrO_2$

11-Bromoundecanoic acid, CA 2834-05-1:   Above 385.93 (1).

$C_{11}H_{21}ClO$

Undecanoyl chloride, CA 17746-05-3:   Above 385.93 (1).

$C_{11}H_{21}ClO_2$

Decyl chloroformate, CA 55488-51-2:   391.45* (4); 393.35 (4).

$C_{11}H_{21}N$

Undecanenitrile, CA 2244-07-7:   383.15 (1).

$C_{11}H_{22}$

1-Undecene, CA 821-95-4:   336.15 (1,3); 344.26 (2).

$C_{11}H_{22}Br_2$

1,11-Dibromoundecane, CA 16696-65-4:   Above 385.93 (1).

$C_{11}H_{22}N_2$

Dipiperidinomethane, CA 880-09-1:   364.82 (1).

$C_{11}H_{22}O$

4-tert-Pentylcyclohexanol, CA 5349-51-9:   373.12 (7).

2-Undecanone, CA 112-12-9:   362.04 (1,6,7).

(continues)

$C_{11}H_{22}O$ *(continued)*

3-Undecanone, CA 2216-87-7: 362.59 (1).

6-Undecanone, CA 927-49-1: 361.48 (1).

omega-Undecylenyl alcohol, CA 112-43-6: 366.48 (1).

Undecylic aldehyde, CA 112-44-7: 369.26 (1); 385.93* (7)

$C_{11}H_{22}OSi$

3-Triethylsiloxy-1,4-pentadiene, CA 62418-65-9: 331.15 (3).

$C_{11}H_{22}O_2$

2-Ethylhexyl glycidyl ether, CA 2461-15-6: 369.82 (1).

Ethyl nonanoate, CA 123-29-5: 367.15 (3); 367.59 (1).

7-Methoxy-3,7-dimethyloctanal, CA 3613-30-7: 371.48 (1).

Methyl decanoate, CA 110-42-9: 363.15 (3); 367.59 (1).

Nonyl acetate, CA 143-13-5: 341.15 (5,6).

Undecanoic acid, CA 112-37-8: Above 385.93 (1).

$C_{11}H_{22}O_2Si$

4-(*tert*-Butyldimethylsilyloxy)-3-penten-2-one, mixed isomers, CA 69404-97-3: 353.71 (1).

$C_{11}H_{23}Br$

1-Bromoundecane, CA 693-67-4: Above 385.93 (1).

$C_{11}H_{23}BrO$

11-Bromo-1-undecanol, CA 1611-56-9: Above 385.93 (1).

$C_{11}H_{23}I$

1-Iodoundecane, CA 4282-44-4: Above 385.93 (1).

$C_{11}H_{23}N$

N-Methylcyclodecylamine, CA 80789-66-8: 376.48 (1).

$C_{11}H_{23}NO_2$

1-Piperidineacetaldehyde diethyl acetal, CA 3616-58-8: 353.71 (1).

$C_{11}H_{24}$

Undecane, CA 1120-21-4: 333.15 (1); 338.15 (2); 338.15* (6,7); 339.15 (3).

$C_{11}H_{24}N_2$

4-Dimethylamino-2,2,6,6-tetramethylpiperidine, CA 32327-90-5: 322.04 (1).

$C_{11}H_{24}N_2O_2$

1,4-Bis(2-hydroxypropyl)-2-methylpiperazine, CA 94-72-4: 422.04* (2).

$C_{11}H_{24}O$

2,4-Diethyl-1-heptanol, CA 80192-55-8: 373.71 (1).

5-Ethyl-2-nonanol, CA 103-08-2: 373.15 (5).

1-Undecanol, CA 112-42-5: Above 385.93 (1).

2-Undecanol, CA 1653-30-1: 370.15 (3); 385.93* (6).

$C_{11}H_{24}O_2$

2,2-Dibutoxypropane, technical, CA 141-72-0: 330.37 (1).

$C_{11}H_{24}O_4$

1,1,3,3-Tetraethoxypropane, CA 122-31-6: 315.37 (1); 361.15 (3,5); 361.15* (6).

Tripropylene glycol, ethyl ether: 405.37* (2).

$C_{11}H_{24}O_6$

Diethylene glycol, monomethyl ether formal, CA 5405-88-9: 427.59* (6).

$C_{11}H_{24}O_6Si$

Tris(2-methoxyethoxy)vinylsilane, CA 1067-53-4: 329.15 (3); Above 385.93 (1).

$C_{11}H_{25}N$

N,N-Diisopropyl-3-pentylamine, CA 68714-10-3: 322.15 (3).

Undecylamine, CA 7307-55-3: 365.37 (1).

$C_{11}H_{25}NO$

1-Dibutylamino-2-propanol, CA 2109-64-0: 369.15 (5); 369.15* (6).

1-Propanamine, 3-[(2-ethylhexyl)], CA 5397-31-9: 372.04* (7).

$C_{11}H_{25}NO_2$

(*N,N*-Dimethylamino)dibutoxymethane, CA 18503-90-7: 326.15 (3).

(*N,N*-Dimethylamino)di-*tert*-butoxymethane, CA 36805-97-7: 308.15 (3).

$C_{11}H_{26}N_2$

3-(Dibutylamino)propylamine, CA 102-83-0: 377.04 (1).

$C_{11}H_{26}O_2Si$

Methyloctyldimethoxysilane: 363.15 (3).

$C_{11}H_{26}O_3Si$

Octyltrimethoxysilane, CA 3069-40-7: 371.15 (3).

Pentyltriethoxysilane, CA 2761-24-2: 341.15 (4).

$C_{11}H_{28}O_3Si_3$

Tris(trimethylsilyloxy)ethylene, CA 69097-20-7: 312.59 (1); 331.15 (3).

$C_{12}F_{27}N$

Perfluorotributylamine, CA 311-89-7: Nonflammable (1).

$C_{12}H_5N_7O_{12}$

2,2',4,4',6,6'-Hexanitrodiphenylamine, technical, CA 131-73-7: 302.59 (1).

$C_{12}H_8BrNO_2$

4-Bromo-3-nitrobiphenyl, CA 27701-66-2: Above 385.93 (1).

$C_{12}H_9Br$

5-Bromoacenaphthene, CA 2051-98-1: Above 385.93 (1).

2-Bromobiphenyl, technical, CA 2052-07-5: Above 385.93 (1).

3-Bromobiphenyl, CA 2113-57-7: 383.15 (1).

4-Bromobiphenyl, CA 92-66-0: 417.04 (6).

$C_{12}H_9BrO$

4-Bromodiphenyl ether, CA 101-55-3: Above 385.93 (1).

$C_{12}H_9ClO$

2-Chloro-4-phenylphenol, CA 92-04-6: 447.04 (6,7).

$C_{12}H_9N$

1-Naphthylacetonitrile, CA 132-75-2: Above 385.93 (1).

$C_{12}H_9NO$

2-Benzoylpyridine, CA 91-02-1: 423.15 (1).

3-Benzoylpyridine, CA 5424-19-1: 423.15 (1).

4-Benzoylpyridine, CA 14548-46-0: 423.15 (1).

$C_{12}H_9NO_2$

2-Nitrobiphenyl, CA 86-00-0: 416.15 (5,6); 452.04 (7).

$C_{12}H_9NO_3$

4-Nitrophenyl phenyl ether, CA 620-88-2: Above 385.93 (1).

$C_{12}H_{10}$

Biphenyl, CA 92-52-4: 386.15 (4,5,6,7).

$C_{12}H_{10}ClN$

2-(4-Chlorobenzyl)pyridine, CA 4350-41-8: Above 385.93 (1).
4-(4-Chlorobenzyl)pyridine, CA 4409-11-4: Above 385.93 (1).
3-Chlorodiphenylamine, CA 101-17-7: Above 385.93 (1)

$C_{12}H_{10}ClOP$

Diphenylphosphinic chloride, CA 1499-21-4: 299.26 (1).

$C_{12}H_{10}ClO_3P$

Diphenyl chlorophosphate, CA 2524-64-3: Above 385.93 (1).

$C_{12}H_{10}ClP$

Chlorodiphenylphosphine, technical, CA 1979-66-9: Above 385.93 (1).

$C_{12}H_{10}Cl_2Si$

Dichlorodiphenylsilane, CA 80-10-4: 415.37 (6); 430.93 (1).

$C_{12}H_{10}Cl_2Sn$

Diphenyltin dichloride, CA 1135-99-5: Above 385.93 (1).

$C_{12}H_{10}N_3O_3P$

Diphenylphosphoryl azide, CA 26386-88-9: Above 385.93 (1).

$C_{12}H_{10}O$

1'-Acenaphthone, CA 941-98-0: Above 385.93 (1).
2'-Acenaphthone, CA 93-08-3: Above 385.93 (1).
Phenyl ether, CA 101-84-8: 385.37 (6); above 385.93 (1); 388.15 (2,4,5,7).
2-Phenylphenol, CA 90-43-7: 397.04 (1,6).
4-Phenylphenol, CA 92-69-3: 438.71 (1).

$C_{12}H_{10}O_2$

4-Methoxy-1-naphthaldehyde, CA 15971-29-6: Above 385.93 (1).
1-Naphthyl acetate, CA 830-81-9: Above 385.93 (1).
3-Phenoxyphenol, technical, CA 713-68-8: 383.15 (1).

$C_{12}H_{10}S$

Phenyl sulfide, CA 139-66-2: Above 385.93 (1).

$C_{12}H_{11}N$

2-Aminobiphenyl, CA 90-41-5: Above 385.93 (1).
4-Aminobiphenyl, CA 92-67-1: Above 385.93 (1).
2-Benzylpyridine, CA 101-82-6: 398.15 (1).
3-Benzylpyridine, CA 620-95-1: 383.15 (1).
4-Benzylpyridine, CA 2116-65-6: 388.15 (1).
Diphenylamine, CA 122-39-4: 425.93 (1,5,6,7).
3-Methyl-2-phenylpyridine, CA 10273-90-2: 369.26 (1).
2-(para-Tolyl)pyridine, CA 4467-06-5: 383.15 (1).

$C_{12}H_{11}NS$

S,S-Diphenylsulfilimine, CA 36744-90-8: Above 385.93 (1).

$C_{12}H_{11}O_3P$

Diphenyl phosphite, CA 4712-55-4: 449.82 (1).

$C_{12}H_{11}P$

Diphenylphosphine, CA 829-85-6: 383.15 (1).

$C_{12}H_{12}$

1,2-Dimethylnaphthalene, technical, CA 573-98-8: Above 385.93 (1).
1,3-Dimethylnaphthalene, CA 575-41-7: 382.59 (1).
1,4-Dimethylnaphthalene, CA 571-58-4: Above 385.93 (1).
1,6-Dimethylnaphthalene, CA 575-43-9: Above 385.93 (1).
Dimethylnaphthalene, mixed isomers, CA 28804-88-8: 374.26 (1).
1-Ethylnaphthalene, CA 1127-76-0: 384.82 (1).
2-Ethylnaphthalene, CA 939-27-5: 377.59 (1).

$C_{12}H_{12}Fe$

Vinylferrocene, CA 1271-51-8: 335.37 (1).

$C_{12}H_{12}N_2O$

4-Aminophenyl ether, CA 101-80-4: 492.04 (1).

$C_{12}H_{12}O$

1-Ethoxynaphthalene, CA 5328-01-8: Above 385.93 (1).

$C_{12}H_{12}O_2$

2-Acetyl-1-tetralone, CA 17216-08-9: 378.15 (1).

$C_{12}H_{12}O_2Si$

Diphenylsilanediol, CA 947-42-2: 327.04 (1).

$C_{12}H_{12}O_3$

Ethyl 3-benzoylacrylate, technical, CA 17450-56-5: Above 385.93 (1).

$C_{12}H_{12}O_6$

Trimethyl 1,2,4-benzenetricarboxylate, CA 2459-10-1: Above 385.93 (1).

$C_{12}H_{12}Si$

Diphenylsilane, CA 775-12-2: 371.48 (1).

$C_{12}H_{13}N$

Citronitrile, mixed isomers: Above 385.93 (1).

N,N-Dimethyl-1-naphthylamine, CA 86-56-6: Above 385.93 (1).

N-Ethyl-1-naphthylamine, CA 118-44-5: Above 385.93 (1).

1-(1-Naphthyl)ethylamine ±, CA 42882-31-5: Above 385.93 (1).

1-Phenyl-1-cyclopentanecarbonitrile, CA 77-57-6: Above 385.93 (1).

$C_{12}H_{13}NO_2$

1-Benzoyl-4-piperidone, CA 24686-78-0: Above 385.93 (1).

Ethyl 3-indoleacetate, CA 778-82-5: Above 385.93 (1).

$C_{12}H_{14}$

1-Phenyl-1-cyclohexene, CA 31017-40-0: 376.48 (1).

$C_{12}H_{14}ClN$

5-Chloro-2-methylene-1,3,3-trimethylindoline, technical, CA 6872-17-9: Above 385.93 (1).

$C_{12}H_{14}Fe$

1,1'-Dimethylferrocene, CA 1291-47-0: 380.93 (1).

$C_{12}H_{14}O$

2-Butylbenzofuran, CA 4265-27-4: 374.26 (1).

Cyclopentyl phenyl ketone, CA 5422-88-8: Above 385.93 (1).

5,7-Dimethyl-1-tetralone, CA 13621-25-5: Above 385.93 (1).

2-Phenylcyclohexanone, CA 1444-65-1: Above 385.93 (1).

$C_{12}H_{14}O_2$

Precocene, CA 17598-02-6: Above 385.93 (1).

$C_{12}H_{14}O_4$

Diethyl phthalate, CA 84-66-2: 390.15 (5); 433.15 (1); 434.15* (4,6); 435.92* (2,7).

Diethyl terephthalate, CA 636-09-9: 390.15 (5,6).

Ethyl (phenylacetoxy)acetate, CA 18801-08-6: 383.15 (1).

Piperonyl isobutyrate, CA 5461-08-5: Above 385.93 (1).

$C_{12}H_{15}ClO$

1-(4-Chlorophenyl)-1-cyclopentanemethanol, CA 80866-79-1: Above 385.93 (1).

4-Pentylbenzoyl chloride, CA 49763-65-7: Above 385.93 (1).

$C_{12}H_{15}ClO_3$

Ethyl 2-(4-chlorophenoxy)-2-methylpropionate, CA 637-07-0: Above 385.93 (1).

$C_{12}H_{15}F_3N_2$

1-[2-(Trifluoromethyl)benzyl]piperazine: 380.37 (1).

$C_{12}H_{15}F_3O_2$

3-(Trifluoroacetyl)camphor, CA 51800-98-7: 355.37 (1).

$C_{12}H_{15}N$

1,2-Dihydro-2,2,4-trimethylquinoline, CA 147-47-7: 374.82 (1).

Julolidine, CA 479-59-4: Above 385.93 (1).

2-Methylene-1,3,3-trimethylindoline, CA 118-12-7: 374.82 (1).

1-Methyl-4-phenyl-1,2,3,6-tetrahydropyridine, CA 28289-54-5: Above 385.93 (1).

$C_{12}H_{15}NO$

1-Benzyl-4-piperidone, CA 3612-20-2: Above 385.93 (1).

$C_{12}H_{15}NO_2$

Acetoacet-*meta*-xylidide, CA 97-36-9: 444.26* (6,7).

4-Methoxymethyl-2-methyl-5-phenyl-2-oxazoline (-), CA 52075-14-6: Above 385.93 (1).

$C_{12}H_{15}NO_3$

Acetoacet-*para*-phenetidide, CA 122-82-7: 435.93 (6); 435.93* (7).

$C_{12}H_{15}N_3O_3$

2,4,6-Triallyloxy-1,3,5-triazine, CA 101-37-1: Above 353.15* (7); 383.15 (1).

Triallyl-1,3,5-triazine-2,4,6(1H,3H,5H)trione, CA 1025-15-6: Above 385.93 (1).

$C_{12}H_{16}$

tert-Butylstyrene, mixed isomers, CA 25338-51-6: 353.71 (6).

Cyclohexylbenzene, CA 827-52-1: 354.15 (3); 372.04 (1,5); 372.04* (6,7).

Methylcyclopentadiene dimer, CA 26472-00-4: 299.82 (1); 326.15 (3).

$C_{12}H_{16}BrNO$

1-[2-(4-Bromophenoxy)ethyl]pyrrolidine, CA 1081-73-8: Above 385.93 (1).

$C_{12}H_{16}N_2O_2$

1-Piperonylpiperazine, CA 32231-06-4: Above 385.93 (1).

$C_{12}H_{16}O$

2-Cyclohexylphenol, CA 119-42-6: 407.04 (6).

Hexanophenone, CA 942-92-7: Above 385.93 (1).

2-Phenyl-1-cyclohexanol, CA 1444-64-0: Above 385.93 (1); 410.93 (7).

$C_{12}H_{16}O_2$

alpha,alpha-Dimethylphenethyl acetate, CA 151-05-3: 369.26 (6).

Isobutyl phenylacetate, CA 102-13-6: Above 373.15 (6).

Methyl 4-tert-butylbenzoate, CA 26537-19-9: Above 385.93 (1).

3-Methyl-2-phenylvaleric acid, CA 7782-37-8: Above 385.93 (1).

$C_{12}H_{16}O_3$

2,2-Diethoxyacetophenone, CA 6175-45-7: 383.15 (1,4); 401.15* (4).

4-(Diethoxymethyl)benzaldehyde, CA 81172-89-6: 347.15 (3).

2',4'-Dimethoxy-3'-methylpropiophenone, CA 77942-13-3: 383.15 (1).

4-Hexanoylresorcinol, CA 3144-54-5: Above 385.93 (1).

Pentyl salicylate, CA 2050-08-0: 405.15 (5,6).

2,4,5-Trimethoxypropenylbenzene, mostly cis, 5273-86-9: Above 385.93 (1).

$C_{12}H_{16}O_4$

Terephthalaldehyde mono-(diethyl acetal), CA 81172-89-6: Above 385.93 (1).

$C_{12}H_{16}O_7$

Tri-O-acetylglucal (+), CA 2873-29-2: Above 385.93 (1).

$C_{12}H_{17}ClO$

2-Chloro-4-*tert*-pentylphenyl methyl ether: 383.15 (6).

$C_{12}H_{17}N$

4-Benzylpiperidine, CA 31252-42-3: Above 385.93 (1).

4-Cyclohexylaniline, CA 6373-50-8: Above 385.93 (1).

$C_{12}H_{17}NO$

1-Benzyl-2-pyrrolidinemethanol (-), CA 53912-80-4: 383.15 (1).

N-Butylacetanilide, CA 91-49-6: 414.26 (6,7).

N,*N*-Diethyl-*meta*-toluamide, CA 134-62-3: Above 385.93 (1).

$C_{12}H_{17}NO_2$

5-Amino-2,2-dimethyl-4-phenyl-1,3-dioxane (+), CA 35019-66-0: 361.15 (3); above 385.93 (1).

4-(3-Dimethylaminopropoxy)benzaldehyde, CA 26934-35-0: Above 385.93 (1).

$C_{12}H_{18}$

1-*tert*-Butyl-3,5-dimethylbenzene, CA 98-19-1: 341.15 (3); 345.37 (1).

1,5,9-Cyclododecatriene, *trans,cis,cis*, CA 2765-29-9: Above 348.15 (1).

1,5,9-Cyclododecatriene, *trans,trans,cis*, CA 706-31-0: 346.15 (3); 360.93 (1).

1,5,9-Cyclododecatriene, *trans,trans,trans*, CA 676-22-2: 346.15 (3); 354.26 (1).

1,5,9-Cyclododecatriene, mixed isomers, CA 4904-61-4: 344.26 (6); 346.15 (3).

1,2-Diisopropylbenzene, CA 577-55-9: 349.82* (7).

1,3-Diisopropylbenzene, CA 99-62-7: 349.82 (1,3).

1,4-Diisopropylbenzene, CA 100-18-5: 349.82 (1).

Diisopropylbenzene, mixed isomers, CA 25321-09-9: 350.15 (5); 350.15* (6).

Hexamethyldewarbenzene, CA 7641-77-2: 308.15 (1).

Hexylbenzene, CA 1077-16-3: 356.48 (1); 366.15 (3).

Pentyltoluene, mixed isomers: 355.15 (5); 355.15* (6).

1,2,4-Triethylbenzene, CA 877-44-1: 349.15 (3); 355.93* (6).

1,3,5-Triethylbenzene, CA 102-25-0: 349.15 (3); 354.26 (1).

Triethylbenzene, mixed isomers, CA 25340-18-5: 356.15 (5,7).

1,2,4-Trivinylcyclohexane, CA 2855-27-8: 342.04 (1).

$C_{12}H_{18}Br_6$

1,2,5,6,9,10-Hexabromocyclododecane, technical, CA 3194-55-6: Nonflammable (1).

$C_{12}H_{18}N_2$

4-Amino-1-benzylpiperidine, CA 50541-93-0: Above 385.93 (1).

2-(2-Piperidinoethyl)pyridine, CA 5452-83-5: 383.15 (1).

$C_{12}H_{18}N_2O$

Oxotremorine, CA 70-22-4: Above 385.93 (1).

$C_{12}H_{18}N_2O_2$

Isophorone diisocyanate, CA 4098-71-9: 436.15* (4).

$C_{12}H_{18}O$

1-Adamantyl methyl ketone, CA 1660-04-4: 381.48 (1).

Benzyl pentyl ether, CA 6382-14-5: 352.59* (2).

6-tert-Butyl-2,4-dimethylphenol, CA 1879-09-0: 384.82 (1).

2-tert-Butyl-5-methylanisole, CA 88-40-0: 363.15 (1).

2,6-Diisopropylphenol, technical, CA 2078-54-8: 385.93 (7); above 385.93 (1).

5,7-Dimethyl-3,5,9-decatrien-2-one: Above 385.93 (1).

1,2-Epoxy-5,9-cyclododecadiene, CA 943-93-1: Above 385.93 (1).

Ethylene glycol, 4-sec-butylphenyl ether: 422.04* (2).

Methyl 4-tert-pentylphenyl ether, CA 2050-03-5: 372.04 (6); 372.04* (2).

6-Pentyl-meta-cresol, CA 53043-14-4: 388.89 (7).

Pentyl tolyl ether, mixed isomers: 363.71* (2).

$C_{12}H_{18}O_2$

Carvyl acetate (-), mixed isomers, CA 97-42-7: 370.93 (1).

2-(4-tert-Butylphenoxy)ethanol, CA 713-46-2: 393.15* (6); 430.37* (2).

4-Hexyloxyphenol, CA 18979-55-0: Above 385.93 (1).

$C_{12}H_{18}O_3$

5-[1-(Acetoxy)-1-methylethyl]-2-methyl-2-cyclohexen-1-one ±, CA 72597-31-0: Above 385.93 (1).

4-Carbethoxy-2-ethyl-3-methyl-2-cyclohexen-1-one, technical, CA 51051-65-1: Above 385.93 (1).

Dipropylene glycol phenyl ether: 430.37* (2).

$C_{12}H_{18}O_4$

1,3-Butanediol dimethacrylate, CA 1189-08-8: 397.15* (4).

1,4-Butanediol dimethacrylate, CA 2082-81-7: Above 385.93 (1).

Butopyronoxyl, CA 532-34-3: Above 385.93 (1).

Di-tert-butyl acetylenedicarboxylate, CA 66086-33-7: Above 385.93 (1).

(continues)

$C_{12}H_{18}O_4$ *(continued)*

3,4-Dihydro-2,2-dimethyl-4-oxo-2*H*-pyran-6-carboxylic acid, butyl ester, CA 532-34-3: 430.37 (7).

1,6-Hexanediol diacrylate, technical, CA 13048-33-4: Above 385.93 (1).

$C_{12}H_{18}O_7$

Diethylene glycol, bis(allylcarbonate), CA 142-22-3: 450.15* (4); 465.38* (6,7).

$C_{12}H_{18}SSi$

1-(Trimethylsilyl)cyclopropyl phenyl sulfide, CA 74379-74-1: Above 385.93 (1).

$C_{12}H_{19}BO_2$

Diisopropoxyphenylborane, CA 1692-26-8: 304.82 (1).

$C_{12}H_{19}Cl$

1-Chloro-3,5-dimethyladamantane, CA 707-36-8: 366.48 (1).

$C_{12}H_{19}N$

4-*tert*-Butyl-*N*,*N*-dimethylaniline, CA 2909-79-7: 382.59 (1).

2,6-Diisopropylaniline, CA 24544-04-5: 397.04 (1).

$C_{12}H_{19}NO$

4-Hexyloxyaniline, CA 39905-57-2: Above 385.93 (1).

$C_{12}H_{20}$

1,3-Dimethyladamantane, CA 702-79-4: 325.93 (1).

$C_{12}H_{20}NO_3P$

Diethyl phenethylamidophosphate, technical, CA 57673-91-3: Above 385.93 (1).

$C_{12}H_{20}O_2$

4,8-Bis(hydroxymethyl)tricyclo[5,2,1,0^{2,6}]decane, CA 26896-48-0: Above 385.93 (1).

Bornyl acetate (-), CA 5655-61-8: 357.59 (1).

Dihydrocarvyl acetate (-), mixed isomers, CA 20777-49-5: 363.15 (1).

Ethyl chrysanthemumate ±, CA 97-41-6: 357.59 (1).

Geranyl acetate, mixed isomers, CA 16409-44-2: 368.15 (3); above 373.15 (6).

Isobornyl acetate ±, CA 17283-45-3: 360.93 (6).

(continues)

$C_{12}H_{20}O_2$ *(continued)*

  Isopulegyl acetate, CA 89-49-6: 358.71 (1).

  Linalyl acetate, CA 115-95-7: 358.15 (6); 363.15 (1); 367.15 (3).

  Terpinyl acetate, CA 80-26-2: 366.48 (6).

$C_{12}H_{20}O_3Si$

  Phenyltriethoxysilane, CA 780-69-8: 315.93 (1); 393.15 (4).

$C_{12}H_{20}O_4$

  Dibutyl fumarate, CA 105-75-9: 410.93 (2); 422.04* (7).

  Dibutyl maleate, CA 105-76-0: 383.15 (1); 408.15 (2); 414.15 (5); 414.15* (6).

$C_{12}H_{20}O_6$

  Glycerol tripropionate, CA 139-45-7: 440.15* (4,6).

$C_{12}H_{20}O_6S_3$

  1,2,6-Hexanetriol trithioglycolate, CA 19759-80-9: 383.15 (1).

$C_{12}H_{20}O_7$

  Triethyl citrate, CA 77-93-0: 383.15 (1); 423.71 (6); 423.71* (7).

$C_{12}H_{20}Sn$

  Tetraallyltin, CA 7393-43-3: 348.15 (1).

$C_{12}H_{21}N$

  Tris(2-methylallyl)amine, CA 6321-40-0: 326.48 (1).

$C_{12}H_{22}$

  Cyclododecene, mixed isomers, technical, CA 1501-82-2: 360.15 (3); 367.04 (1).

  Dicyclohexyl, CA 92-51-3: 347.04 (6); 374.82 (1).

  1,4-Diisopropyl-1-cyclohexene, mixed isomers, CA 39000-66-3: 344.26 (1).

  Dimethyl decalin, mixed isomers, CA 28777-88-0: 357.59 (1).

  1-Dodecyne, CA 765-03-7: 352.59 (1).

$C_{12}H_{22}O$

  Cyclododecane epoxide, CA 286-99-7: Above 385.93 (1).

  Cyclohexylcyclohexanol, mised isomers: 405.37 (6).

  10-Dodecadien-1-ol, *trans,trans*, CA 57002-06-9: 335.37 (1).

$C_{12}H_{22}O_2$

  5-Decen-1-yl acetate: 335.37 (1).

  delta-Dodecanolactone ±, CA 713-95-1: Above 385.93 (1).

  2-Ethylhexyl methacrylate ±, CA 688-84-6: 365.35 (1).

  Menthyl acetate, mixed isomers, CA 16409-45-3: 350.15 (3); 365.35 (1).

  4-Pentylcyclohexanecarboxylic acid, *trans*, CA 38289-29-1: 383.15 (1).

$C_{12}H_{22}O_3$

  Hexanoic anhydride, CA 2051-49-2: Above 385.93 (1).

  (-)-Menthoxyacetic acid, CA 40248-63-3: Above 385.93 (1).

$C_{12}H_{22}O_4$

  Diethyl suberate, CA 2050-23-9: Above 385.93 (1).

  Dimethyl sebacate, CA 106-79-6: 418.15* (1,6); 433.15 (2).

  Dipentyl oxalate, CA 20602-86-2: 389.15 (2); 391.15 (5,6).

$C_{12}H_{22}O_6$

  Dibutyl tartrate, CA 87-92-3: 364.15 (5,6); above 385.93 (1).

$C_{12}H_{23}BrO_2$

  12-Bromododecanoic acid, CA 73367-80-3: Above 385.93 (1).

$C_{12}H_{23}ClO$

  Lauroyl chloride, CA 112-16-3: Above 385.93 (1).

$C_{12}H_{23}N$

  Dicyclohexylamine, CA 101-83-7: 369.26 (1); 372.15 (3,5); above 372.15
  (6,7); 373.15* (2).

  Undecyl cyanide, CA 2437-25-4: Above 385.93 (1).

$C_{12}H_{23}P$

  Dicyclohexylphosphine, CA 829-84-5: 274.82 (1).

$C_{12}H_{24}$

  Cyclododecane, CA 294-62-2: 362.15 (3).

  1-Dodecene, CA 112-41-4: 346.15 (3); 350.93 (1); 352.04 (2).

  2-Methyl-1-undecene, CA 18516-37-5: 345.37 (1).

$C_{12}H_{24}N_2$

  1,2-Diperidinoethane, CA 1932-04-3: 383.15 (1).

$C_{12}H_{24}O$

Cycloundecanemethanol, technical, CA 29518-02-3:   Above 385.93 (1).

Decyl vinyl ether, CA 765-05-9:   358.15* (2).

7-Dodecen-1-ol, *cis*, CA 20056-92-2:   334.26 (1).

Dodecyl aldehyde, CA 112-54-9:   374.82 (1).

1,2-Epoxydodecane, CA 2855-19-8:   317.59 (1).

2-Methylundecanal, CA 110-41-8:   366.48 (1).

2,6,8-Trimethyl-4-nonanone, CA 123-18-2:   360.37 (2); 361.15 (4); 363.15* (4); 364.15 (5); 364.15* (6,7).

$C_{12}H_{24}O_2$

Ethyl decanoate, CA 110-38-3:   Above 373.15 (6); 375.37 (1).

Lauric acid, CA 143-07-7:   Above 385.93 (1).

Methyl undecanoate, CA 1731-86-8:   382.59 (1).

$C_{12}H_{24}O_3$

2,2,4-Trimethyl-1,3-pentanediol, 1-isobutyrate, CA 77-68-9:   393.15* (6); 399.15* (2).

$C_{12}H_{24}O_4Sn$

Dibutyltin diacetate, technical, CA 1067-33-0:   Above 385.93 (1); 416.48* (7).

$C_{12}H_{24}O_6$

18-Crown-6, CA 17455-13-9:   Above 385.93 (1).

$C_{12}H_{25}Br$

1-Bromododecane, CA 143-15-7:   383.15 (1); 417.15 (5,6); 419.26 (2).

2-Bromododecane, CA 13187-99-0:   383.15 (1).

$C_{12}H_{25}BrO$

12-Bromo-1-dodecanol, CA 3344-77-2:   Above 385.93 (1).

$C_{12}H_{25}Cl_3Si$

Dodecyltrichlorosilane, CA 4484-72-4:   Above 385.93 (1).

$C_{12}H_{25}I$

1-Iodododecane, CA 4292-19-7:   Above 385.93 (1).

$C_{12}H_{25}N$

Cyclododecylamine, CA 1502-03-0:   394.26 (1,3).

$C_{12}H_{25}NO_2$

1-Nitrododecane, CA 16891-99-9:   Above 385.93 (1).

$C_{12}H_{26}$

Dodecane, CA 112-40-3:   344.26 (1,2); 347.15 (3,5,6,7).

$C_{12}H_{26}N_2$

4-(Dimethylamino)-1,2,2,6,6-pentamethylpiperidine, CA 52185-74-7:
359.26 (1).

$C_{12}H_{26}N_2O_3Si$

3-(2-Imidazolin-1-yl)propyltriethoxysilane, CA 58068-97-6:   360.15 (3).

$C_{12}H_{26}O$

2-Butyl-1-octanol, CA 3913-02-8:   383.15 (6,7).

1-Dodecanol, CA 112-53-8:   Above 385.93 (1); 399.82 (6,7); 402.59* (2).

Hexyl ether, CA 112-58-3:   349.82* (2,6,7); 350.15 (5); 351.48 (1).

2,6,8-Trimethyl-4-nonanol, CA 123-17-1:   366.15 (5); 366.48* (6,7).

$C_{12}H_{26}O_3$

Diethylene glycol dibutyl ether, CA 112-73-2:   320.93 (1); 371.15 (3);
391.48* (2,4,6,7).

1,1,3-Triethoxyhexane, CA 101-33-7:   372.04* (6,7).

$C_{12}H_{26}O_4$

Tripropylene glycol isopropyl ether: 397.04* (2).

$C_{12}H_{26}O_5$

Tetraethylene glycol diethyl ether, CA 4353-28-0:   Above 385.93 (1).

$C_{12}H_{26}O_7$

Hexaethylene glycol, CA 2615-15-8:   Above 385.93 (1).

$C_{12}H_{26}S$

1-Dodecanethiol, CA 112-55-0:   360.93 (1); 360.93* (2); 400.93* (6,7).

tert-Dodecanethiol, mixed isomers, CA 25103-58-6:   363.71 (1); 369.26*
(2,6,7).

Hexyl sulfide, CA 6294-31-1:   Above 385.93 (1).

$C_{12}H_{27}Al$

Triisobutyl aluminum, CA 100-99-2:   Below 257.59 (7).

$C_{12}H_{27}AlO_3$

Aluminum tri-*sec*-butoxide, technical, CA 2269-22-9: 300.93 (1).

$C_{12}H_{27}B$

Tributyl borane, CA 122-56-5: 237.59 (7).

$C_{12}H_{27}BO_3$

Tributyl borate, CA 688-74-4: Below 343.15 (3); 366.48 (1,5); 366.48* (6,7).

Tri-*sec*-butyl borate, CA 22238-17-1: 347.04* (7).

Tri-*tert*-butyl borate, CA 7397-43-5: 302.59 (1).

Triisobutyl borate, CA 13195-76-1: 358.15 (5); 358.15* (6,7).

$C_{12}H_{27}BrSn$

Tributyltin bromide, technical, CA 1461-23-0: Above 385.93 (1).

$C_{12}H_{27}ClSi$

Chlorotributylsilane, CA 995-45-9: 370.15 (3); 377.59 (1).

Chlorotriisobutylsilane: 360.93 (1).

$C_{12}H_{27}ClSn$

Tributyltin chloride, CA 1461-22-9: Above 385.93 (1).

$C_{12}H_{27}N$

Dihexylamine, CA 143-16-8: 368.15 (1); 377.59* (6,7).

Dodecylamine, CA 124-22-1: 389.93* (2); above 385.93 (1).

Tributylamine, CA 102-82-9: 336.48 (1); 342.04 (2); 358.15 (3); 359.15 (5); 359.26* (2,6,7).

Triisobutylamine, CA 1116-40-1: 323.15 (3).

$C_{12}H_{27}O_3P$

Dibutyl butanephosphonate, CA 78-46-6: 428.15* (7).

Tributyl phosphite, CA 102-85-2: 355.15 (3); 393.15* (6,7); 394.26* (1,2).

$C_{12}H_{27}O_4P$

Tributyl phosphate, CA 126-73-8: 419.15* (2,5,6,7); 466.48 (1).

Triisobutyl phosphate, CA 126-71-6: 408.15 (5,6).

$C_{12}H_{27}P$

Tributylphosphine, CA 998-40-3: 310.37 (1); 313.15 (3).

$C_{12}H_{27}PS_3$

   S,S,S-Tributyl trithiophosphite, CA 150-50-5: 419.26* (7).

$C_{12}H_{28}Cl_2OSi_2$

   1,3-Dichloro-1,1,3,3-tetraisopropyldisiloxane, CA 69304-37-6: 349.82 (1); 383.15 (3).

$C_{12}H_{28}N_2$

   1,12-Diaminododecane, CA 2783-17-7: 428.15 (1).

$C_{12}H_{28}O_4Si$

   Tetraisopropoxysilane, CA 1992-48-9: 333.15 (4).

   Tetrapropoxysilane, CA 682-01-9: 368.15 (1,4).

$C_{12}H_{28}O_4Ti$

   Tetraisopropoxy titanium, CA 546-68-9: 295.37 (1); 333.15 (3).

   Tetrapropoxy titanium, CA 3087-37-4: 315.93 (1); 336.15 (3).

$C_{12}H_{28}O_4Zr$

   Tetrapropoxy zirconium, CA 23519-77-9: 295.15 (3).

$C_{12}H_{28}O_8Si$

   Tetrakis(2-methoxyethoxy)silane, CA 2157-45-1: 413.15 (4).

$C_{12}H_{28}Si$

   Tributylsilane, CA 998-41-4: 356.48 (1).

   Triisobutylsilane, CA 6485-81-0: 382.59 (1).

$C_{12}H_{28}Sn$

   Tributyltin hydride, CA 688-73-3: 313.15 (1).

$C_{12}H_{29}N_3$

   Bis(hexamethylene)triamine, technical, CA 143-23-7: Above 385.93 (1).

$C_{12}H_{30}BP$

   Borane-tributylphosphine complex, CA 4259-20-5: 308.71 (1).

$C_{12}H_{30}N_3P$

   Hexaethylphosphorous triamide, CA 2283-11-6: 332.04 (1).

$C_{12}H_{30}OSi_2$

Hexaethyldisiloxane, CA 994-49-0: 356.15 (3).

$C_{13}H_5F_5O$

2,3,4,5,6-Pentafluorobenzophenone, CA 1536-23-8: Above 385.93 (1).

$C_{13}H_7ClOS$

2-Chlorothioxanthen-9-one, CA 86-39-5: 364.26 (1).

$C_{13}H_7F_5O$

2,3,4,5,6-Pentafluorobenzhydrol, CA 1944-05-4: Above 385.93 (1).

$C_{13}H_8Cl_2O_2$

3-(3,4-Dichlorophenoxy)benzaldehyde, CA 79124-76-8: Above 385.93 (1).

3-(3,5-Dichlorophenoxy)benzaldehyde, CA 81028-92-4: Above 385.93 (1).

$C_{13}H_8CrO_6$

Pentacarbonyl(*alpha*-methoxybenzylidene)chromium, CA 27436-93-7: 374.26 (1).

$C_{13}H_8F_2O$

2,4-Difluorobenzophenone, CA 342-25-6: Above 385.93 (1).

2,4'-Difluorobenzophenone: Above 385.93 (1).

2,5-Difluorobenzophenone, CA 85068-36-6: Above 385.93 (1).

2,6-Difluorobenzophenone, CA 59189-51-4: Above 385.93 (1).

3,4-Difluorobenzophenone: Above 385.93 (1).

$C_{13}H_9ClO$

2-Chlorobenzophenone, CA 5162-03-8: Above 385.93 (1).

$C_{13}H_9ClO_2$

3-(4-Chlorophenoxy)benzaldehyde, CA 69770-20-3: Above 385.93 (1).

$C_{13}H_9FO$

2-Fluorobenzophenone, CA 342-24-5: Above 385.93 (1).

4-Fluorobenzophenone, CA 345-83-5: Above 385.93 (1).

$C_{13}H_9N$

7,8-Benzoquinoline, CA 230-27-3: Above 385.93 (1).

$C_{13}H_{10}Cl_2$

Chloro(4-chlorophenyl)phenylmethane, CA 134-83-8: Above 385.93 (1).

Dichlorodiphenylmethane, CA 2051-90-3: Above 385.93 (1).

$C_{13}H_{10}F_2$

Bis(4-Fluorophenyl)methane, CA 457-68-1: Above 385.93 (1).

$C_{13}H_{10}F_2O$

4,4'-Difluorobenzhydrol, CA 365-24-2: Above 385.93 (1).

$C_{13}H_{10}O$

Benzophenone, CA 119-61-9: Above 385.93 (1).

$C_{13}H_{10}O_2$

2-Hydroxybenzophenone, CA 117-99-7: Above 385.93 (1).

3-Phenoxybenzaldehyde, CA 39515-51-0: Above 385.93 (1).

4-Phenoxybenzaldehyde, CA 67-36-7: Above 385.93 (1).

$C_{13}H_{10}O_3$

Phenyl salicylate, CA 118-55-8: Above 385.93 (1).

$C_{13}H_{11}Br$

Bromodiphenylmethane, CA 776-74-9: Above 385.93 (1).

$C_{13}H_{11}Cl$

Chlorodiphenylmethane, CA 90-99-3: Above 385.93 (1).

$C_{13}H_{11}N$

Benzophenone imine, CA 1013-88-3: Above 385.93 (1).

$C_{13}H_{11}NO$

1-(1-Naphthyl)ethyl isocyanate, CA 88442-63-1: 366.48 (1).

$C_{13}H_{12}$

Diphenylmethane, CA 101-81-5: Above 385.93 (1); 403.15 (5,6,7).

2-Phenyltoluene, CA 643-58-3: 380.37 (1); 410.93* (6).

3-Phenyltoluene, CA 643-93-6: Above 385.93 (1).

4-Phenyltoluene, CA 644-08-6: Above 385.93 (1).

$C_{13}H_{12}O$

2-Biphenylmethanol, CA 2928-43-0: Above 385.93 (1).

2-Hydroxydiphenylmethane, CA 28994-41-4: Above 385.93 (1).

2-Methoxybiphenyl, CA 86-26-0: Above 385.93 (1).

3-Phenoxytoluene, CA 3586-14-9: Above 385.93 (1).

$C_{13}H_{12}O_2$

2-(Benzyloxy)phenol, CA 6272-38-4: Above 385.93 (1).

Methyl 1-naphthaleneacetate, CA 2876-78-0: Above 385.93 (1).

1-Naphthyl propionate, CA 3121-71-9: Above 385.93 (1).

3-Phenoxybenzyl alcohol, CA 13826-35-2: Above 385.93 (1).

$C_{13}H_{12}S$

Benzyl phenyl sulfide, CA 831-91-4: Above 385.93 (1).

$C_{13}H_{12}S_2$

Bis(phenylthio)methane, CA 3561-67-9: Above 385.93 (1).

$C_{13}H_{13}ClSi$

Chlorodiphenylmethylsilane, CA 144-79-6: Above 385.93 (1).

$C_{13}H_{13}N$

Aminodiphenylmethane, CA 91-00-9: Above 385.93 (1).

2-Benzylaniline, CA 28059-64-5: Above 385.93 (1).

3-Methyldiphenylamine, CA 1205-66-7: Above 385.93 (1).

N-Phenylbenzylamine, CA 103-32-2: Above 385.93 (1).

$C_{13}H_{13}NO$

4-Dimethylamino-1-naphthaldehyde, CA 1971-81-9: Above 385.93 (1).

$C_{13}H_{13}OP$

Methyl diphenylphosphinite, CA 4020-99-9: Above 385.93 (1).

$C_{13}H_{13}O_4P$

Diphenyl methyl phosphate, technical, CA 115-89-9: Above 385.93 (1).

$C_{13}H_{13}P$

Methyl diphenylphosphine, CA 1486-28-8: Above 385.93 (1).

$C_{13}H_{14}$

2,3,5-Trimethylnaphthalene CA 2245-38-7: 383.15 (1).

$C_{13}H_{14}N_2$

Bis(4-aminophenyl)methylene, CA 26446-71-9: 493.15 (6).

4,4'-Methylenedianiline, CA 101-77-9: 494.26 (1,4); 499.82 (7).

$C_{13}H_{14}O_6$

Methyl phthalyl ethyl glycolate, CA 85-71-2: 466.48* (6,7).

$C_{13}H_{14}Si$

Diphenylmethylsilane, CA 776-76-1: Above 385.93 (1).

$C_{13}H_{16}O$

Cyclohexyl phenyl ketone, CA 712-50-5: Above 385.93 (1).

$C_{13}H_{16}O_2$

1-(2-sec-Butylphenoxy)-2-propanol: 405.37 (7).

2-(3-Methoxyphenyl)cyclohexanone, CA 15547-89-4: Above 385.93 (1).

Phenylpropiolic aldehyde diethyl acetal, CA 6142-95-6: Above 385.93 (1).

$C_{13}H_{16}O_3$

Ethyl 2-benzylacetoacetate, CA 620-79-1: Above 385.93 (1).

Precocene II, CA 644-06-4: Above 385.93 (1).

$C_{13}H_{16}O_4$

Diethyl phenylmalonate, CA 83-13-6: Above 385.93 (1).

$C_{13}H_{17}ClO$

4-Hexylbenzoyl chloride, CA 50606-95-6: Above 385.93 (1).

$C_{13}H_{17}ClO_2$

4-Hexyloxybenzoyl chloride, CA 39649-71-3: Above 385.93 (1).

$C_{13}H_{17}FeN$

(Dimethylaminomethyl)ferrocene, CA 1271-86-9: Above 385.93 (1).

$C_{13}H_{17}NO$

4-Acetyl-4-phenylpiperidine, CA 34798-80-6: Above 385.93 (1).

$C_{13}H_{17}NO_2$

Ethyl *N*-benzyl-*N*-cyclopropylcarbamate, CA 2521-01-9: Above 385.93 (1).

$C_{13}H_{18}FN$

1-(4-Fluorophenyl)-*N*-methylcyclohexylamine, CA 78987-76-5: Above 385.93 (1).

$C_{13}H_{18}N_2O$

2-(*N*-Propylcarbamoyl)-1,2,3,4-tetrahydroisoquinoline, CA 88630-43-7: Above 385.93 (1).

$C_{13}H_{18}O$

Cyclamen aldehyde, CA 103-95-7: 360.93 (6).

Heptanophenone, CA 1671-75-6: Above 385.93 (1).

$C_{13}H_{18}O_2$

4-*tert*-Butylphenyl 2,3-epoxypropyl ether, CA 3101-60-8: 374.82 (1).

4-*tert*-Pentylphenyl acetate: 388.71 (6).

2-Phenylpropyl butyrate, CA 80866-83-7: Above 385.93 (1).

2-Phenylpropyl isobutyrate, CA 65813-53-8: Above 385.93 (1).

$C_{13}H_{18}O_4$

2-Butoxyethyl salicylate: 430.37 (6).

$C_{13}H_{19}BrO_2$

Methyl 3-bromo-1-adamantaneacetate, CA 14575-01-0: Above 385.93 (1).

$C_{13}H_{19}N$

1-Phenethylpiperidine, CA 332-14-9: 383.15 (1).

$C_{13}H_{19}NO_2S$

Isobornyl thiocyanatoacetate, CA 115-31-1: 355.15 (7).

$C_{13}H_{20}$

1-Phenylheptane, CA 1078-71-3: 368.15 (1).

$C_{13}H_{20}N_2Si$

N-Benzyl-*N*-(trimethylsilylmethyl)aminoacetonitrile, CA 87813-07-8: 383.15 (1).

$C_{13}H_{20}O$

alpha-Ionone, CA 127-41-3: Above 373.15 (6); 377.59 (1).

beta-Ionone, CA 79-77-6: Above 373.15 (3); above 385.93 (1).

Pentyl xylyl ether, mixed isomers: 369.15 (5); 369.26* (2,6).

$C_{13}H_{20}O_2$

Carvyl propionate, CA 97-45-0: 380.93 (1).

2-(4-tert-Pentylphenoxy)ethanol, CA 6382-07-6: 410.93 (6).

Propylene glycol, 4-sec-butylphenyl ether: 419.26* (2).

$C_{13}H_{20}O_3$

Triethyl orthobenzoate, CA 1663-61-2: 369.82 (1).

$C_{13}H_{20}O_4$

Diethyl diallylmalonate, CA 3195-24-2: Above 385.93 (1).

$C_{13}H_{21}N$

2,6-Di-tert-butylpyridine, CA 585-48-8: 344.15 (3); 345.37 (1).

$C_{13}H_{22}$

Perhydrofluorene, CA 5744-03-6: 337.59 (1).

$C_{13}H_{22}N_2$

1,3-Dicyclohexylcarbodiimide, CA 538-75-0: Above 385.93 (1).

$C_{13}H_{22}O$

Geranyl acetone, CA 38237-36-4: Above 385.93 (1).

Neryl acetone, CA 3879-26-3: 338.15 (3).

$C_{13}H_{22}O_2$

Geranyl propionate, CA 105-90-8: Above 373.15 (6).

$C_{13}H_{24}N_2$

3,3,6,9,9-Pentamethyl-2,10-diazabicyclo(4,4,0)dec-1-ene, CA 69340-58-5: 369.15 (3).

$C_{13}H_{24}O$

Cyclotridecanone, technical, CA 832-10-0: Above 385.93 (1).

$C_{13}H_{24}O_2$

Decyl acrylate, CA 2156-96-9: 500.37* (6).

256

$C_{13}H_{24}O_4$

Diethyl azelate, technical, CA 624-17-9: Above 385.93 (1).

$C_{13}H_{26}$

1-Tridecene, CA 2437-56-1: 352.59 (1,2); 365.15 (3).

$C_{13}H_{26}N_2$

1,1'-Methylenebis(3-methylpiperidine), CA 68922-17-8: 383.15 (1).

$C_{13}H_{26}O$

Tridecanal, technical, CA 10486-19-8: Above 385.93 (1).

2-Tridecanone, CA 593-08-8: 380.37 (6).

7-Tridecanone, CA 462-18-0: Above 385.93 (1).

$C_{13}H_{26}O_2$

Ethyl undecanoate, CA 627-90-7: Above 385.93 (1).

Methyl laurate, CA 111-82-0: Above 385.93 (1).

Tridecanoic acid, CA 638-53-9: Above 385.93 (1).

$C_{13}H_{27}Br$

1-Bromotridecane, CA 765-09-3: Above 385.93 (1).

2-Bromotridecane, CA 59157-17-4: Above 385.93 (1).

$C_{13}H_{28}$

Tridecane, CA 629-50-5: 352.59 (1,2); 367.15 (3).

$C_{13}H_{28}O$

1-Tridecanol, CA 112-70-9: 355.37* (6); above 385.93 (1); 394.26* (6,7).

$C_{13}H_{28}O_3$

Tributyl orthoformate, CA 588-43-2: 362.15 (3).

$C_{13}H_{28}O_4$

Tetraisopropyl orthocarbonate, CA 36597-49-6: 330.37 (1).

Tripropylene glycol butyl ether, CA 57499-93-1: 408.15* (2).

$C_{13}H_{29}N$

Tridecylamine, CA 2869-34-3: Above 385.93 (1).

$C_{13}H_{29}NO_2$

(*N*,*N*-Dimethoxy)dineopentoxymethane, CA 4909-78-8: 324.82 (1); 335.15 (3).

$C_{13}H_{29}NO_4$

N,*N*-Bis(2,2-diethoxyethyl)methylamine, CA 6948-86-3: 333.15 (1).

$C_{13}H_{30}OSn$

Tributyltin methoxide, CA 1067-52-3: 372.04 (1).

$C_{14}H_8O_2$

Anthraquinone, CA 84-65-1: 458.15 (1,6); 458.15* (5).

$C_{14}H_9Br$

9-Bromophenanthrene, CA 573-17-1: Above 385.93 (1).

$C_{14}H_9F_3O$

3-(Trifluoromethyl)benzophenone, CA 728-81-4: Above 385.93 (1).

$C_{14}H_9F_3O_2$

3-[3-(Trifluoromethyl)phenoxy]benzaldehyde, CA 78725-46-9: Above 385.93 (1).

$C_{14}H_{10}$

Anthracene, CA 120-12-7: 394.15 (5,6,7).

Phenanthrene, CA 85-01-8: 444.26* (6).

$C_{14}H_{10}Cl_2O$

2-Chloro-2,2-diphenylacetyl chloride, CA 2902-98-9: Above 385.93 (1).

$C_{14}H_{10}O_3$

Benzoic anhydride, CA 93-97-0: Above 385.93 (1).

$C_{14}H_{10}O_9$

Tannic acid, CA 1401-55-4: 472.04* (6).

$C_{14}H_{12}$

9,10-Dihydrophenanthrene, CA 776-35-2: Above 385.93 (1).

1,1-Diphenylethylene, CA 530-48-3: Above 385.93 (1).

Stilbene, *cis*, CA 645-49-8: Above 385.93 (1).

258

$C_{14}H_{12}O$

Deoxybenzoin, CA 451-40-1: Above 385.93 (1).

Diphenylacetaldehyde, CA 947-91-1: Above 385.93 (1).

2-Methylbenzophenone, CA 131-58-8: Above 385.93 (1).

3-Methylbenzophenone, CA 643-65-2: Above 385.93 (1).

$C_{14}H_{12}O_2$

Benzyl benzoate, CA 120-51-4: 421.15 (1,5,6,7).

3-(4-Methylphenoxy)benzaldehyde, CA 79124-75-7: Above 385.93 (1).

4'-Phenoxyacetophenone, CA 5031-78-7: Above 385.93 (1).

$C_{14}H_{12}O_3$

Benzyl salicylate, CA 118-58-1: Above 373.15 (6).

3-(4-Methoxyphenoxy)benzaldehyde, CA 62373-80-2: Above 385.93 (1).

$C_{14}H_{13}ClO$

2-Biphenyl 2-chloroethyl ether: 433.15 (6).

4-Chloro-2-(*alpha*-methylbenzyl)phenol ±, CA 5828-70-6: Above 385.93 (1).

$C_{14}H_{13}N$

N-Benzylidenebenzylamine, CA 780-25-6: 383.15 (1).

$C_{14}H_{13}P$

Diphenylvinylphosphine, CA 2155-96-6: Above 385.93 (1).

$C_{14}H_{14}$

Bibenzyl, CA 103-29-7: Above 385.93 (1); 402.04 (6,7).

3,3'-Dimethylbiphenyl, CA 612-75-9: Above 385.93 (1).

1,1-Diphenylethane, CA 612-00-0: Above 373.15 (6); 402.15 (5).

2-Ethylbiphenyl, CA 1812-51-7: Above 373.15 (5).

4-Ethylbiphenyl, CA 5707-44-8: Above 385.93 (1).

$C_{14}H_{14}N_2O$

2-Methyl-1,2-di-3-pyridyl-1-propanone, CA 54-36-4: Above 385.93 (1).

$C_{14}H_{14}O$

2-Benzoyl-5-norbornene, mixed isomers, technical, CA 6056-35-5: Above 385.93 (1).

2-Benzylbenzyl alcohol, CA 1586-00-1: Above 385.93 (1).

Benzyl ether, CA 103-50-4: 408.15 (1,5,6,7); 408.15* (2).

(continues)

$C_{14}H_{14}O$ *(continued)*

2,2-Diphenylethanol, CA 1883-32-5: Above 385.93 (1).

4-Methylbenzhydrol, CA 1517-63-1: Above 385.93 (1).

$C_{14}H_{14}O_2$

3-Benzyloxybenzyl alcohol, CA 1700-30-7: Above 385.93 (1).

3,3'-Dimethoxybiphenyl, CA 6161-50-8: Above 385.93 (1).

1-Naphthylbutyrate, CA 3121-70-8: 335.37 (1).

$C_{14}H_{14}O_3S$

ortho-Tolyl-*para*-toluenesulfonate, CA 599-75-7: 457.04 (6).

$C_{14}H_{14}O_4$

Diallyl phthalate, CA 131-17-9: 383.15 (1); 438.71 (6,7).

$C_{14}H_{14}S$

Benzyl sulfide, CA 26898-12-4: Above 385.93 (1).

$C_{14}H_{14}S_2$

para-Tolyl disulfide, CA 103-19-5: Above 385.93 (1).

$C_{14}H_{15}F_7O_2$

3-Heptafluorobutyrylcamphor, CA 51800-99-8: 367.04 (1).

$C_{14}H_{15}N$

Dibenzylamine, CA 103-49-1: 416.48 (1).

1,2-Diphenylethylamine, CA 25611-78-3: Above 385.93 (1).

2,2-Diphenylethylamine, CA 3963-62-0: Above 385.93 (1).

$C_{14}H_{15}NO$

N-(2-Phenoxyethyl)aniline, CA 622-18-4: 443.15 (6).

$C_{14}H_{15}OP$

Ethyl diphenylphosphinite, CA 1733-57-9: Above 385.93 (1).

$C_{14}H_{15}O_3P$

Dibenzyl phosphite, technical, CA 17176-77-1: Above 385.93 (1).

$C_{14}H_{16}$

Butylnaphthalene, mixed isomers, CA 31711-50-9: 633.15 (6).

$C_{14}H_{16}N_2O_2$

3,3'-Dimethoxybenzidine, CA 119-90-4: 479.15* (1,5,6,7).

$C_{14}H_{16}O_4$

Diethyl benzalmalonate, CA 5292-53-5: Above 385.93 (1).

$C_{14}H_{16}O_6$

Phthalic acid, ethyl ethylglycolate ester, CA 84-72-0: 458.15* (5,6).

$C_{14}H_{17}N$

N-Butyl-1-naphthylamine, CA 6281-00-1: 419.26 (2).

$C_{14}H_{18}N_2$

1-Benzyl-4-piperidineacetonitrile, CA 78056-67-4: Above 385.93 (1).

1,8-Bis(dimethylamino)naphthalene, N,N,N',N'-tetramethyl-1,8-naphthalene-diamine, CA 20734-58-1: Above 385.93 (1).

$C_{14}H_{18}O$

alpha-Pentylcinnamaldehyde, CA 122-40-7: Above 385.93 (1).

$C_{14}H_{18}O_3$

7-Ethoxy-6-methoxy-2,2-dimethylchromene, CA 65383-73-5: Above 385.93 (1).

$C_{14}H_{18}O_4$

Diethyl benzylmalonate, CA 607-81-8: Above 385.93 (1).

Dipropyl phthalate, CA 131-16-8: Above 385.93 (1).

$C_{14}H_{18}O_6$

Bis(2-methoxyethyl) phthalate, CA 117-82-8: 408.15 (6); 460.37 (6); 483.15* (6).

$C_{14}H_{18}O_7$

Pentaerythritol triacrylate, technical, CA 3524-68-3: Above 385.93 (1).

$C_{14}H_{19}ClO$

4-Heptylbenzoyl chloride, CA 50606-96-7: Above 385.93 (1).

$C_{14}H_{19}ClO_2$

4-Heptyloxybenzoyl chloride, CA 40782-54-5: Above 385.93 (1).

$C_{14}H_{19}N$

5-(4-Pyridyl)-2,7-nonadiene, mixed isomers, technical: 383.15 (1).

$C_{14}H_{19}NO_4$

6-(Carbobenzyloxyamino)caproic acid, CA 1947-00-8: Above 385.93 (1).

$C_{14}H_{20}$

tert-Butyltetralin, mixed isomers: 633.15 (6).

$C_{14}H_{20}N_2O$

2-(*N*-Butylcarbamoyl)-1,2,3,4-tetrahydroisoquinoline, CA 88630-42-6: Above 385.93 (1).

$C_{14}H_{20}O$

Octanophenone, CA 1674-37-9: Above 385.93 (1).

$C_{14}H_{20}O_3$

2-(4-*tert*-Butylphenoxy)ethyl acetate: 435.37* (6).

$C_{14}H_{20}O_4$

Hexyl vanillate: Above 385.93 (1).

$C_{14}H_{21}NO$

6-Ethoxy-1,2,3,4-tetrahydro-2,2,4-trimethylquinoline, CA 16489-90-0: Above 385.93 (1).

$C_{14}H_{22}$

1,3-Di-*tert*-butylbenzene, CA 1014-60-4: 357.04 (1).

1-Phenyloctane, CA 2189-60-8: 380.93 (1).

$C_{14}H_{22}O$

Di-*sec*-butylphenol, mostly 2,6-isomer, technical, CA 5510-99-6: 400.37 (1).

2,4-Di-*tert*-butylphenol, CA 96-76-4: 388.15 (1); 402.59 (7).

2,6-Di-*tert*-butylphenol, CA 128-39-2: 391.48 (1).

Methyl ionone, CA 79-69-4: Above 373.15 (6).

$C_{14}H_{22}O_2$

2,5-Di-*tert*-butylhydroquinone, CA 88-58-4: 488.71* (6).

4-Propylbenzaldehyde diethyl acetal: Above 385.93 (1).

$C_{14}H_{22}O_4$

Glyoxal bis(diallyl acetal), CA 16646-44-9:   Above 385.93 (1).

$C_{14}H_{22}O_6$

Triethylene glycol dimethacrylate, CA 109-16-0:   Above 385.93 (1).

$C_{14}H_{22}O_7$

Diethylene glycol dilevulinate, CA 67385-17-5:   444.15* (5); 444.26 (6).

Tetraethylene glycol diacrylate, technical, CA 17831-71-9:   Above 385.93 (1).

$C_{14}H_{23}N$

N,N-Dibutylaniline, CA 613-29-6:   383.15 (2,6).

2,6-Di-*tert*-butyl-4-methylpyridine, CA 38222-83-2:   357.04 (1).

N-(2-Ethylhexyl)aniline, CA 10137-80-1:   435.93* (6,7).

4-Octylaniline, CA 16245-79-7:   Above 385.93 (1).

$C_{14}H_{23}NO_4$

N-(3-Methoxypropyl)-3,4,5-trimethoxybenzylamine, CA 34274-04-9:   383.15 (1).

$C_{14}H_{24}N_2$

N,N-Di-*sec*-butyl-1,4-phenylenediamine, CA 29467-99-0:   Above 385.93 (1); 413.71* (7).

N,N'-Di-*sec*-butyl-1,4-phenylenediamine, CA 101-96-2:   405.37 (6).

$C_{14}H_{24}O_2$

trans-7,*cis*-9-Dodecadien-1-yl acetate, CA 55774-32-8:   334.26 (1).

trans-8,*trans*-10-Dodecadien-1-yl acetate, CA 53880-51-6:   335.37 (1).

Geranyl butyrate, CA 166-29-6:   Above 373.15 (6).

$C_{14}H_{24}O_4$

Dipentyl maleate, CA 10099-71-5:   405.37 (6).

$C_{14}H_{26}$

Butyldecalin, mixed isomers: 533.15 (6).

tert-Butyldecalin, mixed isomers, CA 27193-30-2:   610.93 (6).

$C_{14}H_{26}N_2O$

N,N'-Dicyclohexyl-*O*-methylisourea, CA 6257-10-9:   Above 385.93 (1).

$C_{14}H_{26}O_2$

7-Dodecen-l-yl acetate, *cis*, CA 14959-86-5: 334.26 (1).

Isodecyl methacrylate, CA 29964-84-9: 394.15* (4).

Myristoleic acid, CA 544-64-9: 335.37 (1).

2,4,7,9-Tetramethyl-5-decyne-4,7-diol, mixed isomers, CA 126-86-3: Above 385.93 (1).

$C_{14}H_{26}O_3$

Heptanoic anhydride, CA 626-27-7: Above 385.93 (1).

$C_{14}H_{26}O_4$

Dibutyl adipate, CA 105-99-7: 389.15* (4).

Diethyl sebacate, technical, CA 110-40-7: Above 385.93 (1).

Diisobutyl adipate, CA 141-04-8: 433.15 (2); 433.15* (4).

$C_{14}H_{26}O_7$

Diethylene glycol, bis(butyl carbonate), CA 73454-84-9: 462.04 (6).

$C_{14}H_{27}BrO_2$

2-Bromotetradecanoic acid, CA 10520-81-7: Above 385.93 (1).

$C_{14}H_{27}ClO$

Myristoyl chloride, CA 112-64-1: Above 385.93 (1).

$C_{14}H_{27}N$

N-Ethyldicyclohexylamine, CA 7175-49-7: 363.15 (3).

$C_{14}H_{28}$

1-Tetradecene, technical, CA 1120-36-1: 383.15 (5,6); 388.71 (1,2).

7-Tetradecene, *trans*, CA 41446-63-3: 372.59 (1).

$C_{14}H_{28}O$

1,2-Epoxytetradecane, technical, CA 3234-28-4: Above 385.93 (1).

7-Tetradecen-l-ol, *cis*, CA 40642-43-1: 334.26 (1).

9-Tetradecen-l-ol,-*cis*, CA 35153-15-2: 335.37 (1).

11-Tetradecen-l-ol, *cis*, CA 34010-15-6: 335.37 (1).

Tetradecyl aldehyde, technical, CA 124-25-4: Above 385.93 (1).

2,6,8-Trimethylnonyl vinyl ether, CA 10141-19-2: 366.48* (7).

$C_{14}H_{28}O_2$

Dodecyl acetate, CA 112-66-3: Above 385.93 (1).   *(continues)*

$C_{14}H_{28}O_2$ (continued)

Methyl tridecanoate, CA 1731-88-0:   Above 385.93 (1).

Myristic acid, CA 544-63-8:   Above 385.93 (1).

$C_{14}H_{28}Sn$

Ethynyltributyltin, CA 994-89-8:   347.04 (1).

$C_{14}H_{29}Br$

1-Bromotetradecane, CA 112-71-0:   Above 385.93 (1).

$C_{14}H_{29}Cl$

1-Chlorotetradecane, CA 2425-54-9:   Above 385.93 (1).

$C_{14}H_{29}N$

N-(2-Ethylhexyl)cyclohexylamine, CA 5432-61-1:   402.59* (6,7).

$C_{14}H_{30}$

Tetradecane, CA 629-59-4:   373.15 (1,5,6,7); 394.26 (2).

$C_{14}H_{30}O$

7-Ethyl-2-methyl-4-undecanol, CA 103-20-8:   413.71* (6).

1-Tetradecanol, CA 112-72-1:   413.71* (6,7); 414.15 (5); 416.48* (2).

2-Tetradecanol, CA 4706-81-4:   Above 385.93 (1).

$C_{14}H_{30}S$

Tetradecyl mercaptan, CA 2079-95-0:   424.82* (2).

tert-Tetradecyl mercaptan, CA 28983-37-1:   394.26 (6).

$C_{14}H_{30}Sn$

Vinyltributyltin, CA 7486-35-3:   Above 385.93 (1).

$C_{14}H_{31}N$

N,N-Dimethyldodecylamine, CA 112-18-5:   383.15 (1).

Monobutyl dipentylamine, CA 79073-34-0:   366.48* (2).

1-Tetradecylamine, technical, CA 2016-42-4:   Above 385.93 (1).

$C_{14}H_{32}N_2O_4$

N,N,N',N'-Tetrakis(2-hydroxypropyl)ethylenediamine, CA 102-60-3:   Above 385.93 (1).

$C_{14}H_{32}N_4$

1,4,8,11-Tetramethyl-1,4,8,11-tetraazacyclotetradecane, CA 41203-22-9: Above 385.93 (1).

$C_{14}H_{32}OSn$

Tributyltin ethoxide, CA 36253-76-6: 313.15 (1).

$C_{14}H_{32}O_3Si$

Octyltriethoxysilane, CA 2943-75-1: 373.15 (4).

$C_{14}H_{32}SSn$

Ethyl tributyltin sulfide, CA 23716-85-0: Above 385.93 (1).

$C_{15}H_{10}N_2O_2$

4,4'-Diphenylmethane diisocyanate, CA 101-68-8: 474.15* (4).

$C_{15}H_{12}O$

Chalcone, CA 94-41-7: Above 385.93 (1).

Dibenzosuberone, CA 1210-35-1: Above 385.93 (1).

beta-Phenylcinnamaldehyde, CA 13702-35-7: Above 385.93 (1).

$C_{15}H_{13}N$

2,2-Diphenylpropionitrile, CA 5558-67-8: Above 385.93 (1).

$C_{15}H_{14}$

9-Ethylfluorene, CA 2294-82-8: Above 385.93 (1).

$C_{15}H_{14}O$

2,4-Dimethylbenzophenone, CA 1140-14-3: Above 385.93 (1).

2,5-Dimethylbenzophenone, technical, CA 4044-60-4: Above 385.93 (1).

3,4-Dimethylbenzophenone, CA 2571-39-3: Above 385.93 (1).

1,3-Diphenylacetone, CA 102-04-5: Above 385.93 (1).

$C_{15}H_{14}O_2$

Benzoin methyl ether ±, CA 3524-62-7: Above 385.93 (1).

2'-Hydroxy-3-phenylpropiophenone, CA 42772-82-7: Above 385.93 (1).

$C_{15}H_{16}$

1,1-Diphenylpropane, CA 1530-03-6: Above 373.15 (5).

2,2-Diphenylpropane, CA 778-22-3: Above 385.93 (1).

(continues)

$C_{15}H_{16}$ *(continued)*

  2-Isopropylbiphenyl, CA 19486-60-3: 413.71 (6).

  4-Isopropylbiphenyl, CA 7116-95-2: Above 385.93 (1).

  2-Propylbiphenyl, CA 20282-28-4: Above 373.15 (5,6).

$C_{15}H_{16}O$

  3,3-Diphenyl-1-propanol, CA 20017-67-8: Above 385.93 (1).

$C_{15}H_{17}N$

  N-Benzyl-2-phenethylamine, CA 3647-71-0: Above 385.93 (1).

  3,3-Diphenylpropylamine, CA 5586-73-2: Above 385.93 (1).

  N-Ethylbenzylaniline, CA 92-59-1: 413.15* (6).

$C_{15}H_{18}$

  Guaiazulene, CA 489-84-9: Above 385.93 (1).

  Pentylnaphthalene, mixed isomers: 380.15 (5); 397.04* (6).

$C_{15}H_{18}O$

  2-Naphthyl pentyl ether, CA 31059-19-5: 427.59* (2).

$C_{15}H_{20}Fe$

  tert-Pentylferrocene, CA 1277-58-3: Above 385.93 (1).

$C_{15}H_{20}O$

  alpha-Hexylcinnamaldehyde, technical, CA 101-86-0: Above 385.93 (1).

$C_{15}H_{20}O_6$

  Trimethylolpropane triacrylate, technical, CA 15625-89-5: Above 385.93 (1); 422.04* (6).

$C_{15}H_{22}O_3$

  Ethyl 3-(1-adamantyl)-3-oxopropionate, CA 19386-06-2: Above 385.93 (1).

$C_{15}H_{23}N$

  2,3-Cyclododecenopyridine, CA 6571-43-3: Above 385.93 (1).

$C_{15}H_{24}$

  alpha-Humulene, CA 6753-98-6: 363.15 (3); 382.04 (1).

  Isolongifolene (-), CA 1135-66-6: 322.15 (3).

  Longicyclene (+), CA 1137-12-8: 365.15 (3).

(continues)

$C_{15}H_{24}$ *(continued)*

Longifolene (+), CA 475-20-7: 371.15 (3); 374.26 (1).

alpha-Longipinene (+), CA 5989-08-2: 365.15 (3).

Nonylbenzene, CA 1081-77-2: 372.04 (5,6); above 385.93 (1).

1,3,5-Triisopropylbenzene, CA 717-74-8: 365.15 (2).

Triisopropylbenzene, mixed isomers, CA 27322-34-5: 370.37* (6).

$C_{15}H_{24}O$

Butyl 4-*tert*-pentylphenyl ether: 408.15 (6).

2,4-Di-*tert*-butyl-5-methylphenol, CA 497-39-2: 400.93* (7).

2,6-Di-*tert*-butyl-4-methylphenol, CA 128-37-0: 400.15 (6) 400.15* (7).

Nonylphenol, mixed isomers, technical, CA 25154-52-3: Above 385.93 (1); 413.71 (6,7).

Santalol, CA 11031-45-1: Above 373.15 (6).

$C_{15}H_{24}O_2$

4-Butylbenzaldehyde diethyl acetal: 372.15 (3); 383.15 (1).

$C_{15}H_{24}O_6$

Tripropylene glycol diacrylate, technical, CA 42978-66-5: Above 385.93 (1).

$C_{15}H_{25}NO_2S$

N,N-Dibutyltoluenesulfonamide, CA 599-65-5: 438.71 (6).

$C_{15}H_{26}O$

Farnesol, mixed isomers, CA 4602-84-0: 369.26 (1).

3-Hydroxy-3,7,11-trimethyl-1,6,10-dodecatriene, CA 7212-44-4: 369.26 (1).

$C_{15}H_{26}O_4S_2$

2,3-Dimercapto-1-propanol tributyrate, CA 58428-97-0: Above 385.93 (1).

$C_{15}H_{26}O_6$

Glycerol tributyrate, CA 60-01-5: 447.04 (1); 453.15* (6).

$C_{15}H_{27}N_3O$

2,4,6-Tris(dimethylaminomethyl)phenol, CA 90-72-2: Above 385.93 (1).

$C_{15}H_{28}$

Isopropyl bicyclohexyl, mixed isomers, CA 28577-51-7: 397.04 (6).

$C_{15}H_{28}O_2$

Dodecyl acrylate, CA 2156-97-0: 356.15 (3).

$C_{15}H_{28}O_4$

Dimethyl brassylate, CA 1472-87-3: Above 385.93 (1).

$C_{15}H_{29}NO_2$

N,N-Dimethylformamide dicyclohexyl acetal, CA 2016-05-9: 363.71 (1).

$C_{15}H_{30}$

1-Pentadecene, CA 13360-61-7: Above 385.93 (1).

1,3,5-Triisopropylcyclohexane, mised isomers, technical, CA 34387-60-5: 367.04 (1).

$C_{15}H_{30}N_2$

4,4'-Trimethylenebis(1-methylpiperidine), CA 64168-11-2: 383.15 (1).

$C_{15}H_{30}N_3OP$

Tripiperidinophosphine oxide, CA 4441-17-2: Above 385.93 (1).

$C_{15}H_{30}O$

8-Pentadecanone, CA 818-23-5: 383.15 (1).

$C_{15}H_{30}O_2$

Ethyl tridecanoate, CA 28267-29-0: Above 385.93 (1).

Methyl myristate, CA 124-10-7: Above 385.93 (1).

Pentadecanoic acid, CA 1002-84-2: Above 385.93 (1).

$C_{15}H_{30}O_3$

Propylene glycol monolaurate, CA 142-55-2: 460.93* (2).

$C_{15}H_{30}O_4$

Glycerol monolaurate, CA 27215-38-9: 491.48* (2).

$C_{15}H_{31}Br$

1-Bromopentadecane, CA 629-72-1: Above 385.93 (1).

$C_{15}H_{32}$

Pentadecane, CA 629-62-9: 405.37 (1,2).

$C_{15}H_{32}O$

1-Pentadecanol, CA 629-76-5: Above 385.93 (1).

$C_{15}H_{32}Sn$

Allyltributyltin, CA 24850-33-7: Above 385.93 (1).

$C_{15}H_{33}BO_3$

Tripentyl borate, CA 621-78-3: 355.15 (5); 355.37* (6).

$C_{15}H_{33}N$

N,N-Diisobutyl-2,4-dimethyl-3-pentylamine, CA 54561-97-6: 366.15 (3).
Triisopentylamine, CA 645-41-0: 350.15 (3).
Tripentylamine, CA 621-77-2: 361.15 (3); 364.26 (2); 375.15 (5); 375.15* (2,6).

$C_{15}H_{33}NO_6$

Tris[2-(2-Methoxyethoxy)ethyl]amine, CA 70384-51-9: Above 385.93 (1).

$C_{16}H_{12}$

1-Phenylnaphthalene, CA 605-02-7: Above 385.93 (1).

$C_{16}H_{12}N_2O_2$

Tolidine diisocyanate, CA 91-97-4: 487.15* (4).

$C_{16}H_{12}N_2O_4$

Dianisidine diisocyanate, CA 91-93-0: 518.15* (4).

$C_{16}H_{13}NO$

2-(3-Benzoylphenyl)propionitrile, CA 42872-30-0: Above 385.93 (1).
2-Methyl-4,5-diphenyloxazole, CA 14224-99-8: Above 385.93 (1).

$C_{16}H_{14}$

1-Phenyl-3,4-dihydronaphthalene, CA 7469-40-1: Above 385.93 (1).

$C_{16}H_{14}O$

1,3-Diphenyl-2-buten-1-one, CA 495-45-4: 449.82* (6).
Methyl styrylphenyl ketone, CA 1322-90-3: 449.82* (7).

$C_{16}H_{14}O_2$

Benzyl cinnamate, CA 103-41-3: Above 385.93 (1).

$C_{16}H_{16}O_2$

4-Benzyloxy-3-methoxystyrene, CA 55708-65-1:   Above 385.93 (1).

$C_{16}H_{18}$

2-Butylbiphenyl, CA 54532-97-7:   Above 373.15 (5,6).
1,1-Diphenylbutane, CA 719-79-9:   Above 373.15 (5,6).

$C_{16}H_{18}O$

4-*tert*-Butyl-2-phenylphenol, CA 577-92-4:   433.15 (6).
alpha-Methylbenzyl ether, CA 93-96-9:   408.15* (6,7).

$C_{16}H_{18}O_2Si$

Methyl (diphenylmethylsilyl)acetate, CA 89266-73-9:   Above 385.93 (1).

$C_{16}H_{19}ClSi$

tert-Butylchlorodiphenylsilane, CA 58479-61-1:   Above 385.93 (1).

$C_{16}H_{19}N$

Bis(*alpha*-methylbenzyl)amine, CA 10024-74-5:   352.59* (7).

$C_{16}H_{19}P$

Butyldiphenylphosphine, CA 6372-41-4:   Above 385.93 (1).

$C_{16}H_{20}N_2$

N,N'-Dibenzylethylenediamine, CA 140-28-3:   Above 385.93 (1).

$C_{16}H_{20}N_2S_2$

2,2'-(3,6-Dithiaoctamethylene)dipyridine, CA 64691-70-9:   Above 385.93 (1).

$C_{16}H_{22}O_4$

Dibutyl isophthalate, CA 3126-90-7:   434.15 (5,6).
Dibutyl phthalate, CA 84-74-2:   430.15 (5,6,7); 444.26* (1,2); 455.37 (2); 463.15* (4).
Diethyl 2-ethyl-2-(*para*-tolyl)malonate, CA 68692-80-8:   Above 385.93 (1).
Diisobutyl phthalate, CA 84-69-5:   447.04 (2); 458.15* (6); 469.26* (7).

$C_{16}H_{22}O_6$

Di(2-ethoxyethyl) phthalate, CA 605-54-9:   446.15 (6,7); 446.15* (5).

$C_{16}H_{24}O$

Decanophenone, CA 6048-82-4: Above 385.93 (1).

$C_{16}H_{25}ClO$

2-Chloro-4,6-di-*tert*-pentylphenol, CA 42350-99-2: 394.26 (6).

$C_{16}H_{26}$

Decylbenzene, mixed isomers, CA 29463-65-8: 380.37 (6).

Dipentylbenzene, mixed isomers: 380.37* (6).

1-Phenyldecane, CA 104-72-3: Above 385.93 (1).

$C_{16}H_{26}O$

2,4-Dipentylphenol, CA 138-00-1: 400.15 (5); 400.15* (6,7).

Pentyl 4-*tert*-phenyl ether: 399.82* (2).

$C_{16}H_{26}O_3$

2-Dodecen-1-ylsuccinic anhydride, CA 25377-73-5: 450.93 (1).

$C_{16}H_{26}Si_2$

1,4-Bis(trimethylsilyl)-1,4-dihydronaphthalene, CA 1085-97-8: 322.04 (1).

$C_{16}H_{27}N$

4-Decylaniline, CA 37529-30-9: Above 385.93 (1).

N,N-Dipentylaniline, mixed isomers, CA 6249-76-9: 399.82 (2).

Di-*tert*-pentylaniline, mixed isomers: 402.59 (2).

$C_{16}H_{28}O_4$

cis-9,*trans*-11-Tetradecadien-1-yl acetate, CA 30562-09-5: 334.26 (1).

cis-9,*trans*-12-Tetradecadien-1-yl acetate, CA 31654-77-0: 335.37 (1).

$C_{16}H_{28}O_4$

Bis(1,3-dimethylbutyl) maleate: 416.48* (6).

Dihexyl maleate, CA 105-52-2: 416.48* (7)

$C_{16}H_{30}O$

9-Hexadecenal, *cis*, CA 56219-04-6: 335.37 (1).

11-Hexadecenal, *cis*, CA 53939-28-9: 334.26 (1).

$C_{16}H_{30}O_2$

Lauryl methacrylate, CA 142-90-5: 383.15 (1); 405.15* (4).

(continues)

$C_{16}H_{30}O_2$ *(continued)*

Palmitoleic acid, CA 373-49-9: 335.37 (1).

7-Tetradecen-1-yl acetate, *cis*, CA 16974-10-0: 335.37 (1).

11-Tetradecen-1-yl acetate, *cis*, CA 20711-10-8: 334.26 (1).

11-Tetradecen-1-yl acetate, *trans*, CA 33189-72-9: 335.37 (1).

Tridecyl acrylate, CA 3076-04-8: 405.37* (6).

$C_{16}H_{30}O_4$

Dibutyl suberate, CA 16090-77-0: Above 385.93 (1).

Diethyl dodecanedioate, CA 10471-28-0: 383.15 (1).

2,2,4-Trimethyl-1,3-pentanediol diisobutyrate, CA 6846-50-0: 394.26* (6,7); 416.15* (4).

$C_{16}H_{31}BrO_2$

2-Bromohexadecanoic acid, CA 18263-25-7: Above 385.93 (1).

$C_{16}H_{32}$

1-Hexadecene, CA 629-73-2: 401.48 (2); 405.37 (1).

$C_{16}H_{32}N_2O_5$

4,7,13,16,21-Pentaoxa-1,10-diazabicyclo(8,8,5)tricosane, CA 31364-42-8: Above 385.93 (1).

$C_{16}H_{32}O$

Di-*tert*-pentylcyclohexanol, mixed isomers: 405.37 (6).

1,2-Epoxyhexadecane, technical, CA 7320-37-8: 366.48 (1).

11-Hexadecen-1-ol, *cis*, CA 56683-54-6: 334.26 (1).

$C_{16}H_{32}O_2$

Ethyl myristate, CA 124-06-1: 383.15 (1).

Methyl pentadecanoate, CA 7132-64-1: Above 385.93 (1).

$C_{16}H_{32}O_4$

Diethylene glycol monolaurate, CA 141-20-8: 416.48 (6).

$C_{16}H_{33}Br$

1-Bromohexadecane, CA 112-82-3: Above 385.93 (1); 449.82 (2).

$C_{16}H_{33}Cl$

1-Chlorohexadecane, CA 4860-03-1: Above 385.93 (1).

$C_{16}H_{33}Cl_3Si$

Hexadecyltrichlorosilane, CA 5894-60-0: 419.26 (6); 419.26* (7).

$C_{16}H_{33}I$

1-Iodohexadecane, CA 544-77-4: Above 385.93 (1).

$C_{16}H_{33}NO$

N,N-Diethyldodecanamide, CA 3352-87-2: Above 385.93 (1).

$C_{16}H_{34}$

2,2,4,4,6,8,8-Heptamethylnonane, CA 4390-04-9: 368.71 (1).

Hexadecane, CA 544-76-3: 408.15 (1,2).

$C_{16}H_{34}O$

2-Ethylhexyl ether, CA 10143-60-9: 385.93 (1).

1-Hexadecanol, CA 124-29-8: Above 385.93 (1); 422.04* (2); as 96% solution, 408.15 (1).

2-Hexadecanol, CA 14852-31-4: Above 385.93 (1).

Octyl ether, CA 629-82-3: Above 385.93 (1).

$C_{16}H_{34}O_4$

2,5-Dimethyl-2,5-di(tert-butylperoxy)hexane, CA 78-63-7: Above 355.37* (7).

$C_{16}H_{34}O_5$

Tetraethylene glycol dibutyl ether, CA 112-98-1: 424.82* (6).

$C_{16}H_{34}S$

Hexadecyl mercaptan, technical, CA 2917-26-2: 374.82 (1); 422.04* (2).

tert-Hexadecyl mercaptan, CA 25360-09-2: 402.59* (6); 416.48 (2).

Octyl sulfide, CA 2690-08-6: Above 385.93 (1).

$C_{16}H_{35}N$

Bis(2-ethylhexyl)amine, CA 106-20-7: 319.15 (3); above 385.93 (1); 405.15 (5); 405.15* (2,6,7).

Dioctylamine, CA 1120-48-5: Above 385.93 (1).

1-Hexadecylamine, technical, CA 143-27-1: 413.71* (1,2).

$C_{16}H_{35}O_3P$

Bis(2-ethylhexyl)hydrogen phosphite, CA 3658-48-8: Above 385.93 (1); 438.71* (2); 444.26* (2).

$C_{16}H_{35}O_4P$

Bis(2-ethylhexyl) hydrogen phosphate, CA 298-07-7: Above 385.93 (1); 469.26* (6).

$C_{16}H_{36}ClN$

Tetrabutylammonium chloride, CA 1112-67-0: Above 385.93 (1).

$C_{16}H_{36}O_4Si$

Tetrabutoxysilane, CA 4766-57-8: 352.04 (1); 383.15 (4).
Tetra(sec-butoxy)silane, CA 5089-76-9: 377.15 (4).

$C_{16}H_{36}O_4Ti$

Tetrabutoxytitanium, CA 5593-70-4: 331.15 (3); 349.82 (1,7).

$C_{16}H_{36}Sn$

Tetrabutyltin, technical, CA 1461-25-2: 380.37 (1).

$C_{17}H_{16}O_4$

Dibenzyl malonate, CA 15014-25-2: Above 385.93 (1).

$C_{17}H_{17}NO_2$

N-(Diphenylmethylene)glycine, ethyl ester, CA 69555-14-2: Above 385.93 (1).

$C_{17}H_{17}N_5O_4$

N^6-Benzoyl-2'-deoxyadenosine hydrate, CA 4546-72-9: Above 385.93 (1).

$C_{17}H_{18}N_2O_2$

N,N'-Bis(salicylidene)-1,3-propanediamine, CA 120-70-7: Above 385.93 (1).

$C_{17}H_{18}O_2$

3-(4-tert-Butylphenoxy)benzaldehyde, CA 69770-23-6: Above 385.93 (1).
tert-Butyl 4-phenoxyphenyl ketone, CA 55814-54-5: Above 385.93 (1).

$C_{17}H_{18}O_3$

4'-Benzyloxy-2'-methoxy-3'-methylacetophenone: Above 385.93 (1).

$C_{17}H_{19}ClN_2$

1-(4-Chlorobenzhydryl)piperazine, technical, CA 303-26-4: Above 385.93 (1).

$C_{17}H_{20}$

1,1-Diphenylpentane, CA 1726-12-1: Above 373.15 (5,6).

4-Pentylbiphenyl, CA 7116-96-3: Above 385.93 (1); 422.04 (7).

$C_{17}H_{20}N_2O$

1,3-Diethyl-1,3-diphenylurea, CA 85-98-3: 423.15 (5,6,7).

$C_{17}H_{20}O_3$

1,3-Dibenzyloxy-2-propanol, CA 6972-79-8: Above 385.93 (1).

$C_{17}H_{22}OSi$

tert-Butyldiphenylmethoxysilane, CA 76358-47-9: 383.15 (1).

$C_{17}H_{28}O_2$

Farnesyl acetate, CA 29548-30-9: 368.15 (3).

4-Hexylbenzaldehyde diethyl acetal: Above 385.93 (1).

$C_{17}H_{32}Sn$

2,4-Cyclopentadien-1-yltributyltin, CA 3912-86-5: Above 385.93 (1).

$C_{17}H_{34}$

1-Heptadecene, CA 6765-39-5: Above 385.93 (1).

$C_{17}H_{34}O$

2-Heptadecanone, CA 2922-51-2: 393.15 (6).

9-Heptadecanone, technical, CA 540-08-9: Above 385.93 (1).

$C_{17}H_{34}O_2$

Isopropyl myristate, technical, CA 110-27-0: Above 385.93 (1); 424.82* (2).

Methyl palmitate, CA 112-39-0: Above 385.93 (1).

Octyl isononanoate, CA 71566-49-9: 399.82* (2); 402.59* (2).

Pentyl dodecanoate, CA 5350-03-8: 422.04 (6,7).

$C_{17}H_{36}$

Heptadecane, CA 629-78-7: 422.04 (1,2).

$C_{17}H_{36}N_2O$

1,1,3,3-Tetrabutylurea, CA 4559-96-8: 366.48 (1).

276

$C_{17}H_{36}O$

3,9-Diethyl-6-tridecanol, CA 123-24-0: 427.59* (6).

1-Heptadecanol, CA 1454-85-9: Above 385.93 (1); 427.15 (5).

Heptadecanol, mixed primary isomers, CA 52783-44-5: 427.59* (7).

$C_{17}H_{36}O_3S$

Hexadecyl methanesulfonate, CA 20779-14-0: 383.15 (1).

$C_{17}H_{37}N$

N-Methyldioctylamine, CA 4455-26-9: Above 385.93 (1).

$C_{18}H_5F_{10}P$

Bis(pentafluorophenyl)phenylphosphine, CA 5074-71-5: Above 385.93 (1).

$C_{18}H_{14}$

ortho-Terphenyl, CA 84-15-1: 383.15 (1); 436.15 (5); 436.15* (6); 444.15 (4).

meta-Terphenyl, CA 92-06-8: 464.15* (5,6); 479.15 (4).

para-Terphenyl, CA 92-94-4: 480.15* (5,7); 483.15 (4).

$C_{18}H_{14}O$

2-Phenoxybiphenyl, CA 6738-04-1: Above 385.93 (1).

$C_{18}H_{15}O_3P$

Triphenyl phosphite, CA 101-02-0: 491.48* (1,2,6,7).

$C_{18}H_{15}O_4P$

Triphenyl phosphate, CA 115-86-6: 493.15 (6,7); 493.15* (5); 497.04 (1).

$C_{18}H_{15}P$

Triphenylphosphine, CA 603-35-0: 453.15* (6,7); 454.82 (1).

$C_{18}H_{15}Sb$

Triphenylantimony, CA 603-36-1: Above 385.93 (1).

$C_{18}H_{16}Si$

Triphenylsilane, CA 789-25-3: 349.26 (1).

$C_{18}H_{18}BO_3P$

Borane-triphenyl phosphite complex, CA 55811-33-1: Above 385.93 (1).

$C_{18}H_{18}O_3$

Ethyl 2-benzylbenzoylacetate, CA 56409-75-7: Above 385.93 (1).

$C_{18}H_{18}O_5$

Diethylene glycol dibenzoate, CA 120-55-8: 458.15 (7); 505.15 (6); 505.15* (4).

$C_{18}H_{18}O_6$

Triallyl 1,3,5-benzenetricarboxylate, CA 17832-16-5: Above 385.93 (1).

$C_{18}H_{18}O_7$

Diethylene glycol, bis(phenylcarbonate): 510.93 (6).

$C_{18}H_{20}$

1,3-Dimethyl-1,3-diphenylcyclobutane, CA 597-28-4: 415.93 (6).

$C_{18}H_{20}O_2$

Benzoin isobutyl ether, CA 22499-12-3: 358.15 (1).

$C_{18}H_{20}O_3$

4'-Benzyloxy-2'-methoxy-3'-methylpropiophenone: Above 385.93 (1).

$C_{18}H_{21}NO$

N-(4-Methoxybenzylidene)-4-butylaniline, CA 26227-73-6: Above 385.93 (1).

$C_{18}H_{22}$

1,1-Bis(3,4-dimethylphenyl)ethane, technical, CA 1742-14-9: Above 385.93 (1).

4-Hexylbiphenyl, CA 59662-31-6: Above 385.93 (1).

$C_{18}H_{24}O$

(1S)-(-)-Nopol benzyl ether, CA 74851-17-5: 383.15 (1).

$C_{18}H_{24}O_6$

Phthalic acid, butyl, butyl glycolate ester, CA 85-70-1: 469.15* (5); 472.04* (6).

$C_{18}H_{25}NO_3$

2-Isobutoxy-1-isobutoxycarbonyl-1,2-dihydroquinoline, CA 38428-14-7: Above 385.93 (1).

$C_{18}H_{26}O_4$

Dipentyl phthalate, CA 131-18-0: 391.15 (5,6).

$C_{18}H_{26}O_6$

Trimethylolpropane trimethacrylate, technical, CA 3290-92-4: Above 385.93 (1); above 422.15* (4).

$C_{18}H_{27}NO_4$

Diethyl 3,3'-(phenethylimino)dipropionate, technical: Above 385.93 (1).

$C_{18}H_{28}O$

Dodecanophenone, CA 1674-38-0: Above 385.93 (1).

$C_{18}H_{30}$

1-Phenyldodecane, CA 123-01-3: Above 385.93 (1); 413.71 (7).

$C_{18}H_{30}O$

Dodecylphenol, mixed isomers, CA 27193-86-8: 435.93* (6,7).

$C_{18}H_{30}O_2$

2-(Dipentylphenoxy)ethanol, mixed isomers: 422.15 (5).

$C_{18}H_{31}N$

4-Dodecylaniline, CA 104-42-7: Above 385.93 (1).

$C_{18}H_{32}O_2$

cis-7,cis-11-Hexadecadien-1-yl acetate, CA 52207-99-5: 335.37 (1).
Linoleic acid, CA 60-33-3: Above 385.93 (1).

$C_{18}H_{32}O_4$

Dimethylpentyl maleate: 416.48 (2).

$C_{18}H_{32}O_7$

Tributyl citrate, CA 77-94-1: 430.15 (5,6).

$C_{18}H_{32}SSn$

Phenyl tributyltin sulfide, CA 17314-33-9: Above 385.93 (1).

$C_{18}H_{33}ClO$

Oleoyl chloride, technical, CA 112-77-6: Above 385.93 (1).

$C_{18}H_{34}O$

13-Octadecenal, *cis*, CA 58594-45-9: 334.26 (1).

$C_{18}H_{34}O_2$

Elaidic acid, CA 112-79-8: Above 385.93 (1).
cis-11-Hexadecen-1-yl acetate, CA 34010-21-4: 334.26 (1).
Oleic acid, CA 112-80-1: 457.59 (6); 462.15 (7); 462.15* (5).
Vaccenic acid, *cis*, CA 506-17-2: Above 385.93 (1).
Vaccenic acid, *trans*, CA 693-72-1: Above 385.93 (1).

$C_{18}H_{34}O_4$

Dibutyl sebacate, CA 109-43-3: 449.82 (2); 451.15* (5,6,7).
Diethyl tetradecanedioate, CA 19812-63-6: 379.26 (1).

$C_{18}H_{34}O_6$

Di(2-butoxyethyl) adipate, CA 141-18-4: 461.15* (4); 463.71 (2).
Triethylene glycol, di(2-ethylbutyrate), CA 95-08-9: 469.26* (7).

$C_{18}H_{34}O_9$

Diethylene glycol, bis(2-butoxyethyl carbonate): 465.93 (6).

$C_{18}H_{35}ClO$

Stearoyl chloride, technical, CA 112-76-5: Above 385.93 (1).

$C_{18}H_{35}N$

Stearonitrile, technical, CA 638-65-3: Above 385.93 (1).

$C_{18}H_{36}$

1-Octadecene, technical, CA 112-88-9: 416.48 (2); 422.04 (1).

$C_{18}H_{36}B_2O_6$

Triethylene glycol diborate, CA 100-89-0: 447.04 (7).

$C_{18}H_{36}O$

1,2-Epoxyoctadecane, technical, CA 7390-81-0: 383.15 (1).
Hexadecyl vinyl ether, CA 822-28-6: 435.93* (2).
Oleyl alcohol, technical, CA 143-28-2: Above 385.93 (1).

$C_{18}H_{36}O_2$

Ethyl palmitate, CA 628-97-7: Above 385.93 (1).

(continues)

$C_{18}H_{36}O_2$ *(continued)*

Methyl heptadecanoate, CA 1731-92-6: Above 385.93 (1).

Stearic acid, CA 57-11-4: 469.15 (4,5,6,7).

$C_{18}H_{36}O_3S_6$

1,5,9,13,17,21-Hexathiacyclotetracosane-3,11,19-triol, mixed isomers: 383.15 (1).

$C_{18}H_{36}O_4$

Diethylene glycol myristate: 416.48 (7).

$C_{18}H_{37}Br$

1-Bromooctadecane, CA 112-89-0: Above 385.93 (1).

$C_{18}H_{37}Cl$

1-Chlorooctadecane, CA 3386-33-2: Above 385.93 (1).

$C_{18}H_{37}Cl_3Si$

Octadecyltrichlorosilane, CA 112-04-9: 362.59 (1,6).

$C_{18}H_{37}I$

1-Iodooctadecane, CA 629-93-6: 383.15 (1).

$C_{18}H_{37}N$

Oleylamine, technical, CA 112-90-3: 427.59 (1).

$C_{18}H_{38}$

Octadecane, CA 593-45-3: 438.71 (1,2).

$C_{18}H_{38}O$

1-Octadecanol, CA 112-92-5: 452.59* (2).

$C_{18}H_{38}S$

Octadecyl mercaptan, CA 2885-00-9: 458.15 (1); 466.48* (2).

$C_{18}H_{39}BO_3$

Trihexyl borate, CA 5337-36-0: 422.04 (7).

$C_{18}H_{39}ClSi$

Chlorotrihexylsilane, CA 3634-67-1: 383.15 (1).

$C_{18}H_{39}N$

Octadecylamine, CA 124-30-1: 383.15 (1).

Trihexylamine, CA 102-86-3: Above 385.93 (1).

$C_{18}H_{39}NO$

Bis(2-ethylhexyl)ethanolamine, CA 101-07-5: 410.93 (6).

$C_{18}H_{39}O_3P$

Trihexyl phosphite, CA 6095-42-7: 433.15* (6).

$C_{18}H_{39}O_7P$

Tris(2-butoxyethyl) phosphate, CA 78-51-3: 497.04 (7).

$C_{18}H_{40}N_2$

N,N-Bis(1-methylheptyl)ethylenediamine, CA 5064-47-1: Above 477.59 (6).

$C_{18}H_{40}Si$

Trihexylsilane, CA 2929-52-4: Above 385.93 (1).

$C_{18}H_{45}O_4PSi_2Sn$

Bis(trimethylsilyl) tributylstannyl phosphate, CA 74785-85-6: 383.15 (1).

$C_{19}H_{16}$

Triphenylmethane, CA 519-73-3: Above 373.15 (5,6).

$C_{19}H_{16}S_3$

Tris(phenylthio)methane, CA 4832-52-4: Above 385.93 (1).

$C_{19}H_{17}O_4P$

Diphenyltolyl phosphate, mixed isomers, CA 26444-49-5: 503.15* (5).

$C_{19}H_{20}O_4$

Benzyl butyl phthalate, CA 85-68-7: 472.04 (6,7); 491.48 (2).

$C_{19}H_{23}NO$

N-(4-Ethoxybenzylidene)-4-butylaniline, CA 29743-08-6: Above 385.93 (1).

$C_{19}H_{25}NO$

4-Dimethylamino-1,2-diphenyl-3-methyl-2-butanol (+), CA 38345-66-3: Above 385.93 (1).

$C_{19}H_{26}$

Nonylnaphthalene, mixed isomers, CA 27193-93-7: Above 366.48 (6).

$C_{19}H_{28}O_4$

2,2,4-Trimethyl-1,3-pentanediol, isobutyrate benzoate: 435.93* (6).

$C_{19}H_{30}O_5$

Piperonyl butoxide, technical, CA 51-03-6: 444.26 (7).

$C_{19}H_{32}$

1-Phenyltridecane, CA 123-02-4: Above 385.93 (1).

$C_{19}H_{32}O_2$

Methyl linolenate, CA 301-00-8: Above 385.93 (1).

Methyl *gamma*-linolenate, CA 16326-32-2: 335.37 (1).

$C_{19}H_{34}O_2$

Methyl linoleate, CA 112-63-0: Above 385.93 (1).

$C_{19}H_{36}O_2$

Methyl oleate, CA 112-62-9: Above 385.93 (1).

Vaccenic acid, methyl ester, CA 52380-33-3: 335.37 (1).

$C_{19}H_{36}O_3$

Methyl 7-oxooctadecanoate, CA 2380-22-5: 383.15 (1).

Methyl 12-oxooctadecanoate, CA 2380-27-0: 383.15 (1).

$C_{19}H_{37}NO$

Octadecyl isocyanate, CA 112-96-9: 457.15* (4); 458.15 (1).

$C_{19}H_{38}$

1-Nonadecene, CA 18435-45-5: Above 385.93 (1).

$C_{19}H_{38}O$

cis-7,8-Epoxy-2-methyloctadecane ±, technical, CA 29804-22-6: 334.26 (1).

2-Nonadecanone, CA 629-66-3: 397.04 (6).

10-Nonadecanone, CA 504-57-4: Above 385.93 (1).

$C_{19}H_{38}O_2$

Isopropyl palmitate, technical, CA 142-91-6: 435.93* (2).

(continues)

$C_{19}H_{38}O_2$ *(continued)*

Methyl stearate, CA 112-61-8: 426.15 (4,6,7).

$C_{19}H_{40}$

Nonadecane, CA 629-92-5: 441.48 (1,2).

2,6,10,14-Tetramethylpentadecane, CA 1921-70-6: Above 385.93 (1).

$C_{19}H_{41}N$

N-Methyloctadecylamine, CA 2439-55-6: Above 385.93 (1).

$C_{20}H_{14}O_4$

Diphenyl phthalate, CA 84-62-8: 497.15* (5,6).

$C_{20}H_{20}O_7P$

Cresyl diphenyl phosphate, mixed isomers: 505.37 (6).

$C_{20}H_{22}O_5$

Dipropylene glycol dibenzoate, CA 27138-31-4: 485.15* (4).

$C_{20}H_{23}ClO$

4-Butyl-*alpha*-chloro-4'-ethoxystilbene, *trans*, CA 33468-13-2: Above 385.93 (1).

$C_{20}H_{26}O_4$

Dicyclohexyl phthalate, CA 84-61-1: 480.37 (2).

$C_{20}H_{28}$

Decylnaphthalene, mixed isomers: 449.82 (6).

Dipentylnaphthalene, mixed isomers, CA 71784-99-1: 430.15 (5,6).

$C_{20}H_{30}O_2$

5,8,11,14,17-Eicosapentaenoic acid, *cis*, CA 10417-94-4: 366.48 (1).

$C_{20}H_{30}O_4$

Butyl octyl phthalate, CA 84-78-6: 460.93 (2).

Di(2-ethylbutyl) phthalate, CA 7299-89-0: 467.04* (6).

Dihexyl phthalate, CA 84-75-3: 449.82 (7).

Diisohexyl phthalate, CA 3068-96-0: 466.48 (2).

$C_{20}H_{30}O_6$

Di(2-butoxyethyl) phthalate, CA 117-83-9: 481.48* (2,6).

$C_{20}H_{30}O_8$

Bis[2-(ethoxyethoxy)ethyl] phthalate, CA 117-85-1: 481.15 (5,6).

$C_{20}H_{31}N$

Dehydroabietylamine, CA 1446-61-3: Above 385.93 (1).

$C_{20}H_{32}O$

Tetradecanophenone, CA 4497-05-6: Above 385.93 (1).

$C_{20}H_{32}O_2$

Arachidonic acid, CA 506-32-1: Above 385.93 (1).

$C_{20}H_{34}O_2$

cis-8,11,14-Eicosatrienoic acid, CA 1783-84-2: 335.37 (1).
Ethyl linolenate, CA 1191-41-9: Above 385.93 (1).

$C_{20}H_{34}O_4$

Di(methylcyclohexyl) adipate, CA 41544-42-7: 463.15* (4).

$C_{20}H_{34}O_8$

Acetyl tributyl citrate, CA 77-90-7: 477.60* (7).

$C_{20}H_{35}N$

4-Tetradecylaniline, CA 91323-12-5: Above 385.93 (1).

$C_{20}H_{36}N_2$

N,N'-Bis(1,4-dimethylpentyl)-1,4-phenylenediamine, CA 3081-14-9:
448.15* (6).

$C_{20}H_{36}O_2$

Linoleic acid, ethyl ester, CA 544-35-4: Above 385.93 (1).

$C_{20}H_{36}O_4$

Di(2-ethylhexyl) fumarate, CA 141-02-6: 466.48 (2).
Di(2-ethylhexyl) maleate, CA 142-16-5: 452.59 (2); 458.15 (6,7).
Diisooctyl fumarate: 466.48 (2).
Diisooctyl maleate, CA 1330-76-3: 452.59 (2).
Dioctyl fumarate, CA 2997-85-5: 458.15* (7).

$C_{20}H_{36}O_6$

   Dicyclohexano-18-crown-6, mixed isomers, CA 16069-36-6: Above 385.93 (1).

$C_{20}H_{38}O_2$

   11-Eicosenoic acid, *cis*, CA 5561-99-9: Above 385.93 (1).

   Ethyl oleate, CA 111-62-6: Above 385.93 (1).

$C_{20}H_{38}O_3$

   Decanoic anhydride, CA 2082-76-0: Above 385.93 (1).

$C_{20}H_{38}O_4$

   Bis(2-ethylhexyl) succinate, CA 2915-57-3: 430.37 (6).

$C_{20}H_{40}$

   1-Eicosene, technical, CA 3452-07-1: Above 385.93 (1).

$C_{20}H_{40}O$

   Octadecyl vinyl ether, CA 930-02-9: 449.82 (6); 449.82* (2).

   Phytol, mostly *trans*, CA 150-86-7: Above 385.93 (1).

$C_{20}H_{40}O_2$

   Ethyl stearate, CA 111-61-5: Above 385.93 (1).

   Isobutyl palmitate, CA 110-34-9: 435.93* (2).

   Methyl nonadecanoate, CA 1731-94-8: Above 385.93 (1).

$C_{20}H_{40}O_3$

   Ethylene glycol monostearate, CA 111-60-4: 472.04* (2).

$C_{20}H_{41}NO_2$

   Ethylene glycol amido stearate, CA 10287-60-2: 455.37* (2).

$C_{20}H_{42}$

   Eicosane, CA 112-95-8: 455.37 (2).

$C_{20}H_{43}ClSi$

   Chlorodimethyloctadecylsilane, CA 18643-08-8: Above 385.93 (1).

$C_{20}H_{43}N$

   Didecylamine, CA 1120-49-6: Above 385.93 (1).

286

$C_{20}H_{44}Si$

Dimethyloctadecylsilane, CA 3295-58-7: Above 385.93 (1).

$C_{21}H_{19}ClO_2$

3,4-Dibenzyloxybenzyl chloride, CA 1699-59-8: Above 385.93 (1).

$C_{21}H_{21}BO_6$

Tri-*ortho*-cresol borate, CA 2665-12-5: 447.04* (7).

$C_{21}H_{21}N$

Tribenzylamine, CA 620-40-6: 338.71 (1).

$C_{21}H_{21}O_4P$

Tri(2-tolyl) phosphate, CA 78-30-8: 498.15 (6,7).
Tritolyl phosphate, mixed isomers, CA 1330-78-5: 383.15 (1); 511.15* (5).

$C_{21}H_{23}NO_2$

4-Cyanophenyl 4-heptylbenzoate, CA 38690-76-5: Above 385.93 (1).

$C_{21}H_{26}O_3$

4-Octylphenyl salicylate, CA 2512-56-3: 488.71* (6).

$C_{21}H_{32}O_2$

Methyl abietate, CA 127-25-3: 453.15* (6,7).
Methyl *cis*-5,8,11,14,17-eicosapentaenoate, technical, CA 2734-47-6: 366.48 (1).

$C_{21}H_{32}O_4$

Benzyl octyl adipate, CA 3089-55-2: Above 473.15* (4).

$C_{21}H_{32}O_{11}$

Glucose pentapropionate, CA 65709-94-6: 538.15 (6).

$C_{21}H_{34}O_2$

Methyl dihydroabietate, CA 30968-45-7: 455.93 (6).

$C_{21}H_{36}$

Tripentylbenzene, mixed isomers: 405.15 (5); 405.15* (6).

$C_{21}H_{36}O$

3-Pentadecylphenol, technical, CA 501-24-6: Above 385.93 (1).

$C_{21}H_{38}BrN$

Benzyldimethyldodecylammonium bromide, CA 7281-04-1: Above 385.93 (1).

$C_{21}H_{40}O_2$

Methyl 11-eicosenoate, *cis*, CA 2390-09-2: Above 385.93 (1).

$C_{21}H_{40}O_3$

2-Methoxyethyl oleate, CA 111-10-4: 469.82 (7).

$C_{21}H_{40}O_4$

Glycerol monooleate, CA 25496-72-4: 497.04* (2).

$C_{21}H_{42}$

9-Uneicosene, *cis*, CA 39836-21-0: 335.37 (1).

$C_{21}H_{42}O_2$

Methyl eicosanoate, CA 1120-28-1: Above 385.93 (1).

$C_{21}H_{42}O_3$

Propylene glycol monostearate, CA 1323-39-3: 472.04* (2).

$C_{21}H_{42}O_4$

Glycerol monostearate, CA 31566-31-1: 483.15* (2); 503.15* (4).

$C_{21}H_{44}$

Uneicosane, CA 629-94-7: Above 385.93 (1).

$C_{21}H_{45}N_3$

Hexetidine, mixed isomers, CA 141-94-6: 343.15 (1).

$C_{22}H_{30}$

Dipentylbiphenyl, mixed isomers: 444.26 (6).

$C_{22}H_{30}O_2S$

3-*tert*-Butyl-4-hydroxy-5-methylphenyl sulfide, CA 96-66-2: 513.71 (1).

$C_{22}H_{31}O_3P$

Diphenylisodecyl phosphate, CA 26544-23-0: 491.48* (2,6).

$C_{22}H_{32}O_2$

4,7,10,13,16,19-Docosahexaenoic acid, CA 6217-54-5: 335.37 (1).

$C_{22}H_{34}O_2$

Ethyl abietate, CA 631-71-0: 450.93* (6).

$C_{22}H_{34}O_4$

Butyl decyl phthalate, CA 89-19-0: 474.82 (2).

$C_{22}H_{39}O_3P$

Dioctyl phenylphosphonate, CA 1754-47-8: Above 385.93 (1).

$C_{22}H_{42}O_2$

Butyl oleate, CA 142-77-8: 453.15* (5,6,7); 466.48* (2).

Erucic acid, CA 112-86-7: Above 385.93 (1).

Stearyl methacrylate, CA 32360-05-7: Above 422.15* (2,4).

$C_{22}H_{42}O_3$

Butyl ricinoleate, CA 151-13-3: 383.15 (6); 493.15 (5).

$C_{22}H_{42}O_4$

Bis(2-ethylhexyl) adipate, CA 103-23-1: 466.48 (2); 469.15 (4,6); 469.15* (4); 479.15* (4,6).

Bis(isooctyl) adipate, mixed isomers, CA 1330-86-5: 477.59 (2); 468.15 to 483.15* (4).

$C_{22}H_{42}O_8$

Dibutoxyethoxyethyl adipate, CA 141-17-3: 439.15* (4).

$C_{22}H_{44}O_2$

Butyl stearate, CA 123-95-5: 433.15 (4,5,6); 460.93* (2).

Isobutyl stearate, CA 646-13-9: 455.37* (2).

Methyl uneicosanoate, CA 6064-90-0: Above 385.93 (1).

$C_{22}H_{44}O_4$

Diethylene glycol monostearate, CA 106-11-6: 474.82* (2).

$C_{22}H_{45}NO$

N,N-Diethylstearamide, CA 7446-68-6: 463.71 (6).

$C_{22}H_{45}NO_2$

Butylamine oleate, CA 26094-13-3: 339.15 (5); 339.15* (6).

$C_{22}H_{46}$

Docosane, CA 629-97-0: Above 385.93 (1).

$C_{23}H_{34}O_2$

Methyl 4,7,10,13,16,19-Docosahexaenoate, CA 301-01-9: 366.48 (1).

$C_{23}H_{42}O_3$

Tetrahydrofurfuryl oleate, CA 5420-17-7: 472.04 (6).

$C_{23}H_{42}O_5$

Ethylene glycol, monomethyl ether acetylricinoleate, CA 140-05-6: 503.15 (7).

$C_{23}H_{44}O_2$

Pentyl oleate, CA 142-57-4: 458.71 (6).

$C_{23}H_{46}$

9-Tricosene, *cis*, CA 27519-02-4: Above 385.93 (1).

$C_{23}H_{46}O_2$

Methyl docosanoate, CA 929-77-1: Above 385.93 (1).
Pentyl stearate, CA 6382-13-4: 458.15* (5,6).

$C_{23}H_{48}$

Tricosane, CA 638-67-5: Above 385.93 (1).

$C_{24}H_{19}O_4P$

Diphenyl 2-biphenyl phosphate, CA 132-29-6: 498.15 (6).

$C_{24}H_{20}F_{24}O_4$

Bis(2,2,3,3,4,4,5,5,6,6,7,7-dodecafluoroheptyl)camphorate ±, technical: Above 385.93 (1).

$C_{24}H_{20}Sn$

Tetraphenyltin, CA 595-90-4: 383.71 (1); 505.15* (5,6).

290

$C_{24}H_{31}ClO$

trans-4-Octyl-alpha-chloro-4'-ethoxystilbene, CA 33468-15-4: Above 385.93 (1).

$C_{24}H_{38}O_4$

Di(2-ethylhexyl) phthalate, CA 117-81-7: 472.15* (5); 480.37 (1); 488.71* (2,4,6).

Di(2-ethylhexyl) terephthalate, CA 6422-86-2: 511.15* (4).

Dicapryl phthalate, CA 117-84-0: 474.26 (2,6).

Diisooctyl phthalate, CA 27554-26-3: 491.48 (2); 505.37 (6).

$C_{24}H_{44}O_4$

Butylacetyl ricinoleate, CA 140-04-5: 383.15 (5,6).

$C_{24}H_{44}O_8$

Dicyclohexano-24-crown-8, CA 17455-23-1: Above 385.93 (1).

$C_{24}H_{46}O_2$

Nervonic acid, CA 506-37-6: Above 385.93 (1).

$C_{24}H_{46}O_3$

Butoxyethyl oleate, CA 109-39-7: 474.82* (2).

Lauric anhydride, CA 645-66-9: 383.15 (1).

$C_{24}H_{46}O_4$

Decyloctyl adipate, CA 110-29-2: 473.15 to 508.15* (4); 477.59 (2).

Dinonyl adipate, CA 151-32-6: 475.15 to 505.15* (4); 491.48 (2).

Isodecylisooctyl adipate, CA 31474-57-4: 477.59 (2).

$C_{24}H_{48}O_2$

Isohexyl stearate, CA 80749-85-5: 505.37* (2).

Methyl tricosanoate, CA 2433-97-8: Above 385.93 (1).

Octyl palmitate, CA 16958-85-3: 474.82* (2).

$C_{24}H_{48}O_3$

Butoxyethyl stearate, CA 109-38-6: 466.48* (2).

$C_{24}H_{50}$

Tetracosane, CA 646-31-1: Above 385.93 (1).

$C_{24}H_{50}S$

  Dodecyl sulfide, CA 2469-45-6: Above 385.93 (1).

$C_{24}H_{51}BO_3$

  Tri(2-ethylhexyl) borate, CA 2467-13-2: 449.82* (7).

  Trioctyl borate, CA 2467-12-1: 460.93* (7).

  Tri(2-octyl) borate, CA 24848-81-5: 438.71* (7).

$C_{24}H_{51}N$

  Triisooctylamine, technical, CA 25549-16-0: Above 385.93 (1).

  Trioctylamine, CA 1116-76-3: Above 385.93 (1).

$C_{24}H_{51}OP$

  Trioctylphosphine oxide, CA 78-50-2: Above 385.93 (1).

$C_{24}H_{51}O_3P$

  Dilauryl phosphite, technical, CA 21302-90-9: Above 385.93 (1).

  Triisooctyl phosphite, mixed isomers, CA 25103-12-2: 469.26* (2).

  Tris(2-ethylhexyl) phosphite, CA 301-13-3: 444.26* (6); 458.15* (2).

$C_{24}H_{51}O_4P$

  Tris(2-ethylhexyl) phosphate, CA 78-42-2: 480.37* (2).

$C_{24}H_{52}O_4Si$

  Tetra(2-ethylbutoxy)silane, CA 78-13-7: 389.15 (4); 441.48* (6,7).

$C_{24}H_{54}OSn_2$

  Bis(tributyltin) oxide, CA 56-35-9: Above 385.93 (1).

$C_{24}H_{54}Sn_2$

  Bis(tributyltin), CA 813-19-4: Above 385.93 (1).

$C_{25}H_{30}NP$

  4-(Diisopropylaminomethyl)triphenylphosphine: 383.15 (1).

$C_{25}H_{34}O_3$

  4-Dodecyloxy-2-hydroxybenzophenone, CA 2985-59-3: 532.04 (6).

$C_{25}H_{38}O_6$

  Bis(2-ethylhexyl) trimellitate, CA 63468-10-0: 536.15* (4).

292

$C_{25}H_{42}O_3$

2-(4-*tert*-Pentylphenoxy)ethyl dodecanoate: 483.15 (6).

$C_{25}H_{48}O_4$

Di(2-ethylhexyl) azelate, CA 103-24-2: 499.82* (6).
Di(isooctyl) azelate, technical, CA 26544-17-2: Above 385.93 (1).

$C_{25}H_{52}$

Pentacosane, CA 629-99-2: Above 385.93 (1).

$C_{26}H_{31}O_4P$

Bis(4-*tert*-butylphenyl)phenyl phosphate, CA 115-87-7: 523.15 (6).

$C_{26}H_{42}O_4$

Decyl octyl phthalate, CA 119-07-3: 502.59 (2); 508.15* (7).
Isodecyl isooctyl phthalate, CA 42343-35-1: 485.93 (2).

$C_{26}H_{46}$

Tetrapentylbenzene, mixed isomers: 419.26 (6).

$C_{26}H_{47}O_3P$

Didecyl phenyl phosphite, CA 1254-78-0: 491.48* (6).
Diisodecyl phenyl phosphite, CA 25550-98-5: 491.48* (2).

$C_{26}H_{50}O_4$

Bis(2-ethylhexyl) sebacate ±, CA 122-62-3: 483.15 (7); 499.82 (2).
Bis(isodecyl) adipate, CA 27178-16-1: 380.37* (6); 491.15* (4); 498.82 (2).
Bis(isooctyl) sebacate, CA 27214-90-0: 499.82 (2).

$C_{26}H_{53}NO$

N,N-Dibutylstearamide, CA 5831-88-9: 488.71 (6).

$C_{26}H_{24}Sn_2$

Bis(tributylstannyl)acetylene, CA 994-71-8: Above 385.93 (1).

$C_{27}H_{42}O_6$

Decyl octyl trimellitate, CA 34870-88-7: 538.71 (2).

$C_{27}H_{52}O_5$

Glycerol dilaurate, CA 27638-00-2: 522.04* (2).

$C_{28}H_{27}O_9$

Bis(2,2,4-Trimethylpentanediolisobutyrate) diglycolate: 468.15* (6).

$C_{28}H_{46}O_4$

Didecyl phthalate, CA 84-77-5: 505.37* (7).
Diisodecyl phthalate, CA 26761-40-0: 505.37* (2,6).

$C_{29}H_{30}P_2$

1,5-Bis(diphenylphosphino)pentane, CA 27721-02-4: Above 385.93 (1).

$C_{29}H_{50}O_2$

Vitamin E, CA 10191-41-0: 383.15 (1).

$C_{30}H_{23}O_4P$

Di(2-biphenyl) phenyl phosphate, CA 587-79-5: 523.15 (6).

$C_{30}H_{50}$

Squalene, CA 111-02-4: Above 385.93 (1).

$C_{30}H_{54}O_8$

Bis(2,2,4-trimethyl-1,3-pentanediolmonoisobutyrate) adipate, CA 31869-96-2: 508.15* (4).

$C_{30}H_{56}O_4$

Ditridecyl maleate, CA 61791-92-2: 474.82 (2).

$C_{30}H_{62}$

Squalane, CA 111-01-3: 490.93 (1).

$C_{30}H_{63}O_3P$

Tridecyl phosphite, CA 2929-86-4: 508.15* (6).
Triisodecyl phosphite, CA 25448-25-3: 508.15* (2).

$C_{31}H_{46}O_2$

Vitamin K1, CA 84-80-0: Above 385.93 (1).

$C_{31}H_{52}O_3$

Vitamin E acetate, CA 7695-91-2: Above 385.93 (1).

$C_{31}H_{52}O_4$

Isodecyl tridecyl phthalate: 513.71 (2).

$C_{32}H_{64}O_4Sn$

Dibutyltin dilaurate, CA 77-58-7: 508.15* (7).

$C_{32}H_{68}O_4Si$

Tetra(2-ethylhexyl) silicate, CA 115-82-2: 461.15 (4); 472.04* (2,6).

$C_{33}H_{54}O_6$

Tri(2-ethylhexyl) trimellitate, CA 3319-31-1: 533.15 (2).

Tri(isooctyl) trimellitate, CA 27251-75-8: 533.15 (2).

$C_{34}H_{58}O_4$

Ditridecyl phthalate, CA 119-06-2: 516.48* (6,7); 522.04 (2).

$C_{34}H_{62}O_{12}$

Bis[(12-Crown-4)-2-ylmethyl] 2-dodecyl-2-methylmalonate, CA 80403-59-4: Above 385.93 (1).

$C_{36}H_{62}O_3$

Linoleic anhydride, CA 24909-68-0: Above 385.93 (1).

$C_{36}H_{66}O_3$

Oleic anhydride, CA 24909-72-6: Above 385.93 (1).

$C_{36}H_{70}O_4Zn$

Zinc stearate, CA 557-05-1: 550.15* (5).

$C_{36}H_{75}BO_3$

Tri(dodecyl) borate, CA 2467-15-4: 513.71* (7).

$C_{36}H_{75}PS_3$

Trilauryl trithiophosphite, CA 1656-63-9: 476.48* (6); 494.26* (2).

$C_{36}H_{78}O_7Si_2$

Hexakis(2-ethylbutoxy)disiloxane, CA 1476-03-5: 488.15 (4).

$C_{37}H_{84}O_{12}Si_4$

Methyltris(tri-sec-butoxysiloxy)silane, CA 60711-47-9: 477.15 (4).

$C_{38}H_{43}N_3O_{18}$

Quin 2 AM, CA 83104-85-2: Above 385.93 (1).

$C_{38}H_{74}O_4$

Ethylene glycol, distearate, CA 627-83-8:   472.04* (2).

$C_{39}H_{72}O_5$

Glycerol dioleate, CA 25637-84-7:   544.26* (2).

$C_{39}H_{76}O_4$

Propylene glycol distearate, CA 6182-11-2:   494.26* (2).

$C_{39}H_{76}O_5$

Glycerol distearate, CA 1323-83-7:   516.48* (2).

$C_{40}H_{78}O_5$

Diethylene glycol distearate, CA 109-30-8:   455.37* (2).

$C_{45}H_{76}O_2$

Cholesteryl linoleate, CA 604-33-1:   Above 385.93 (1).

$C_{45}H_{78}O_2$

Cholesteryl oleate, CA 303-43-5:   Above 385.93 (1).

$C_{46}H_{80}O_3$

Cholesteryl oleyl carbonate, CA 17110-51-9:   Above 385.93 (1).

$C_{76}H_{52}O_{46}$

Tannic acid, CA 1401-55-4:   472.04* (7).